C0-ATH-318

Changing Landscapes: An Ecological Perspective

Isaak S. Zonneveld
Richard T.T. Forman
Editors

Changing Landscapes: An Ecological Perspective

Contributors

J. Baudry, F. Burel, R.T.T. Forman, J.F. Franklin, E.R. Fuentes, W. Haber, W.B. Harms, G. Merriam, L. Miklos, E.P. Odum, P. Opdam, P.G. Risser, M. Ruzicka, K.-F. Schreiber, J.R. Sedell, F.J. Swanson, M.G. Turner, R.G. Woodmansee, I.S. Zonneveld

Springer-Verlag
New York Berlin Heidelberg
London Paris Tokyo Hong Kong

I.S. Zonneveld
International Institute
for Aerospace Survey
and Earth Sciences
7511-AL Enschede
The Netherlands

R.T.T. Forman
Harvard University
Cambridge, MA 02138
U.S.A.

Library of Congress Cataloging-in-Publication Data
Changing landscapes: an ecological perspective/Isaak S.
Zonneveld and Richard T.T. Forman, editors; contributors, J. Baudry
. . . [et al.].
p. cm.—
Includes index.
ISBN 0-387-97102-5 (alk. paper)
1. Landscape protection. 2. Ecology. I. Zonneveld, Isaak
Samuel. II. Forman, Richard T.T.
QH75.C43 1989
333.73' 16—dc20 89-11523

Printed on acid-free paper

Typeset by Spectrum Publisher Services
Printed and bound by Edwards Brothers, Inc., Ann Arbor, Michigan.
Printed in the United States of America.

9 8 7 6 5 4 3 2 1

ISBN 0-387-97102-5 Springer-Verlag New York Berlin Heidelberg
ISBN 3-540-97102-5 Springer-Verlag Berlin Heidelberg New York

Preface

The concept of landscape ecology developed over the past few decades is gaining momentum in the scientific and planning-management worlds. Generated in Central Europe and becoming prosperous via application, it has recently spread to the New World. Its state of maturity was signaled by an invitation to IALE, the International Association of Landscape Ecology, to organize a symposium and plenary lecture at the 1986 International Congress of Ecology in Syracuse, USA. The idea for this volume exploring the ecology of actively changing landscapes emerged at that congress.

The chapters, here arranged in four parts, represent a diverse array, though not as wide as the papers presented at ongoing landscape ecology meetings.

Part I, *Evolving Approaches,* provides an introduction to the field and a history of landscape ecology. It shows the roots in biology and geography, as well as portraying the recent surge of interest in developing both theory and its applications. The focus on people and topics within landscape ecology emphasizes the positive feedback operating in the early growth phase of a discipline.

Part II, *Energy, Nutrient, and Species Fluxes in a Mosaic,* delves directly into many of the functional processes and mechanisms of nature operating at a landscape scale. These range from wind- and water-transported energy and materials to the dispersal of plants and animals by means of animal locomotion. The studies show that landscape structure, especially the configuration of patches and corridors, exerts a major control on these flows and movements. Understanding of these fluxes is gained through creative experiments, spatially explicit measurement, hierarchical analysis, and simulation models.

Part III, *Natural and Human Processes Interacting to Cause Landscape Change,* explores how natural processes and human activities interact to change the structure of landscapes over time. Linking several time scales with cyclic and non cyclic natural events is especially revealing. The human dependence on, effect on, and productivity of spatially juxtaposed portions of the land are also critical. These models or case studies of relatively unplanned (at the landscape level) human-land patterns and processes provide sound bases for understanding and decision making.

Part IV, *Planning and Management of Landscapes,* illustrates how landscape change, still mediated by human-land interaction, is organized and directed by planning. While long- and short-term planning focus on somewhat different variables, both maintain spatial configuration of the landscape as the centerpiece. The results of planning and management approaches based on landscape ecology, despite contrasting premises, spatial models, and detailed procedures, show some striking similarities and generalizations.

By focusing on general concepts and principles the field is applicable for ecological understanding of any landscape, from primeval forest to irrigated plain to suburbia. With this approach landscape ecology has attracted the interest of theoreticians, planners and managers. People have been drawn to it from a diversity of disciplines; no one should be surprised that the approaches to understanding the ecology of a landscape are highly diverse. Indeed this hybrid vigor provides power to the field. Moreover, when landscape ecologists from Europe recently came in contact with the people of the New World, an infusion of intriguing ideas provided a stimulus to both. Thus was this book born.

Gazing into the future is both hazardous and our responsibility. Where will or should landscape ecology be in the near future? We sense that a top priority is to develop and consolidate the core body of theory, principles, and concepts. Expansion into new areas will doubtless continue, though perhaps at a decreasing rate; the field cannot do everything nor should it be spread too thinly. Building the central body of theory avoids the pitfall of some interdisciplinary areas of thought that never really flourished as more than the overlap area of established disciplines. This core also provides the essential impetus for, as well as solidifies, landscape ecology applications.

Society gradually ponders larger areas, such as the globe, and longer time frames, as in sustainability. We can no longer view a local ecosystem or land use in isolation from the horizontal and vertical configuration of the surrounding landscape mosaic. Similarly we can no longer view a region or the globe without focusing on the heterogeneous pattern of landscapes within it.

Humans require resources, while nature provides resources, as well as constraints on their use. Flows and movements in a landscape both respond to, and create, spatial patterns on the land. Understanding the linkages between (a) humans, resources, and constraints, and (b) flows, patterns, and land is to understand the ecology of a landscape. Putting this landscape ecology to work is to act in the long-term interest.

I.S. Zonneveld
R.T.T. Forman

Contents

Contributors

Baudry, J. Chargé de Recherche, Institut National de la Recherche Agronomique, Département de Recherches sur les Systémes Agraires et le Développement, Lieury, 14170 Saint-Pierre-sur-Dives, France. Landscapes of major influence include the bocages of western France; New Jersey farmland; eastern Ontario; and the plateau of central France.

Burel, F. Chercheur contractuel, Université de Rennes, Laboratoire d'Evolution des Systémes Naturels et Modifiés du Muséum National d'Histoire Naturelle, Campus de Beaulieu, 35042 Rennes CEDEX, France. Landscapes of major influence include the Brittany coastline; the bocages of western France; the Canadian boreal forest; and the Alps.

Forman, R.T.T. PAES Professor of Landscape Ecology, Graduate School of Design, Harvard University, Cambridge, Massachusetts 02138, USA Landscapes of major influence include the New England mountains; the Central American tropics; New Jersey Pine Barrens and farmland; vineyards of southern France; English countryside; and arid New Mexico lands.

Franklin, J.F. Professor of Forest Ecosystems, College of Forest Resources, University of Washington, Seattle, Washington 98195, USA, and Chief Plant Ecologist, U.S.D.A. Forest Service, Pacific Northwest Research Station, Forestry Sciences Laboratory, 3200 Jefferson Way, Corvallis, Oregon 97331, USA. Landscapes of major influence are the extensive coniferous forests, subalpine meadow-tree parklands, large wildfire areas, fragmented patterns of dispersed clearcuts, and tall volcanic peaks, all in the Cascade Range of northwestern North America.

FUENTES, E.R. Professor of Biological Sciences, Facultad de Ciencias Biologicas, Pontificia Universidad Católica de Chile, Casilla 114-D, Santiago, Chile. Landscapes of major influence include the lakes, volcanoes, *Nothofagus* trees and farmlands of southern Chile; Mediterranean-type landscapes near Santiago; similar landscapes in California, the Mediterranean Basin, South Africa, and Australia; central Europe; and wet American tropics.

HABER, W. Dr. rer. nat., Professor of Landscape Ecology, Munich University of Technology, Lehrstuhl für Landschaftsökologie, Weihenstephan, D-8050 Freising 12, Federal Republic of Germany. Landscapes of major influence include the hedgerow area of Westphalia and the Luneburg Heath near Hamburg; the coast and mountains of central Italy; alpine areas of Austria and Yugoslavia; the countryside of Bavaria and of eastern China; Honshu, Japan urban landscape; and tropics of Java.

HARMS, W.B. Department of Landscape Development, The Winand Staring Centre for Integrated Land, Soil and Water Research, P.O. Box 125, 6700 AC Wageningen, The Netherlands. Landscapes of major influence include the Sawah landscape of Java; the mosaic of the eastern part of The Netherlands; the Negev Desert of Israel; and the Alps.

MERRIAM, G. Professor of Biology and Director, Ottawa-Carleton Institute of Biology, Department of Biology, Carleton University, Ottawa K1S 5B6, Canada. Landscapes of major influence include the farmland in the Great Lakes-St. Lawrence lowlands of Ontario and the Finger Lakes region of New York; the boreal forest of northern Ontario and Saskatchewan; aspen parklands of Alberta; and the Edwards Plateau region of central Texas.

MIKLOS, L. Institute of Experimental Biology and Ecology CBES, Slovak Academy of Sciences, Bratislava 814 34, CSSR (Czechoslovakia). Landscapes of major influence include different parts of Slovakia, and especially the agricultural areas of the East Slovakian Lowland; northern Bulgaria; and the southern part of the German Democratic Republic.

ODUM, E.P. Calloway Professor Emeritus and Director Emeritus, Institute of Ecology, University of Georgia, Athens, Georgia 30602, USA. Landscapes of major influence include Georgia and the Southeast; Illinois where V. Shelford took us all over; and the Illinois River valley.

OPDAM, P. Chairman, Department of Landscape Ecology, Research Institute for Nature Management, P.O. Box 46, 3956 ZR Leersum, The Netherlands. Landscapes of major influence include Western European fragmented landscapes and Dutch hedgerow landscapes.

RISSER, P.G. Professor of Biology and Vice President for Research, University of New Mexico, Albuquerque, New Mexico 87131, USA. Landscapes of major influence include the farmlands and rangelands of northern Oklahoma; fragmented landscapes of central Illinois; and the enchanting lands of New Mexico.

RUZICKA, M. Former Director, Institute of Experimental Biology and Ecology CBES, Slovak Academy of Sciences, Bratislava 814 34, CSSR (Czechoslovakia). Landscapes of major influence include different parts of Slovakia and especially the agricultural areas of the East Slovakian Lowland, northern Bulgaria, and the southern part of the German Democratic Republic.

SCHREIBER, K.-F. Lehrstuhl Landschaftsökologie, Institut für Geographie, Westfälische Wilhelms-Universität Münster, Robert Koch Strasse 26, 44 Münster, Federal Republic of Germany. Landscapes of major influence include: southwestern and northwestern Germany; southeastern France; Yugoslavia; northern Israel; and the Negev Desert.

SEDELL, J.R. Research Ecologist, U.S.D.A. Forest Service, Pacific Northwest Research Station, Forestry Sciences Laboratory, 3200 Jefferson Way, Corvallis, Oregon 97331, USA. Landscapes of major influence include the rivers and streams in the Pacific Northwest, Southeast Alaska, French Pyrenees, central Idaho, and Pennsylvania.

SWANSON, F.J. Research Geologist, U.S.D.A. Forest Service, Pacific Northwest Research Station, Forestry Sciences Laboratory, 3200 Jefferson Way, Corvallis, Oregon 97331, USA. Landscapes of major influence include the: coast and mountains of the Pacific Northwest; the landscape beautifully and incredibly modified by the 1980 eruptions of Mount St. Helens; Beartooth Mountains of Montana; and Galapagos.

TURNER, M.G. Research Staff Scientist, Environmental Sciences Division, Oak Ridge National Laboratory, Oak Ridge, Tennessee 37831, USA. Landscapes of major influence include the: southeastern United States; Georgia barrier islands; and the Rocky Mountains, especially Yellowstone National Park.

WOODMANSEE, R.G. Director, Natural Resource Ecology Laboratory, Colorado State University, Fort Collins, Colorado 80523, USA.

ZONNEVELD, I.S. Professor of Vegetation Surveys, Department of Land Resource Surveys and Rural Development, International Institute for Aerospace Survey and Earth Sciences (ITC), 350 Boulevard 1945, P.O. Box 6, 7511 AL Enschede, and V.P.O. Department, Agricultural University Wageningen, 6708 PD Wageningen, The Netherlands. Landscapes of major influence include the river deltas in The Netherlands and Southeast Asia; savanna and sahelian landscapes in Africa; tropical forests in Asia, Africa, and America; diverse cultural landscapes; and arctic landscapes at Spitsbergen.

Part I Evolving Approaches

Change, as a universal law, is highlighted in landscapes. From historians to paleoecologists and from demographers to geologists we have gained insights into changing landscapes. This volume presents examples and new understanding from the perspective of landscape ecology. Part I explores some roots and evolving approaches of the field, and includes the following themes or ideas.

Landscape ecology is fundamentally a science, with a paradigm or body of theory and a network of researchers and practitioners. The field focuses on the land or landscape as an object, especially utilizing spatial and ecosystematic (and to a lesser extent, aesthetic) perspectives. Terminology remains fluid, characteristic of the early stages of a field, though work may be recognized as morphological, classificational, chorological (spatial patterns and variation), chronological, relationship study, or geospheric in approach. Holism (understanding wholes or systems without knowing all internal details), vertical and horizontal heterogeneity (or Greek "topology" and "chorology"), and systems analysis provide a philosophical and operational framework. Theories and concepts, and then their application, are the goals and products of landscape ecology.

Key early roots of the field in Germany and the USA are outlined to be illustrative as well as to pinpoint important developments. Naturally major contributions have been made from many other areas, such as Czechoslovakia, The Netherlands, Canada, etc. The direct origins in Germany are in 19th and early 20th century biology and geography, leading to Carl Troll who in the

1930's used the term landscape ecology in linking patterns of aerial photographs to the then developing concepts of ecology. Landscape ecology in the USA began in the 1980s, building on an active period of ecosystem science and biogeography, and stimulated by contact with Europeans at international meetings. The meetings of the Ecological Society of America pinpointed the promise of the field to the wider scientific community, and began the diversification and sifting process of subjects and people. Separate annual landscape ecology meetings then catalyzed a linkage within a unique set of disciplines, and provided a ready forum for presenting work in progress or completed.

Theoretical development of landscape ecology in Germany is closely linked to geography, soil science, vegetation analysis, cartography, geomorphology and biology, whereas in the USA it is more linked with ecology, geography, forestry, landscape architecture, and wildlife biology. In application, the concepts are highly integrated into official planning and landscape architecture in Germany; in America the concepts are informally penetrating several fields, with forestry containing the first government agency to officially recognize landscape ecology to provide solutions for land fragmentation and biodiversity problems. Publication of textbooks (handbooks) and the establishment of a journal have helped solidify the field of landscape ecology, and provide a significant framework for understanding landscape change.

1. Scope and Concepts of Landscape Ecology as an Emerging Science

Isaak S. Zonneveld

Why Is Landscape Ecology a Science?

Landscape ecology is an emerging science, with a complex character and heterogeneous content, but with a clearly scientific philosophical (epistemological) background. Before we describe the subject, it is worth explaining why land ecology or landscape ecology is a science, rather than just a "state of mind" or a mix of social activities and attitudes (I. Zonneveld, 1982; Theorie Werkgroep, 1986).

A human activity (institutional or otherwise) can be called *science* if it can be described as "methodical channeled knowledge" (Theorie Werkgroep, 1986). That knowledge must be based on theory, on empirically observed facts and phenomena, and on understanding. Understanding is acquired by systematically ordering the observed data, and then connecting and associating them with each other via mental activity.

Knowledge is further channeled via that understanding and transmitted through words and images to knowledge storage systems to be used later either by the same scientist or for communication with others. Through this communication, understanding can be considerably enlarged, because one knowledge source can be combined with others and thus disseminated to many minds. Thus, the chances of generating more and better understanding and ideas are improved. Where more than one mind is active in certain knowledge channels, there is a possibility of consensus and also of stimulating differences in opinion.

A considerable amount of consensus is considered to be an indication of fact. Thus, it appears that transmittability is a firm prerequisite for a science. This book contains such transmittable knowledge.

Science is subdivided into many areas. The boundaries between these areas are more often vague than discontinuous, but for practical reasons a subdivision in scientific fields is necessary. However, this science continuum results in a rather strong overlap between sciences. One disadvantage of this is financial– when the money suppliers of two overlapping science fields each want the other to pay.

The overlap is sometimes illusory, however, because scientists studying the same subject may differ methodologically, or also according to their considered dimensions and scales in space and time. Apparent overlaps of landscape ecology with other disciplines should be interpreted in this way. Nevertheless, there is a demand for the recognizability of a certain field of science as distinct from others.

The two main criteria for this are the existence of:

1. A paradigm, that is, the whole set of subjects, theories, methods and assumptions applied by the scientists active in that field. (The term *theory* in science and philosophy may point to a set of current hypotheses that still have to be verified and that provide the continued incentive for action; it can also be used to include all the common concepts, facts, understandings and assumptions—making it similar to a paradigm.)
2. A network of researchers having contact with each other via at least a written tool (articles in journals, preferably a specialized journal), but especially by personal contact through congresses and symposia. (Symposium, literally "to drink together," implies true social interaction—in addition to scientific exchange.) This social contact is especially important in creating a firm paradigm—often under the influence of strong personalities who may affect the science's direction (Cramer et al., 1984; see also Theorie Werkgroep, 1986; I. Zonneveld, 1985).

Landscape ecology is supported by such a network of scientists who are joining together all over the world, forming groups and associations and calling themselves landscape ecologists. Many of these are united in the International Association of Landscape Ecology, which contributes to publishing the scientific journal, *Landscape Ecology*. They hold symposia, communicate, and encourage social interaction that provides the impetus for the development and definition of their science—creating and cultivating an emerging paradigm (see, among others, Brandt and Agger, 1984; Tjallingii and de Veer, 1982; Rickler and Schönfelder, 1986; Ruzicka and Miklos 1982; Slovak Academy of Sciences, 1985).

The Subject of Study

The subject of study in landscape ecology is the land or landscape, its form, function, and genesis (change). The English word *landscape* seems to have evolved from the Dutch word *landschap,* introduced via Dutch landscape painting *(landschappen).* The terms in German and Dutch, *Landschaft* and *landschap,* have gradually developed from a mere indication of an area somewhere in space (the Greek *chore*—an area according to its place) to the character of an area according to its contents (Greek *topos*) (see Schmithüsen, 1963; 1974; Neef, 1967).

It goes without saying that *landscape* is sometimes identical in meaning to *land* and sometimes only narrowly related to it. As with any common word used in science, only a few of the definitions of the terms *land* and *landscape* may be used within the scope of landscape ecology.

It appears that landscape ecologists are concerned about land or landscape from three points of view that cannot be completely separated. The first is the visual aspect of landscape (the German *Landschaftsbild*), e.g., landscape scenery. This is the oldest meaning of *landscape* and the most common, especially in *landscape architecture,* which clearly contains an esthetic element. Modern landscape architecture goes well beyond the visual—and certainly the esthetic—aspect. Moreover, this visual aspect has become an important diagnostic of land (see Bartkowski, 1984; I. Zonneveld, 1984).

The second perspective is the chorological aspect (a conglomerate of land attribute units or map patterns). (See Figure 1.) In geology, geomorphology, soil science, and vegetation science, the word *landscape* is used to indicate the pattern of individual surface patches belonging to each of the land attributes that are the subjects of these sciences—i.e., the rocks, landforms (relief types), soil bodies, and vegetation patches. Surveyors used to speak about geologic landscapes, soil landscapes, and vegetation landscapes, indicating the units on legends of relatively small-scale maps (i.e., generalized or broad areas such as whole nations). The geologist, geomorphologist, soil scientist, and vegetation scientist usually use the concept for more than just horizontal patterns, however. They also often tend to incorporate the concept of the ecosystem.

The third point of view sees the landscape as an ecosystem (e.g., the many figures in J. Zonneveld, 1985b). This is the most comprehensive concept, and includes the two preceding ones. It expresses the open system at the Earth's surface formed by all factors acting there—including the physical, biological, and noospherical. These factors form complex three-dimensional phenomena that can be recognized visually as a horizontal pattern of mutually related elements (units of land), and as vertical, mutually related strata, the so-called land attributes (Fig. 1) that may be intensively intermingled. These land attributes include atmosphere/climate, rock, landform, soil, water, vegetation, fauna, and the noospheric aspect of humans. These three-dimensional units may change in time, and thus, they also have a fourth dimension. While each separate relevant science (geology, soil science, etc.) selects a stratum for study and considers the others as "forming factors" for its own selected attribute, *landscape ecology* takes the vertical heterogeneity formed by all land attributes as a holistic object of study. This is in fact one of its main characteristics. The other and equally important characteristic is that landscape ecology also considers the chorological pattern as a whole. The heterogeneity within that whole

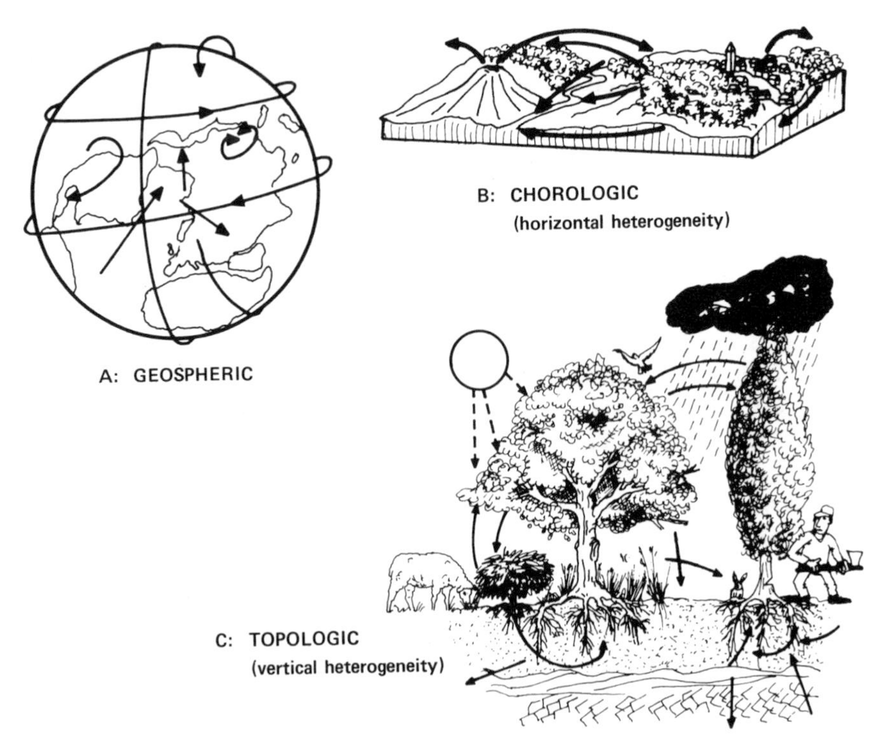

Figure 1. Landscape heterogeneity. Vertical landscape heterogeneity is expressed by *land attributes;* horizontal heterogeneity by *land units*. The latter can be distinguished by chorologic classification at various scales.

object—horizontally as well as vertically—is the major subject of landscape ecology.

This holistic approach can be viewed as a translation of the way farmers, hunters, herdsmen, and other outdoor people perceive the land. Their land concept is synonymous with landscape-as-ecosystem and represents a very old idea that has developed gradually since early humans became aware of their total dependence on the environment. Only in the last century have we learned to study it following modern, analytical, separate sciences.

The early geographer von Humbolt defined landscape very concisely as *Der totale Character einer Erdgegend* [the total character of a patch of the Earth] (cited by Troll, 1950). Landscape is the subject of the applied discipline *land evaluation,* which determines the value or capability or suitability of land for a certain use, and at the same time the need for its conservation or improvement as a human environment.

One of the important tasks of landscape ecology is therefore the "selection of the proper differentiating characteristics of each landscape as a regional unit" (Vink, 1982; see also Haase, 1984; I. Zonneveld, 1979) as a basis for that evaluation, and also for fundamental study.

The landscape units recognized from a prescientific era can still be used by modern science; they may already bear names predating recorded history. Many of the older toponyms in all languages refer to land units as defined by a combination of landform, vegetation, soil condition, and often land use. (Toponymic studies may therefore reveal striking landscape ecological information.) Toponyms can also date from peoples living in an area whose language disappeared long ago. From the many thousands of land-unit toponyms, examples of some typical river landscape terms include *dambo* (Swahili), *fadama* (Hausa), *waard* (Dutch). Dutch topography has inspired a number of more discriminatory toponyms, including *schor* or *kwelder* for a marine foreland with a closed halophytic vegetation, and *haar* for a relatively dry area between peat land (often a former land dune ridge that now has a humuspodzol soil profile). *Alp* in German is a grazed mountain meadow, etc. Many of these are short words, even monosyllabic, suggesting that they have been in use for a very long time (see, e.g., Troll, 1950; I. Zonneveld, 1979).

The Scientists

The network of scientists developing the emerging paradigm of landscape ecology is quite diverse. In the beginning, these scientists often had contact mainly because of the term *landscape ecology*. They usually had rather different ideas about the concept of *landscape*, as well as about the concept of *ecology*. Thus one could speak of a polygenetic origin of landscape ecology.

Nevertheless, the origin of landscape ecology as an emerging scientific discipline goes back to C. Troll (1950). It can be described as a marriage of geography (land and landscape) and biology (ecology). After studying biology, Troll became a geographer; he was impressed by the ecosystem concept as defined by the biologist A. Tansley, and by the comprehensive view of land-(scape) units depicted on aerial photographs (C. Troll, personal communication). Since Troll, geographers and biologists can be found in the network, as well as landscape architects, soil scientists, regional planners, foresters, agronomists, and conservationists of all kinds. Each necessarily brings his or her own perspectives on what constitutes land, landscape, and ecology (and thus also landscape ecology). Some analytically oriented scientists see this heterogeneity as a drawback, but many scholars find that a comprehensive view of the subject is stimulated by this variety, provided that the scientists share a similar basic concept—a holistic view of the landscape as an ecosystem.

The vertical and horizontal heterogeneity of landscapes splits the study of land into many disciplines served by a variety of sciences. In landscape ecology, however, this vertical and horizontal heterogeneity within landscapes is the principal subject of study—indeed, the essence of landscape ecology (see Risser et al., 1984). Thus zoologically oriented landscape ecologists study the influence of chorological heterogeneity on animal populations (see Merriam, 1984; Forman, 1982). This is also one of the essential questions facing humanity: Is the survival of groups of people (up to humankind as a whole) essentially dependent on landscape heterogeneity? Agriculture, recreation, and other basic human activities seem to indicate this. On the other hand, each of the smallest units of land is a tangible ecosystem characterized by structure, function, and change. In a horizontal sense, these basic units are relatively homogeneous, but they are spatially defined, and study requires a holistic approach and cooperation among soil science, vegetation science, hydrology, and other disciplines.

The landscape-ecology approach also leads to such applications as land evaluation. To the zoologist, attention to horizontal (chorological) relationships predominates, but the vertical (topological, site-specific) aspect is also important because the organisms moving between landscape patches are at each location sensing its specific conditions. The topologic study of a single land unit (including relationships among soil, vegetation, and other land attributes) also requires comparison with other land units for the sake of classification. Without mapping, such studies are of less value and quality, as every soil and vegetation surveyor will agree.

Landscape systems, moreover, are open systems, so a single unit is well described only if the influences of other horizontal components are also considered. Soil scientists, vegetation scientists, and certain physical geographers became aware of this long ago, and among them are many landscape ecologists *avant la lettre* (before they realized that they were doing landscape ecology). A clear example of this is the European school of the French-Swiss phytosociologist Braun-Blanquet, with important exponents such as Tuxen and Ellenberg (see also Doing, 1974). The same holds for Edelman's school of soil science in Wageningen, The Netherlands.

Similar developments can be observed among geomorphologists and others—e.g., in the land system concept (Christian and Stewart, 1964; Leser, 1984; and others), in holistic land surveys in Australia and by the British DOS (Brunt, 1967; Perry, 1967; Poore and Robertson, 1964; I. Zonneveld, 1979), in the practical ecological land surveys in Canada by the Canadian Lands Directorate (Thie and Ironside, 1976; Rowe, 1988) and in similar actions elsewhere. Geomorphology, soil science, and vegetation science integrated their knowledge and survey methods, and are progenitors of landscape-ecology study and application.

The Science: Basic Concepts and Theory

Transdisciplinary Science

From the preceding, it is clear that landscape ecology must be multidisciplinary. The study of a tract of land requires many disciplines. It is better to call it a *transdisciplinary* science (Naveh and Lieberman, 1984), because it is not just a combination of the methods of various sciences but is an integration on a higher level that in turn influences—even embraces—other disciplines in basic philosophy and application. The wide and complex field means that at present hardly anyone is just a landscape ecologist per se without being a specialist in one of the component sciences, although this may change as the core of landscape ecology develops.

Holism and Systems Theory

The basic assumption of landscape ecology, and what makes it different from other disciplines, is that it considers a specific tract of landscape as a holistic entity, including all its hetereogeneous components. The term *holistic* deserves some further explanation, because it is considerably misused and misunderstood. Holism is essentially a philosophy formulated by Smuts (1926) and continued by many ecologically minded scientists and philosophers. It states that reality consists of wholes in a hierarchical structure in the sequence: atoms, molecules, minerals, organisms, human society, the world as a total ecosystem, the galaxy, and the cosmos. Each whole is a system, that is, an organized set of relationships in a relatively steady state. That steady state may break up, however, and change or develop into a different steady state.

The mechanism of maintaining the steady state is called *homeostasis,* self-regulation by a set of positive and negative feedback factors that keep the system in a dynamic equilibrium. Closely related, and in a brief time scale scarcely to be distinguished, is the concept of *homeorhesis* (Theorie Werkgroep, 1986). This points to steady movement. There is change in the long run, but it is protected from strong fluctuations by feedback mechanisms like those in homeostasis.

The essential aspect of holism as a scientific assumption is that it provides the basis for studying certain wholes or systems (for example, an organism) without knowing all the details of their internal functions. It removes the necessity of first defining all elements and their relationships before defining the whole. Successes in biology, and especially many practical biological applications in agronomy and medicine, have proved the usefulness of this approach.

The principles of classification are based on holism. We are able to recognize and characterize a whole or an entity by a limited number of its abstracted properties called *diagnostic characteristics*. For many purposes, the entities as described in the classification can be used as "black boxes" and studied only according to their inputs and outputs without knowing details of the processes occurring inside. This has enabled scientists to develop more comprehensive knowledge—for example, by formulating theories, or at least valuable empirical generalizations, as a framework for fruitful research and application.

Thus, holism permits the simplification of scientific activity by reducing analytic observations to better understand very complex structures and processes. At the same time it warns against attempting to study wholes by analyzing them in separate pieces without connecting them with each other. Criticism of holists would be justified if they denied the usefulness of gradually making the black boxes more transparent by analytical study. The importance of holism is that in many cases the objects of study—like life and landscapes—are so complex that real understanding gained by working from the basic elements upwards would be extremely difficult, time-consuming, and hence expensive—if it were even possible. Odum and Polunin (1986 lecture, IV INTECOL Congress, Syracuse, USA) recently pointed this out in relation to regional planning.

Holism is sometimes misinterpreted as indicating a need to study all details, which would be prohibitive in terms of both human and financial resources.

Another misunderstanding occurs when the term is related in a nonscientific context to a metaphysical connection between all components of the universe (see Capra, 1982). We must carefully exclude these uses of the word when discussing methodological problems in landscape ecology. The term (landscape) ecology is also being used to indicate a philosophical point of view, a state of mind (see I. Zonneveld, 1982), a basic philosophy: the perception of land as a holistic entity that must be considered, studied and treated (managed) as a system, and cannot—without danger to humanity—be treated or studied in pieces. It includes a warning against narrow analyses and action in science as well as in management.

From general philosophy, however, it is just a small step to religion. Discussion within environmental movements often takes on an emotional intensity similar to religion, with doomsday prophets versus the merry faithful, the dogmatic precisionists versus the moderates, and it is obvious that scientific and philosophic reason has given way to almost religious fervor. This is not difficult to understand. In the preface to *Le Phénomène Humain* (1955), Teilhard de Chardin warned against mixing these three human categories, stating that the more we look to wholes and the wider fields in science, the more science, philosophy, and religion tend to come together, "as the meridians towards the Pole," without, however, touching each other before the supreme point of the pole itself. It is clear, though, that philosophy, and possibly also metaphysics, stimulates our science of landscape ecology.

Holism gave an impetus to the development of general systems theory and the more mechanical branch of that has, thanks to modern computers, provided an important tool—modeling—to bridge the gap between pure analyst and holist. Although the most complicated model is still a very rough and oversimplified vague imitation of reality, it is possible to at least visualize major cybernetic loops and other mechanisms that form the ecologic system. Thus the complex relationships between, and also inside, black boxes can be studied more effectively.

Recent developments in the hard- and software for geographic information systems offer a very important new opportunity to extend the models with cartographic input and output (Burrough, 1986). They allow the integration of different land attributes, and are also magnificent tools for analyzing holistically surveyed landscape data. Phipps (1981) and Kwakernaak (1982; 1986) introduced the use of the term *information* (in the cybernetics context) as a tool for quantitative analysis and classification of landscape units. Thus a thesis derived from philosophy has stimulated not only basic concepts but also important means of study.

An important element of landscape-ecology theory is the basic hypothesis that specific landscapes (landscape units on various scales) are holistic entities. This is what keeps together a network of scientists working in the emerging transdisciplinary science of landscape ecology. Landscape, in its various hierarchical structures, was not treated as a whole in general systems theory, in Smuts's holism, or in Capra's or Lovelock's philosophies. In general ecology, in describing hierarchies, we jump from the organism, population, and association levels to the world as a whole (Gaia, see Lovelock, 1979). Smuts takes human society in between as an important holistic entity, but the concept of land or landscape as a whole does not appear.

Traditional ecology, since Tansley used the term *ecosystem,* tends to avoid defining boundaries. Ecosystems are sometimes described as environments around special organisms of interest within a sufficiently homogeneous minimum area: the ecosystem of man, of a certain animal, and so on. Soil science, vegetation science, and geomorphology have stimulated classification and thinking in specific units, and so paved the way for landscape ecology. Through these disciplines, we learned to speak of ecosystems as specific and mappable units. From that point on, we deal with landscape ecology.

The holistic paradigm, at the borderline between science and philosophy, is the main challenge for pure scientific landscape ecologic research. Thus, we must study the mechanisms of homeostasis and homeorhesis, that is, the relationships among all factors acting at the Earth's surface resulting in the horizontal and vertical heterogeneity of landscape features. For practical applications, cause-and-effect relationships in the system are important (especially in the context of agronomic and other landscape planning). An important aspect of application is conservation, from sustained production to pure nature protection.

Theories and Concepts

To fulfill these theoretical and practical needs in a field as complex as landscape ecology, we need theories in the sense of verified hypotheses or at least firm sources of generalized empirical knowledge. The difficulty of doing inductive studies (from the measured details to the whole) often demands deductive reasoning based on limited (diagnostic) characteristics of the situation and using basic theory (see Theorie Werkgroep, 1986). When studying systems, moreover, it is characteristic that not only can we not separate clear causal reasoning (linear cause-effect assessment) from functional assessment, but the two tend to merge. Because of the emerging character of landscape ecology, such clear cause-effect theories are still embryonic in this field.

Van Leeuwen's *relation theory* (van Leeuwen, 1982) can also be considered as a language (hence a tool for formulating, storing, handling, and transmitting generalized empirical knowledge) applicable in landscape ecology. An important element of his theory is the critical treatment of the complementary relationship between variation in space (diversity, heterogeneity) and variation in time (dynamic versus constant). The theory also touches on the crucial thesis of homeostasis and homeorhesis.

Qualitative and quantitative aspects of general systems theory (holism, cybernetics, and information theory) have been mentioned as important tools (also via modeling) for developing and ordering landscape-ecological knowledge (see also Naveh and Lieberman, 1984).

Landscape ecology will continue developing hypotheses, leading to new concepts and theories and extending existing ones as a basis for the science and its applications. An example is *connectivity* (Merriam, 1984), an essential concept in the study of the chorological (horizontal) relationships between landscape patches. It is a significant elaboration from the classical concept of accessibility in phytosociology and population dynamics, which is linked to MacArthur and Wilson's island theory (1967).

Connectivity is a genuine landscape ecologic concept that explains the feedback phenomena in an important law of ecology: Beierink Baasbecking's "everything is everywhere, but the environment selects." It is an important tool in studying and describing the predominant regulation and selection functions (van Leeuwen, 1982) at a landscape scale.

Main Subfields and Applications

The preceding indicates that the main task of landscape ecology is to gain knowledge about the relationships among the building blocks of the landscape and, from these, about the functioning of the landscape as a system. This knowledge, in turn, can be used as a basis for planning, managing, improving, and conserving land.

Before we can study relationships in a system, however, we should describe and order the elements and structure of that system. This leads to the following areas of landscape study, each with its own specific set of methods:

- morphology (description of the structure and its elements)
- classification (systematic ordering)
- chorology (study of spatial patterns and variation)
- chronology (study of temporal variation)
- relationships (landscape ecology in the narrow sense—the study of the relationships as such between all land attributes as a scientific interest and as a basis for land evaluation)

Readers familiar with soil science, vegetation science, or geomorphology may recognize analogies with research areas within those sciences.

The important role of the first four descriptive fields would seem to be a valid reason to call the science as a whole *landscape science* rather than landscape ecology, as is sometimes done (see I. Zonneveld, 1979; J. Zonneveld 1981; 1985a; Theorie Werkgroep, 1986). Within the paradigm described here, however, it has become common practice to use the term landscape ecology for all five areas of study together. A valid reason for this is that the five areas are quite interdependent. Study of relationships needs description of form and structure, and usually also classification. But classification and mapping in turn require reasoning about relationships between the attributes and elements. This mutual interdependence especially includes the study of landscape change.

One of the most promising fields of application of landscape ecology must be land evaluation, in which a tract of land (or a combination of tracts) is described in terms of its suitability for various land uses (see Beek, 1978; FAO, 1976; Vink, 1980; Haber, chapter 12; Ruzicka, chapter 13; Ruzicka and Miklos, 1982; van der Maarel and Dauveillier, 1978; Vink, 1982; Siderius, 1984; I. Zonneveld, 1979; 1986). Mainly topological information (see the preceding paragraphs) is used now in the evaluation process, but modern landscape ecology can and should also contribute invaluable chorological data. The relationship of social, technological, and economic aspects in land evaluation is shown in Figure 2.

Landscape Morphology (Structure and Its Components)

It has become common practice to call the vertically overlying components *land attributes* (rock, soil, landform, vegetation, atmosphere [climate], animals, and humans including artifacts). The component patches of the horizontal landscape mosaic are indicated by the term *element*. These can be single trees, or patches or clumps of plants, or parcels, but on a wider scale complex areas composed of mosaics of components can also be elements of mosaics of a higher order of magnitude, as units of land (land units). (See Figure 2.)

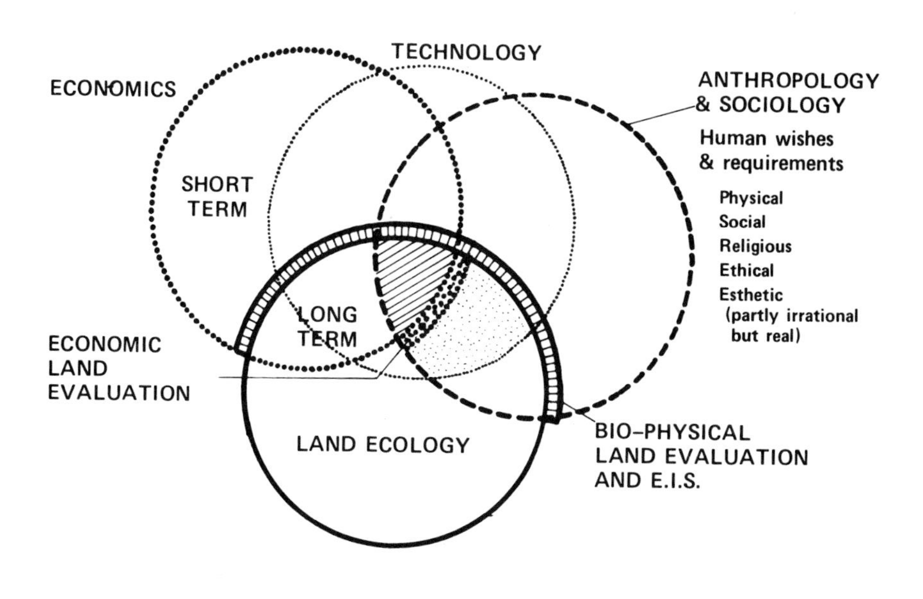

Optimal land use planning arrives at

(Short-term economic reasoning and a too heavy technological impact usually violate land economic and human values)

Figure 2. The four circles determining optimal land use planning. Land(scape) ecology helps in assessing the sustainable biophysical potential (biophysical land evaluation). Anthropology and sociology assist in formulating and analyzing human wishes and responses. Realization is enhanced and modified via technology and economics, which help to determine the final optimal suitability (economic land evaluation).

The methods of landscape description are derived partly from the sciences dealing with land attributes, with the addition of visual structural-spatial aspects. The latter are usually the provenance of landscape architecture. An important theoretical question is to determine at what minimum scale (in size as well as complexity) a specific area (tract of land) can be considered to be the smallest unit of land.

In practice, this question of minimum area is as simply solved as in soil and vegetation science. It will not be a single tree or a small waterhole of a few square meters. The smallest patches that can be described as a horizontally homogeneous (eco)system are recognized as such by farmers, herdsman, hunters, and so on, and—as noted above—often already have characteristic toponyms. In cultivated areas, farm parcels or combinations of such parcels may be considered as such. Hedgerows and other linear patches acting as corridors are special forms of such minimum land units.

Since Troll (1950), these smallest homogeneous pieces of land have been called *ecotopes* in landscape ecology literature. Other terms used, but not recommended, include *site, tessera, landscape cell, biotope,* and *landscape element*. Ecotope is a logical and not much misused term; it expresses the crucial character of landscape ecology: the study of the contents of a patch of land (Greek *topos*) as an *eco*system, hence *ecotope*. The ecotope is the smallest land element that deserves the name *landscape*.

The size of these smallest relatively homogeneous landscape units usually exceeds 100 m^2 or even a square kilometer. In general, we can map ecotopes at map scales larger than 1:25,000. However, in complex hilly areas, even on a scale of 1:10,000, the mapping units may represent complexes of ecotopes. Mapping (and classifying) ecotopes is a fundamental activity of landscape ecology.

Although I have used ecotope as the smallest patch of land that constitutes a landscape, others define a landscape as a complex of ecotopes. This emphasizes the character of a landscape as a chorologically heterogeneous complex, with the ecotopes and other point elements being the basic components (tangible ecosystems) of the landscape.

The definition of ecotope, as either a landscape or just a landscape element, depends on the weight given within landscape ecology to the chorological versus the topological approach. It is my conviction that the most characteristic aspect of landscape ecology, as a field of science, is the integration of topological and chorological research. Relationships between land elements (chorological units) can be understood only if the topological relationships (between land attributes) are known (Figure 1). A practical circumscription of an ecotope is: a tract of land where at least one attribute—landform, soil, or vegetation—is homogeneous. The visual structure of such a tract may still show some—but only minor—heterogeneity in vegetation (e.g., by single trees or other point elements). Seasonal changes or episodic changes over various years may cause a cyclic alternation of vegetation phases or even types within one ecotope. The same may be caused by annual crops in cultivated areas. Permanent crops, however, change the site in such a way that the original ecotope is changed into a new (culturally defined) one.

Landscape Classification

Landscape classification is a major connection point for applications. The contributions of Haber and Ruzicka to this book illustrate this. Classification units form the basis for land evaluation, which in turn is the basis for land use planning and management (see also I. Zonneveld, 1979 and 1986).

In classification, soil surveyors and vegetation surveyors clearly distinguish between a general abstract typology on the one hand, and the legend of a map on the other. A map legend is a description of the patterns of a land attribute or of land as a whole that uses an abstract typology. For soils and vegetation, and to a certain degree also for landforms, such general abstract typologies exist or can be designed for the purpose (see I. Zonneveld, 1979; I. Zonneveld in Küchler and Zonneveld, 1988). For landscapes, such general typologies are rarely described. The commonly used land classifications—as in the Australian and Commonwealth (Christian and Stewart, 1964) and Canadian land surveys (Thie and Ironside, 1976; see also I. Zonneveld, 1979)—are in essence map legends with a hierarchy that distinguishes ecotopes (i.e., sites), combinations of ecotopes (i.e., land facets), combinations of land facets (i.e., land systems), and combinations of land systems (main landscapes) (Figure 1).

Theoretically, these legend units—which Neef (1967) called ecotope, microchore, mesochore, macrochore, megachore, and so on—could be described by an abstract typology based on ecotopes. Vinogradov (in I. Zonneveld, 1979) gave an example of this (see also Pedroli, 1983), but in practice we use the typologies of the land attributes—which in most cases is very satisfactory.

Landscape Chorological Studies (Spatial Patterns and Variation)

Chorological studies require mapping, which is the description of patterns. The resulting maps can be used at least as a basis for studying the chorological relationships. These in turn can be used for a better understanding of landscape functions or for management and land use planning and conservation. An indispensible tool for this is remote sensing, varying from systematic reconnaissance flights and direct observation from small aircraft, via aerial photography, to electromagnetic sensing of light and heat, or radar, using aircraft or satellites as platforms (see Naveh and Lieberman [1984] for a short summary of some remote sensing literature).

The comprehensive overall perspective of remote sensing has considerably stimulated the holistic view of the landscape, as had mapping before it. It should be mentioned here that for mapping—even of single attributes or single values of attributes, especially on a reconnaissance scale—aerial photographs and even satellite remote-sensing images are essential and efficient tools (see I. Zonneveld, 1979; I. Zonneveld in Küchler and Zonneveld, 1988).

A most important aspect of chorological description in the context of relationships is the study of the configuration of the structure of land elements, patches, and corridors, in relation to the flow of information through the landscape (in minerals, animals, plant diaspores, etc). The studies of Forman and Godron (1986; see also Forman, 1982; Merriam, 1984 and chapter 8; Vos et al., 1982; Risser et al., 1984; Risser, chapter 4.) demonstrate important perspectives for understanding landscape function, management, and planning. The important concept of connectivity in this context has been mentioned above.

Landscape Change and Relationships

The study of landscape change is an important component of landscape ecology, as it has been for geomorphologists, soil scientists, and vegetation scientists. The practical questions put to landscape ecologists almost always deal with either stimulating or preventing change—i.e., conservation. Studies of landscape change and landscape relationships are narrowly related. An interesting compilation of graphic representations of relationship models is given in J. Zonneveld (1985b). The fields of study mentioned are mainly descriptive in character, but experimental research is also possible (e.g., Merriam, chapter 8). Landscape architecture in its various forms, from agricultural engineering to environmental care and building esthetics, is in fact often an experiment on a scale of one-to-one. If monitored, this will provide landscape-ecological knowledge about the functioning of the landscape and the character of landscape change.

Neef (1967) distinguished three dimensions in which such studies are carried out (Figure 3).

1. Study in the *topologic dimension** emphasizes ecological understanding of a relatively homogeneous area (ecotope), and involves cooperation among the traditional disciplines of soil science, geomorphology, vegetation science, zoology, and so on. The difference from other ecological studies is that biology is not the dominant discipline, but is of equal importance with the others (Figure 3). Topological study is the basis for understanding the chorological relationships and also most important in land evaluation applications.
2. In the *chorological dimension,* we study especially the landscape-ecological relationships among the ecotopes of a mosaic, without neglecting topological knowledge. The combination of topological and chorological studies is the most characteristic aspect of landscape ecology (Figure 3).
3. Finally, in the *geospheric dimension,* we study the main lines of relationships on continental and global scales. These tend to overlap the general ecological studies of the Gaia type (see Lovelock, 1979) but, again, with equal attention to all attributes and not just mainly human life as the subject (Figure 3).

An enormous impulse for studies of landscape ecology since the advent of remote sensing is the development of sophisticated calculation machines that are

*Not to be confused with the mathematical field of topology.

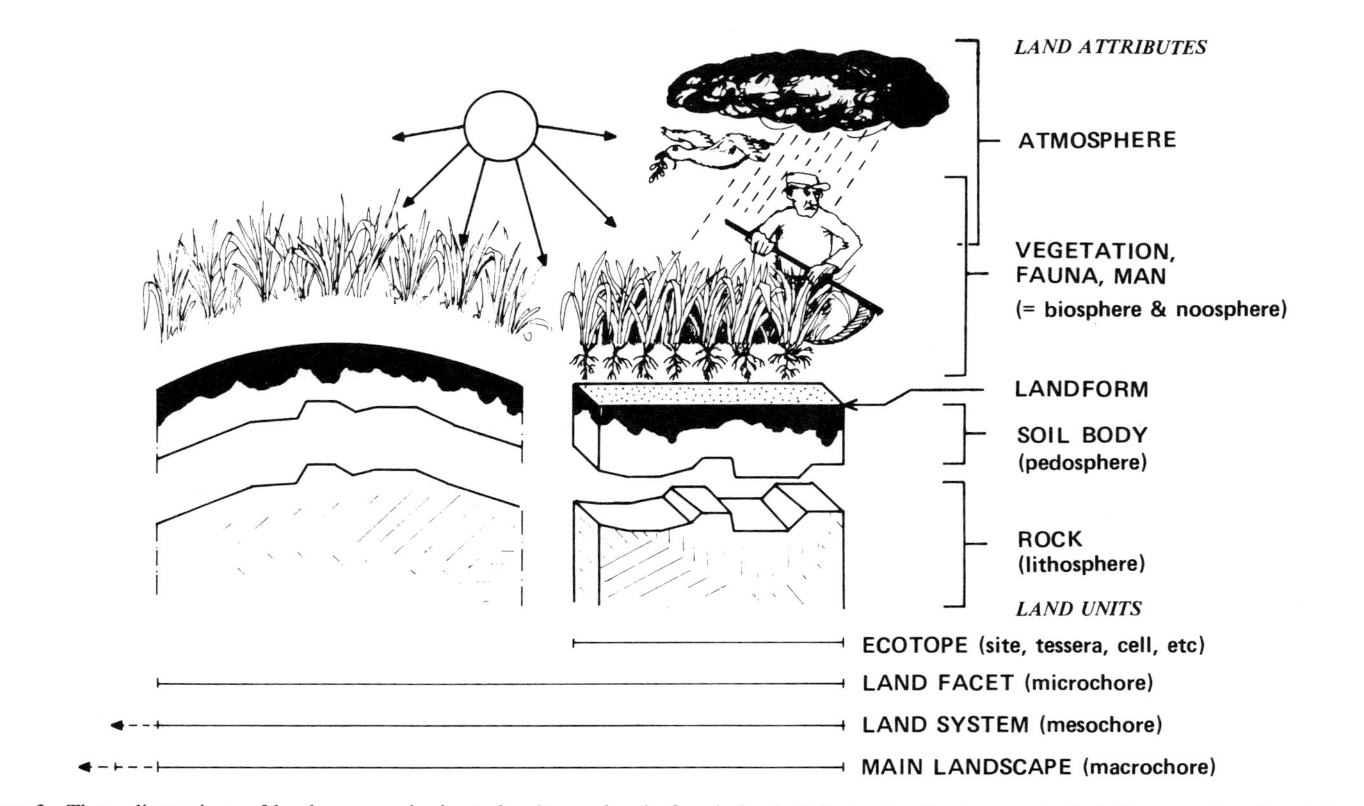

Figure 3. Three dimensions of landscape ecologic study. At any level of scale land attributes (vertical or topological dimension) and land elements (horizontal or chonologic dimension) play important roles, in addition to the specific landscape characteristics under study.

able to handle intricate models of reality, in combination with computerized geographic information systems (see Burrough, 1986). Provided that we do not forget that reality consists mostly of unmeasured operational factors, and that only a few conditional factors can be used as parameters, these models can bring us much farther than can merely qualitative philosophies about wholes and relationships.

Cooperation between applied and more fundamental research in the future must enlarge our knowledge about landscape functioning and about the crucial questions for the maintenance of all our environments—questions of homeostasis and homeorhesis. These crucial themes are too complex to be studied by the classical single-attribute sciences. Only by transdisciplinary research conducted by new generations of scientists—trained to think in wholes rather than compartments divided by walls of envy, financial competition, and narrowmindedness—can these questions be answered.

The emerging science of landscape ecology is a tool for such studies and will be the cradle for advanced studies in the future.

Acknowledgments

I am very grateful to Ann Stewart who helped so much in compiling this paper, to Gert Lutke Schipholt and Benno Masselink for their artistic contributions, to Richard Forman for his useful comments and to Anny Lefers for typing the manuscript.

References

Bartkowski, T. 1984. Introduction to the workshop landscape ecological concepts and the relation between landscape ecology and landscape perception. In *Proceedings of the first international seminar on methodology in landscape ecological research and planning*. Roskilde University Center, Denmark, Vol. 1, pp. 109–16

Beek, K. J. 1978. *Land evaluation for agriculture development*. ILRI Publication Number 3. Wageningen, The Netherlands.

Brandt, J. and P. Agger, eds. 1984. *Proceedings of the first international seminar on methodology in landscape ecological research and planning*. 5 vol. Roskilde University Center, Denmark.

Brunt, M. 1967. The methods employed by the Directorate of Overseas Surveys in the assessment of land resources. Etude de Synthèse VI, Proceedings of the DOS Symposium. *International Archives of Photogrammetry* 16:2–10.

Burrough, P. A. 1986. *Principles of geographic information systems for land resources assessment*. Clarendon Press. Oxford.

Capra, F. 1982. The turning point—science, society and the rising culture. Fontana, London.

Christian, C. S. and G. A. Stewart. 1964. Methodology of integrates surveys. *Proceedings of the UNESCO conference on principles and methods of integrated aerial surveys of natural resources for potential development. UNESCO Recherches sur les Resources Naturelles* 6:233–80.

Cramer, S., M. Kuiper, and C. Vos. 1984. Landschapsecologie: Een nieuwe onderzoeksrichting? *Landschap* 1:176–83

Doing, H. 1974. *Landschapsecologie van de duinstreek tussen Wassenaar en IJmuiden [Landscape ecology of the sand dune areas between Wassenaar and IJmuiden (Netherlands)]*. Med LH, Wageningen, The Netherlands.

FAO. 1976. *A framework for land evaluation*. Soils Bulletin 32. FAO, Rome.

Forman, R. T. T. 1982. Interaction among landscape elements: A core of landscape ecology. In S. P. Tjallingii and A. A. de Veer, eds., *Perspectives in landscape ecology, Proceedings of the International Congress of the Netherlands Society of landscape ecology*. PUDOC, Wageningen, The Netherlands. pp. 35–48.
Forman, R. T. T. and M. Godron. 1986. *Landscape ecology*. Wiley, New York.
Haase, G. 1984. The development of a common methodology of inventory and survey in landscape ecology. In J. Brandt and P. Agger, eds., *Proceedings of the first international seminar on methodology in landscape ecological research and planning*. Roskilde University Center, Denmark, Vol. 5, pp. 68–107.
Küchler, A. W., and I. S. Zonneveld, eds. 1988. *Handbook of vegetation science and vegetation mapping*. Junk, The Hague.
Kwakernaak, C. 1982. Landscape ecology of a prealpine area. A contribution to the unifying concept of landscape ecology based on investigations in the la Berra–Schwarzsee area (Fribourg, Switzerland). Ph. D. Dissertation, University of Amsterdam.
Kwakernaak, C. 1986. Informatie als begrip in de landschapsecologie (Information as a concept in landscape ecology). *Landschap* 3:182–9, 248.
Leser, H. 1984. Zum Oekologie, Oekosystem und Oekotopbegriff. *Natur und Landschaft* 59(9):301–7.
Lovelock, J. E. 1979. *Gaia, a new look at life on earth*. (Dutch ed.) Bruna, Utrecht/Antwerp, The Netherlands.
MacArthur, R. H. and E. O. Wilson. 1967. *The theory of island biogeography*. Princeton University Press, NJ.
Merriam, G. 1984. Connectivity: A fundamental ecological characteristic of landscape pattern. In J. Brandt and P. Agger, eds., *Proceedings of the first international seminar on methodology in landscape ecological research and planning*. Roskilde University Center, Denmark, Vol. 1, pp. 5–17.
Naveh, Z. and A. S. Lieberman. 1984. *Landscape ecology: Theory and application*. Springer-Verlag, New York.
Neef, E. 1967. *Die theoretischen Grundlagen der Landschaftslehre*. Haack, Gotha/Leipzig.
Pedroli, B. 1983. Landscape concept and landscape and rangeland surveys in the Soviet Union. *ITC Journal* 1983–1984:307–21.
Perry, R. A. 1967. Integrated surveys of pastoral areas. In *Integrated survey of natural grazing areas*. ITC/UNESCO Center for Integrated Surveys, Publication S20, Delft, The Netherlands.
Phipps, M. 1982. Information theory and landscape analysis. *Proc. int. congr. neth. soc. landscape ecol., Veldhoven*. PUDOC, Wageningen, The Netherlands, pp. 57–64.
Poore, M. E. D. and V. C. Robertson. 1964. *An approach to the rapid description and mapping of biological habitats*. The Nature Conservancy, London.
Rickler, H. and G. Schönfelder, eds. 1986. Landscape synthesis. Kongress und tagungsbericht. M. Luther Universität, Halle-Wittenberg, German Democratic Republic.
Risser, P. G., J. R. Karr, and R. T. T. Forman. 1984. *Landscape ecology:—directions and approaches*. Illinois Natural History Survey Special Publication Number 2, Champaign, Illinois.
Rowe, S. 1988. Landscape ecology: the study of terrain in ecosystems. In R. Moss, ed. *Landscape ecology and Management*. Polyscience, Montreal.
Ruzicka, M. and L. Miklos. 1982. Methodology of ecological landscape evaluation for optimal development of territory. In S. P. Tjallingii and A. A. de Veer, eds., *Perspectives in landscape ecology, Proceedings of the international congress Netherlands society of landscape ecology*. PUDOC, Wageningen, The Netherlands, pp. 99–109.
Schmithüsen, J. 1963. Der wissenschaftliche Landschaftsbegriff. *Mitt. der floristisch-soziologischen Arbeitsgemeinschaft* NF 10:9–19.
Schmithüsen, J. 1974. *Landschaft und vegetation. Gesammelte Aufsätze von 1934 bis 1971*. Georgr. Inst. Univ. Saarlandes, Saarbrücken, German Democratic Republic.

Siderius, W., ed. 1984. *Proceedings of the workshop on land evaluation for extensive grazing,* Addis Ababa. ILRI, Wageningen, The Netherlands.
Slovak Academy of Sciences. 1985. *Proceedings seventh international symposium on problems of landscape ecological research.* Pezinok, Bratislava, Czechoslovakia.
Smuts, J. C. 1926. *Holism and evolution.* MacMillan, London.
Teilhard de Chardin, P. 1955. *Le phénomène humain.* Editions du Seuil, Paris.
Theorie Werkgroep WLO. 1986. Methodes der begrippen in de landschapsecologie. Netherlands Society for Landscape Ecology. *Landschap* 3:172–81, 248.
Thie, J. and G. Ironside, eds. 1976. *Ecological (biophysical) land classification in Canada.* Ecological Land Classification Series Number 1. Lands Directorate, Environment Canada, Ottawa, Canada.
Tjallingii, S. P. and A. A. de Veer, eds. 1982. *Perspectives in landscape ecology, Proceedings of the international congress of the Netherlands Society for landscape ecology.* PUDOC, Wageningen, The Netherlands.
Troll, C. 1950. *Die geografische Landschaft und ihre Forschung.* Studium Generale 3, 45. Springer-Verlag, Berlin. pp. 163–81.
van Leeuwen, C. G. 1982. From ecosystem to ecodevice. In S. P. Tjallingii and A. P. de Veer, eds., *Perspectives in landscape ecology, Proceedings of the international congress of the netherlands society of landscape ecology.* PUDOC, Wageningen, The Netherlands. pp. 29–36
van der Maarel, E. and P. L. Dauvellier. 1978. *Naar een globaal ecologisch model voor de ruimtelijke ontwikkeling van nederland* (GEM, 3 Vols). Summary, General ecological model. General Physical Planning Outline, part 3. Min Ruimt Ordening, The Hague.
Vink, A. P. A. 1980. *Landschapecologie en landgebruik.* Bohn, Scheltema and Holkema, Amsterdam.
Vink, A. P. A. 1982. Anthropocentric landscape ecology in rural areas. In S. P. Tjallingii and A. F. de Veer, eds., *Perspectives in landscape ecology, Proceedings of the international congress of the netherlands society for landscape ecology.* PUDOC, Wageningen, The Netherlands: pp. 87–98.
Vos, W., W. B. Harms, and A. H. F. S. Storfelder, eds. 1982. *Voor onderzoek naar landschap ecologische relaties tussen ecosystemen.* Rijksinstituut voor Onderzoek in Bos- en Landschapsbouw, Rep. 246. De Dorschkamp, Wageningen, The Netherlands.
Zonneveld, I. S. 1979. *Land evaluation and land(scape) science.* ITC Textbook VII.4, 2nd ed. ITC, Enschede, The Netherlands.
Zonneveld, I. S. 1982. Land(scape) ecology, a science or a state of mind. In S. P. Tjallingii and A. f. de Veer, eds. *Perspectives in landscape ecology, Proceedings of the international congress of the Netherlands society of landscape ecology.* PUDOC, Wageningen, The Netherlands, pp. 9–15.
Zonneveld, I. S. 1984. Landschapsbeeld en landschapsecologie (Landscape physiognomy and landscape ecology). *Landschap* 1:5–9, 72.
Zonneveld, I. S. 1985. Conclusions and outlook of the 1st IALE seminar 1984, Roskilde. *IALE Bulletin.* 3:7–18.
Zonneveld, I. S. 1986. A systematic approach to the evaluation of rangeland inventory data. In *Rangelands, a resource under siege, Proceedings of the 2nd international rangeland congress,* Australian Academy of Science, Canberra, pp. 515–16.
Zonneveld, J. I. S. 1981. The ecological backdrop of regional geography. In S. P. Tjallingii and A. f. de Veer, eds., *Perspectives in landscape ecology, Proceedings of the international congress of the Netherlands society of landscape ecology,* Veldhoven. PUDOC, Wageningen, The Netherlands. pp. 67–9
Zonneveld, J. I. S. 1985a. The landscape, our environment. *Geograficky Casopis, Roch'k* 37:277–86.
Zonneveld, J. I. S. 1985b. *A collection of graphical models used in landscape ecology.* Department of Geography, State University of Utrecht, The Netherlands.

2. The History of Landscape Ecology in Europe

Karl-Friedrich Schreiber

From Ecology to Ecosystem Research

It was quite obviously an expression of the bioscientific currents of his time when, in January 1869, the young Ernst Haeckel used the term *Oecologie* during a lecture on the developments and tasks of zoology. At the beginning of the century A. von Humboldt (1808) had already given these currents new impulses. Haeckel meant a new understanding of a field of work developing in biology, which considered the organism not only on its own, but also as embedded in a relationship between the nonliving (abiotic) and living (biotic) components of the environment (Haeckel, 1870). Some years earlier, in the treatise "General Morphology of Organisms" he had already defined the term *Oecologie*, which he linked exclusively with research into life-space interactions governed by natural laws, as the *Naturhaushaltslehre* (science of the nature household) (Haeckel, 1866).

It should not be surprising, therefore, if the term Oekologie or oecology is to be found in the scientific correspondence of that period. For example, Oehser (1959, based on Harding and Bode, 1958) presumed he had found the first use of the word in the work of the nineteenth-century naturalist and writer, Henry David Thoreau. However, Harding (1965) later corrected this by pointing out that Thoreau's handwritten word was undoubtedly "geology," not "ecology."

More than ten years earlier, Sendtner (1854) and de Candolle (1855) had already analysed the environmental factors influencing the incidence and growth of plants. They used the almost synonymous terms *Standörtlichkeit* and *station,* respectively, for the entirety of the environmental influences. Later, at the International Botanical Congress, Flahault and Schröter (1910) proposed the terms *Standort-station-habitat* meaning a cause-effect complex of climatic, edaphic, and biotic environmental factors. In the English literature, as a rule the term *site* is used. In zoology (Haeckel, 1866; Möbius, 1877) and especially in botany, the ecological approach to the investigation of the arrangement and distribution of plants and animals became more and more generally accepted (e.g., Grisebach 1838; 1872; 1884). Ellenberg (1980) pointed out the exemplary contributions of Grisebach as a vegetation ecologist, which are still of interest today. In addition, Schimper (1898), Warming (1896), and Clements (1905) were also key contributors to the developing science of ecology at this time. On both a local and global scale they dealt with the conditionality and interdependence of living communities (*Biocoenose* according to Möbius, 1877) and site conditions, that is the nonliving environmental conditions. From a zoological point of view, Dahl (1908) called these environmental conditions *Biotop*. This is nearly identical to the *Standort* (site, habitat) of plants as far as the contents are concerned, but does not always include the spatial delimitation implied by *Standort*. The link between living communities and site conditions is especially relevant for animals, because the influence of vegetation structure can hardly be separated from the influence of abiotic factors.

In the first half of the twentieth century, one began to consider this relationship between living communities and environment as a compound system. Woltereck (1928, cf. Müller, 1981) speaks of ecological systems and of ecological *configuration systems (Gestalt-Systeme)*. In doing so, he understands Gestalt as the result of states of equilibrium which remain constant in their integral lawfulness *(Ganzheitsgesetzlichkeit)* year in and year out. This gives an internal and external configuration to the system. Only a few years later, Tansley (1935) examined the different vegetational concepts and terms of his time, and created the term *ecosystem,* which is still in use today. He meant by it:

> the whole system (in the sense of physics), including not only the organism-complex, but also the whole complex of physical factors forming what we call environment of the biome—the habitat factors in the widest sense . . . It is the systems so formed which, from the point of view of the ecologist, are the basic units of nature on the surface of the earth . . . Our natural human prejudices force us to consider the organism (in the sense of the biologist) as the most important part of these systems, but certainly the inorganic "factors" are also parts—there could be no system without them, and there is constant interchange of the most various kinds within each system, not only between the organism but between the organic and inorganic . . . In an ecosystem the organism and the inorganic factors . . . are components, which are in relatively stable dynamic equilibrium.

This forms a direct relation between Woltereck the zoologist and Tansley the botanist. In the year 1939 the latter stated: "The ecosystem consists of both the organic and inorganic components, which may be conveniently grouped under the heads of climate, physiography and soil, animals and plants." Ellenberg (1973a) added to this definition the following idea, one which Woltereck had already started to work on: "An ecosystem is a cause-effect network constituted by organisms and their inorganic environment, although an open system, with the ability to regulate itself to a certain degree" (translation).

Up until the 1970s, ecology was regarded as a subdiscipline of biology (cf. for example Odum, 1971; Stugren, 1978). More recently however, (cf. Ellenberg, 1977; Ellenberg et al., 1986; and others) ecosystem research has given a new dimension to ecology by the intensified and integrated study of abiotic processes, specifically material cycles and energy flows. Thus, ecology is firmly established both in the biological sciences and in the geosciences (Figure 1).

In the intricate cause-effect network between living communities and the abiotic environment it is possible only in theory to draw a clear-cut line between biological and geoscientific research. The merging of biotic processes with abiotic processes in the metabolism of organic substances in the soil is an eloquent example of this fuzziness. This specifically applies to the studies of complete material cycles and energy flows, in which the substance under consideration—or the energy accumulated in it—often has to change form during transformation from the living to the dead state (i.e., from biotic to abiotic compartments in the ecosystem). Thus, ecology has emerged from biology and has become a new inter-discipline, combining parts of biology as well as parts of the different geosciences.

Carl Troll's Landscape Ecology

In the late 1930s, the fascination and possibilities of aerial photo interpretation led the geographer and ecologist Carl Troll to form the concept of landscape ecology. Inspired especially by the papers of Robbins (1934), he writes: "From completely different sides, from the science of forest vegetation and biological aerial photo interpretation and from geography as 'landscape science' and 'ecology', all the methods of natural science meet here" (Troll, 1939 [translation]). Elsewhere he writes: "Aerial photo research is to a great extent landscape ecology, even if it is used, for instance, for archaeology or soil science. In reality it is the consideration of the geographical landscape and of the ecological cause-effect network in the landscape" (Troll, 1968a). At the International Association for Vegetation Science meeting in 1963 he defined *landscape ecology* in accordance with Tansley's concept of the ecosystem as follows:

> Landschafts-ökologie (ist) das Studium des gesamten, in einem bestimmten Landschaftsausschnitt herrschenden komplexen Wirkungsgefüges zwischen den Lebensgemeinschaften (Biozönosen) und ihren Umweltbedingungen. Dieses äussert sich räumlich in einem bestimmten Verbreitungsmuster oder einer naturräumlichen Gliederung verschiedener Grössenordnungen. [Landscape ecology (is) the study of the entire complex cause-effect network between the living communities (biocoenoses)

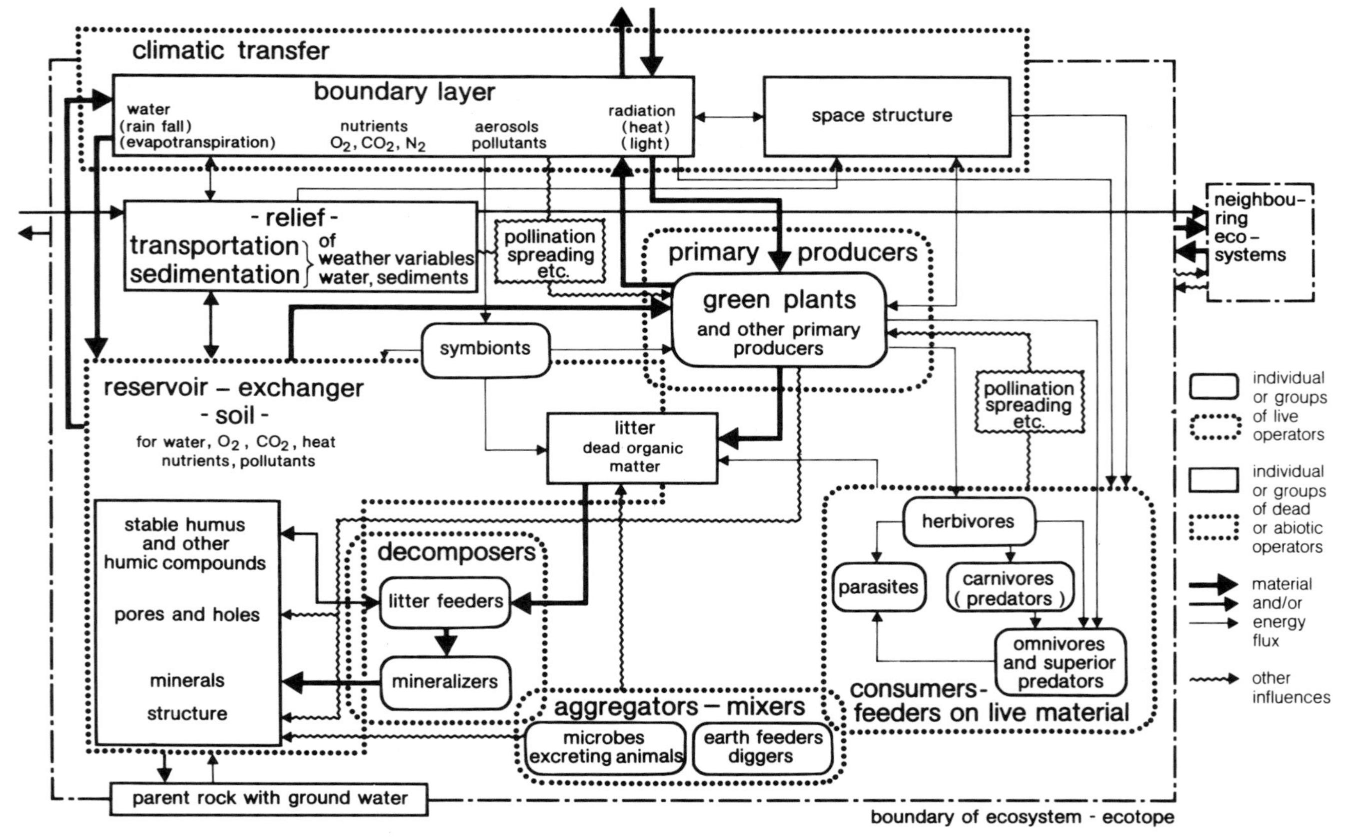

Figure 1. Functional diagram of a terrestrial ecosystem (after Ellenberg, 1978), but placing equal emphasis on the abiotic and biotic compartments, which are closely linked in a functional network.

and their environmental conditions which prevails in specific section of the landscape. This becomes apparent in a specific landscape pattern or in a natural space classification of different orders of size.] (Troll, 1968a).

For the smallest landscape units, which are the result of this ecosystematic consideration of the landscape, Troll (1966) used the term ecotope *(Ökotop)*. In a more recent terminological study Troll remarks that he had overlooked the fact that Tansley (1939) had already used the word *ecotope* (Troll, 1970). Troll wished to use the term ecotope, in the sense of the geographers' topographical dimension, for the smallest spatial unit only. This is in contrast to Tansley (1939) who characterized the ecotope as the whole complex of organisms and factors of environment in an "ecological unit of any rank." Ellenberg (1973b) took this into account by differentiating between nano-, micro-, meso-, and macro-ecosystems. Finally, Troll (1966, 1968a) indicated that landscape ecology at that point was not a new science, but merely a special viewpoint for understanding complex natural phenomena.

At about the same time, the Russian forest botanist Sukachew (1944; 1945; 1949 [quoted in Troll, 1970]; 1953) developed the concept of *biogeocoenology* in the Soviet Union (Sukachew and Dylis 1964). This is, also in Troll's opinion (1970), identical to his own approach to landscape ecology. *Biogeocoenose,* a term subsequently used by European ecologists, too (e.g., Walter, 1976), is the smallest, ecological, indivisible spatial unit with a specific biotic community, and corresponds to the ecotope as a spatial representation of ecosystems (Troll, 1970).

Landscape Ecology and Ecosystem Research: Two Sides of a Coin

As can be seen above, in both ecosystem research and landscape ecology the ecosystem is the main object of research (Figure 2). Ecosystem research is mainly concerned with material cycles and energy flows within an ecosystem, as well as with the roles of internal compartments in affecting turnover and processes. In contrast, landscape ecology mainly focuses on the spatial dimensions, regularities of arrangement, distribution, and contents of ecosystems in a landscape sector, and the roles of spatial configuration in affecting function, (i.e., fluxes, interactions, and changes). In essence, on the one hand there is an intensive but mainly vertical research of processes in an ecosystem without a clear spatial delimitation, and on the other hand, a mostly horizontal, boundary-crossing spatial research in and between ecosystems. This is naturally only in theory, and without anyone being obliged to draw clear-cut lines between "horizontal" and "vertical" research.

An ecotope is the spatial-temporal manifestation of a specific ecosystem in a natural or cultural landscape. Troll (1950) defined it as an ecological, mostly homogeneous section of the global sphere. The homogeneity refers mainly to the abiotic factors, but even those are subjected to change. These changes or

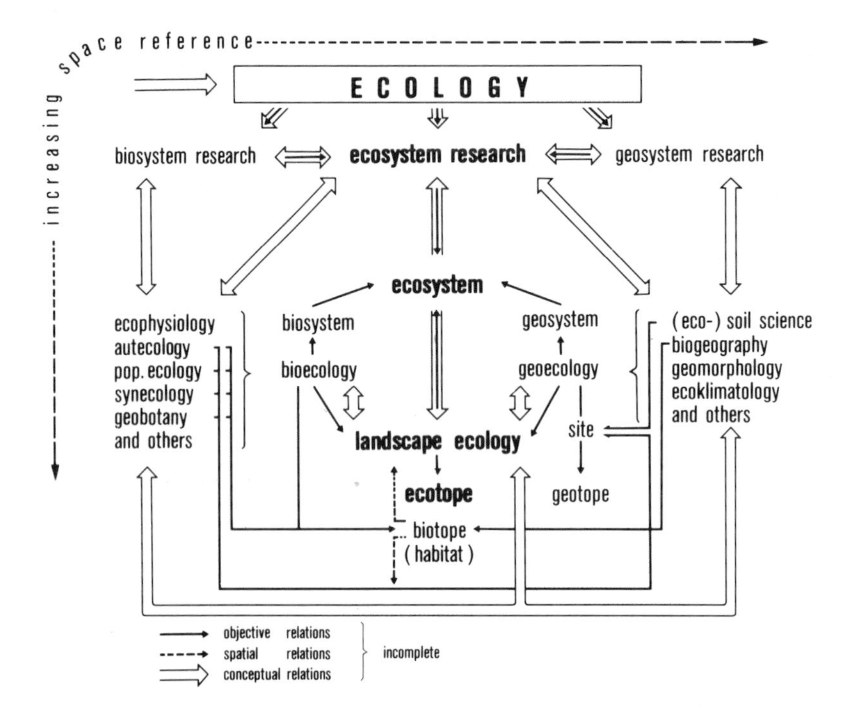

Figure 2. Simplified presentation of the variety of terms that have developed around *landscape ecology* and *geoecology,* mainly in geographical publications in German-speaking countries (modified from Schreiber, 1985). The individual terms are grouped either in the biotic or the abiotic field; Troll's landscape ecology concept combines both. The ecosystem as a research subject common to both ecosystem science and landscape ecology is placed in the center. The ecotope is the spatial, mainly structurally limited manifestation of the ecosystem. The biotope, as the living area of plants and animals (Schmithüsen, 1961), is theoretically more or less identical with the ecotope, but with a highly variable spatial dimension according to the species or group of species considered (e.g., a biotope of birds of prey includes several ecotopes). Altering the present meaning of the term *biotope* to regard it as a spatial expression of the biosystem (Leser, 1984) merely for the sake of symmetry (Haase, 1967; Figure 1) does not seem appropriate. For plants and animals, specific biological structures are often no more than abiotic elements, and therefore can be considered as equivalent to abiotic compartments (e.g., a tree's bark as substrate for epiphytic lichens). The term *biosystem* is generally understood as a plant or animal organism or part thereof, rather than a community of organisms or the entirety of the biotic compartments and resulting cause-and-effect network. Thus a biosystem concept proposed as essentially a counterpart to geosystem (Leser, 1981; 1984) has not been used here, nor the combination of terms such as *geo-ecosystem,* since the geosystem is merely a subsystem of the ecosystem. Note that Stugren (1978) considers animal interrelations, e.g., trophodynamic relations, as a biosystem, and Thienemann (1941) uses it in the sense of the term ecosystem today. Also compare Schramm's (1985) contribution to the discussion.

variations often occur in small spatial dimensions, especially if one considers soil conditions. The demand for homogeneity in an ecotope, just as for the term *site,* is more a convention based on theory, and can at best mean quasi-homogeneity. With reference to ecological conditions in an ecotope we often have gradients with centripetal or centrifugal dynamic processes. In a sloping ecotope we commonly find predominant processes in one direction only, such as air-mass turnover, water and material transport, a catena of different soil conditions, and levels of phytomass production or composition of biotic communities.

Thus, the spatial investigation of ecological relationships and gradients with reference to the landscape mosaic builds on and goes well beyond existing ecosystem research. Troll's concept of landscape ecology may also be understood in this sense. In connecting biological-ecological and geographical-regional methods together with Sukachew's biogeocoenology, Troll sees a comprehensive research into abiotic and biotic systems that deserves the name *ecoscience* (Troll, 1970).

Landscape Ecology and Geoecology

Since the late 1960s Troll (e.g., 1968b; 1970; 1972a,b) has used the term *geoecology* in English texts (and for English usage) as a synonym for landscape ecology, especially in connection with the International Geophysical Union (IGU) Commission for Geography. He assumed that this newly coined word would better correspond to the Anglo-American linguistic instinct and be accepted more easily. However, recent developments in the International Association for Landscape Ecology did not prove him right.

Leser (1981), who has always been committed to the physico-geographic method, which he understood per se as ecological, recommended the pair of words bioecology and geoecology (but see the critique by Hard [1982]). In this way, the close connection between the methods and concepts in ecology and geography intended by Troll's concept of landscape ecology is, in principle, resolved.

In the first edition of his book *Landschaftsökologie,* Leser (1976) focused on the methods and examples of physico-geographic landscape research, with brief coverage of geobotanic and soil science approaches. They had fallen through the grid and sieve, respectively, of the spatial-geographic perspective chosen. Zonneveld's analysis (e.g., 1975) covered the spatial aspect in the Troll tradition of aerial photo interpretation and vegetation research. The use of potential natural vegetation (Tüxen, 1957; Trautmann, 1966) as an ecological aid had also been previously utilized as a classification criterion (Klink, 1964).

Troll (1968a) especially pinpointed Klink's work (1964; 1966), originating in the Czaika school, as being the most thorough and well-balanced ecological study of a space, one that interrelated morphology, climate, soil, and plant communities. In this context Troll also mentioned the work of the Neef school

(Neef et al., 1961; Haase, 1961; 1964; Hubrich, 1964; 1967), with which Leser was associated, in regard to terminology. However, these investigations and mappings, mostly of agricultural land, were more strongly abiotically oriented (Haase, 1967) than the agro- and landscape-ecological studies carried out by the Ellenberg school (e.g., Ellenberg and Zeller, 1951; Ellenberg et al., 1956; Schreiber, 1968; 1969; Weller et al., 1978). The studies by Ellenberg and colleagues also fell through the above-mentioned grid. The *Geoökologische Praktikum* by Späth (1976) encourages the tendency within geography to drift to the purely abiotic field.

This development within the geography of German-speaking countries that I have briefly discussed was not, however, supported by all geographers (cf. among others, Klink, 1975). In fact, at the 1977 meeting of German Geographers, Finke and Fiolka (1978) already regarded geoecology as the abiotic branch, and bioecology as the biotic branch of landscape ecology (Figure 2).

Finally, Leser published an extensive hierarchy of terms in 1984, in which he finally separates geoecology, as a section confined to the abiotic geosystem, from landscape ecology. This present state of terminology is summarized in Schreiber (1985) and in Figure 2.

Back to Troll's Landscape Ecology

The discussion about the congruence of landscape ecology and geoecology will doubtless continue, even though there is no longer a compelling reason to have a synonym for the term *landscape ecology,* especially one so heavily encumbered. Although Troll's concept of landscape ecology is derived from natural landscapes, it can be applied to cultural landscapes and even to urban ones (Finke, 1986).

I have described the concept of landscape ecology, although as yet have said nothing about the development and meaning of the word *landscape*. Many geographers may regard this as a deadly sin. Yet, in the German literature alone, there are about a thousand terminological studies that deal with the linguistic and scientific meaning of landscape. Thus, a discussion of the term seemed to me to be dispensable, as Troll (1966; 1968a) has explained it quite clearly.

Indispensable, however, were supplements to Leser's landscape ecology book (1976), and Naveh and Lieberman's (1984) brief review of European landscape ecology, introduction to holistic study, and examples of applications. Forman and Godron (1986), without doubt more bioecologically oriented, presented a successful handbook especially for teaching at the university, with two main chapters on landscape structure and landscape dynamics as important fields within landscape ecology. These focus on function and process in a spatial mosaic. In addition, the German booklet by Finke (1986) offers a balanced but critical survey of the foundations and applications of landscape ecology.

Theory and Practice in European Landscape Ecology

During the later period, planning practice doubtlessly gave strong impetus to Troll's concept of landscape ecology. This is especially true for The Netherlands, where as early as 1972 a Society of Landscape Ecology had been founded. Zonneveld (1982) noted that the Society consisted of very different groups of scientists and practitioners, who work in and with the landscape. Therefore, it is not surprising that the opinions on what landscape ecology is, or what it should do, vary considerably. They range from the approaches described above to the idea that landscape ecology is conservation or landscape planning. This is probably the result of the natural challenges of this small country, which force it to proceed and plan practically on a rigorous and broad scientific basis (cf. Vink, 1975; 1980; and Burggraaff et al., 1979). Thus, it is no wonder that Zonneveld (1982) writes: "I doubt whether pure holistic ecologists exist in The Netherlands."

In Germany, on the one hand, landscape ecology is tightly tied to planning and landscape architecture. For example, in a compilation of terms, Reichholf (1983) formulates under the key word landscape ecology:

> Ein weites Anwendungsfeld ökologischer Grundprinzipien eröffnet sich für Planer und Gestalter der Landschaft, da sie mit ihren Planungen auf die Vorgänge in der Landschaft Rücksicht nehmen müssen. Die Anwendung der Ökologie in der Praxis der Gestaltung und Nutzung des Raumes bildet daher das Arbeitsfeld der Landschaftsökologie. Sie bezieht in besonderem Masse auch den Menschen in ihre Forschung ein. [A broad field of application of basic ecological principles presents itself to landscape planners and architects, as they are obliged to show consideration for ecological processes in the landscape. The application of ecology in the practical design and use of space, therefore, constitutes the field of work of landscape ecology. It especially includes man in its research.]

On the other hand, the field embraces a range of theoretical and practical problems and methods (cf., the list of topics of the 1981 international congress "Perspectives in Landscape Ecology," [Tjallingii and de Veer, 1982]). At their 1978 annual meeting the Deutsche Gesellschaft für Ökologie (Association of German-speaking Ecologists) attempted to form a link between basic research and application, between theory and practical requirements (Schreiber, 1979) without, however, viewing ecology and landscape ecology, respectively, as planning. The richness of theoretical and applied topics in landscape ecology, and their rapid development, are striking (cf. Ruzicka, 1970; Ruzicka and Miklos, 1982; Brandt and Agger, 1984; Risser et al. 1984; Schreiber, 1988).

Only ecological study and rigorous use of the methods of landscape ecology will form the basis for useful, ecologically founded, practical guidelines for the planner. Although of course, this is not planning as such, it is expected that the rapidly developing body of landscape ecology theory and methods will increasingly provide a foundation for good planning.

Two studies based on this form of applied landscape ecology are representative: (a) *Entwicklungsmöglichkeiten im Andenhochland in ökologischer Sicht* [Potentials of the Andean Highlands from an Ecological Point of View] (Beck and Ellenberg, 1977), and (b) an ecologist's answers to key questions regarding rural development in the tropics and subtropics in *Entwicklung ohne Rückschläge* [Development Without Setbacks] (Ellenberg, 1984). These studies, containing extremely practical information, are based on extensive geobotanical, agroecological, and landscape ecological investigations. They offer the planner the necessary ecological foundation for sound and concrete planning development.

Not only the linkage of ecological and geographical principles, but also basic research and application, constitute the breadth of landscape ecology, and it is this that makes it so attractive.

References

Beck, S. and H. Ellenberg. 1977. *Entwicklungsmöglichkeiten im Andenhochland in ökologischer Sicht*. Lehrstuhl Geobot., Göttingen, Federal Republic of Germany.

Brandt, J. and P. Agger, eds. 1984. *Proceedings of the first international seminar on methodology in landscape ecology research and planning*. Vol. 2. Roskilde University Center, Denmark.

Burggraaff, M., L. van Deijl, G. Laeijendecker, H. A. Meester-Broertjes, and A. H. P. Stumpel. 1979. *Milieukartering, methoden, toepassing en perspectief*. PUDOC, Wageningen, The Netherlands.

Clements, F. T. 1905. *Research methods in ecology*. Lincoln, Nebraska.

Dahl, F. 1908. Grundsätze und Grundbegriffe der biozönologischen Forschung. *Zoologischen Anzeiger* 33:349–53.

de Candolle, A. 1855. *Géographie botanique raisonée*. Paris.

Ellenberg, H., ed. 1973a. *Ökosystemforschung*. Parey, Berlin.

Ellenberg, H. 1973b. Versuch einer klassifikation der ökosysteme nach funktionalen gesichtspunkten. In H. Ellenberg, ed., *Ökosystemforschung*. Parey, Berlin, pp.235–65.

Ellenberg, H. 1978. *Vegetation mitteleuropas mit den alpen in ökologischer sicht*. 2 Auft. Ulmer, Stuttgart, Federal Republic of Germany.

Ellenberg, H. 1980. August grisebach as vegetationsökologie. *Georgia Augusta* 5:15–20.

Ellenberg, H. 1984. *Entwicklung ohne Rückschläge: Antworten eines Ökologen auf 20 Fragen im Hinblick auf die ländliche Entwicklung in den Tropen und Subtropen*. Deutsche Gesellschaft für Technische Zusammenarbeit, Eschborn, Federal Republic of Germany.

Ellenberg, H. and O. Zeller. 1951. Die Pflanzenstandortskarte, am Beispiel des Kreises leonberg. *Forschungs und Sitzungsbericht der Akademie für Raumforschung und Landesplanung* 2:11–49.

Ellenberg, H., K.-F. Schreiber, R. Silbereisen, F. Weller, and F. Winter. 1956. Grundlagen und Methoden der Obstbau-Standortskartierung. *Obstbau* 75:75–7, 90–2, 107–10.

Ellenberg, H., R. Mayer, and J. Schauermann, eds. 1986. *Ökosystemforschung. Ergebnisse des Sollingprojekts 1966–1986*. Ulmer, Stuttgart Federal Republic of Germany.

Finke, L. 1986. *Landschaftsökologie*. Westermann, Braunschweig.

Finke, L. and J. Fiolka, 1978. Ökologische Kriterien für die Verkehrsplanung. *Verhandlungen Deutsche Geographentags* 41:580–93.

Flahault, C. and C. Schröter. 1910. *Phytogeographische Nomenklatur*. Zürich, Switzerland.

Forman, R. T. T. and M. Godron. 1986. *Landscape ecology*. Wiley, New York.
Grisebach, A. 1838. Über den Einfluβ des Klimas auf die Begrenzung der natürlichen floren. *Linnaea* 12:159–200.
Grisebach, A. 1872. Die Vegetation der Erde. Vol. 1 and 2. Wilhelm Engelmann, Leipzig, Germany.
Grisebach, A. 1884. *Die Vegetation der Erde nach ihrer Klimatischen Anordnung*. Vol. 1 and 2. Wilhelm Engelmann, Leipzig, Germany.
Haase, G. 1961. Hanggestaltung und ökologische Differenzierung nach dem Catenaprinzip. *Petermanns Geographische Mitteilungen* 105:1–8.
Haase, G. 1964. Landschaftsökologische Detailuntersuchung und naturräumliche Gliederung. *Petermanns Geographische Mitteilungen* 108:8–30.
Haase, G. 1967. Zur Methodik groβmaβstäbiger landschaftsökologischer und naturräumlicher Erkundung. *Wissenschaftliche Abhandlungen der Geographischen Gesellschaft der* DDR 5:35–128.
Haeckel, E. 1866. *Generelle Morphologie der Organismen*. G. Reimer, Berlin.
Haeckel, E. 1870. Über Entwicklungsgang und Aufgabe der Zoologie. *Jenaische Zeitschrift für Medizin und Naturwissenschaften* 5:352–370.
Hard, G. 1982. *Ökologie/Landschaftsökologie/Geoökologie*. Metzler Handbuch für den Geographieunterricht. J. B. Metzlersche, Stuttgart. pp. 232–6.
Harding, W. 1965. Thoreau and "Ecology": Correction. *Science* 149:707.
Harding, W. and C. Bode. 1958. *The correspondence of Henry David Thoreau*. University Press, New York.
Hubrich, H. 1964. Die Bedeutung äolischer Decken für die ökologische Differenzierung von Sandstandorten in Nordwest-Sachsen. *Petermanns Geographische Mitteilungen* 108: 31–44.
Hubrich, H. 1967. Die landschaftsökologische Catena in reliefarmen Gebieten, dargestellt an Beispielen aus dem nordwestsächsischen Flachland. *Petermanns Geographische Mitteilungen* 118:167–72.
Klink, H.-J. 1964. Landschaftsökologische Studien im südniedersächsichen Bergland. *Erdkunde* 18:267–84.
Klink, H.-J. 1966. *Naturräumliche Gliederung des Ith-Hils-Berglandes: Art und Anordnung der Physiotope und Ökotope*. Forschungen zur Deutschen Landeskunde 159, Bad Godesberg, Federal Republic of Germany.
Klink, H.-J. 1975. *Geoökologie–zielsetzung, Methoden und Beispiele. Verhandlungen der Gesellschaft für* Ökologie, Erlangen 1974. The Hague, pp.211–23.
Leser, H. 1976. *Landschaftsökologie*. UTB 521. 2nd ed., 1978. Ulmer, Stuttgart, Federal Republic of Germany.
Leser, H. 1981. *Fazit zum BGC: "Naturgefahren und geoökologische Hochgebirgsforschung."* Geomethodica 6, Basel, Switzerland. pp. 175–83.
Leser, H. 1984. Zum Ökologie-, Ökosystem- und Ökotopbegriff. *Natur und Landschaft* 59:351–7.
Möbius, K. 1877. *Die Auster und die Austernwirtschaft*. Berlin.
Müller, P. 1981. *Arealsysteme und Biogeographie*. Ulmer, Stuttgart.
Naveh, Z. and A. S. Lieberman. 1984. Landscape ecology. Springer-Verlag, New York.
Neef, E., G. Schmidt, and M. Lauckner. 1961. Landschaftsökologische Untersuchungen an verschiedenen Physiotopen in Nordwestsachsen. *Abhandlungen der Sächsidchen Akademie der Wissenschaften* 47(1).
Odum, E. P. 1971. Ecosystem theory in relation to man. In *Ecosystems: structure and function*. 31st Biological Colloquium, Oregon State University Press, Corvallis.
Oehser, P. H. 1959. The word "Ecology." *Science* 129:992.
Reichholf, J. 1983. Erläuterungen einiger ökologischer Begriffe. In W. Engelhardt, ed., *Ökologie im Bau- und Planungswesen. Wissenschaftliche Verlagsgesellschaft,* Stuttgart, Federal Republic of Germany. pp. 181–6.
Risser, P. G., J. R. Karr, and R. T. T. Forman. 1984. *Landscape ecology—directions*

and approaches. Illinois Natural History Survey Special Publication Number 2, Champaign, Illinois.
Robbins, C. R. 1934. Northern Rhodesia: An experiment in the classification of land with the use of aerial photography. *Journal of Ecology* 22:88–105.
Ruzicka, M., ed. 1970. *Theoretische Probleme der biologischen Landschaftsforschung*. Questiones geobiologicae 7, Bratislava, Czechoslovakia.
Ruzicka, M., and L. Miklos. 1982. Methodology of ecological landscape evaluation for optimal development of territory. In S. P. Tjallingii, and A. A. Veer, eds. *Perspectives in landscape ecology, Proceedings of the international congress of the Netherlands society of landscape ecology*. PUDOC: Wageningen, The Netherlands, pp.99–107.
Schimper, A. F. W. 1898. Planzengeographie auf physiologischer Grundlage. Fischer, Jena, Germany.
Schmithüsen, J. 1961. *Allgemeine Vegetationsgeographie: Lehrbuch der allgemeinen Geographie. IV*. de Gruyter, Berlin.
Schramm, E. 1985. Theoretische und ökologiehistorische Bemerkungen zum Ökosystem- und Biotopbegriff. *Natur und Landschaft* 60:112–14.
Schreiber, K.-F. 1968. *Ecologie appliquée à l'agriculture dans le nord vaudois*. Beitrage zur Geobotanischen Landesaufnature der Schweiz 50 Huber, Bern, Switzerland.
Schreiber, K.-F. 1969. *Landschaftsökologische und standortskundliche untersuchungen im nördlichen Waadtland als Grundlage für die Orts- und Regionalplanung*. Arbeiten der Universität Hohenheim 45. Ulmer, Stuttgart, Federal Republic of Germany.
Schreiber, K.-F., ed. 1979. *Verhandlungen der Gesellschaft für Ökologie*. Verh. GfÖ 7, Münster 1978. Göttingen, Federal Republic of Germany.
Schreiber, K.-F. 1985. *Was leistet die Landschaftsökologie für eine ökologische Planung?* Schriften reihe für Orts-, Regional-und Landesplanung 34. Zürich, Switzerland. pp.7–28.
Schreiber, K.-F., ed. 1988. Connectivity in landscape ecology. *Münstersche Geographische Arbeiten* 29, Münster, Federal Republic of Germany.
Sendtner, O. 1854. *Die Vegetations-verhältnisse Südbayerns*. Munich, Germany.
Späth, H.-J. 1976. *Geoökologisches Praktikum*. UTB 607. Schöningh-verlag, Paderborn, Federal Republic of Germany.
Stugren, B. 1978. *Grundlagen der allgemeinen ökologie*. Fischer, Stuttgart/New York.
Sukachew, V. N. 1944. Principles of genetical classifikation in biogeocoenology (russ). *Zh. obshch. Biol. (USSR)* 6.
Sukachew, V. N. 1945. *Biogeocoenology and phytocoenology*. Readings of the Akademy of Sciences USSR 4:g (russ.)
Sukachew, V. N. 1949. On the relation of the notions of geographical landscape and biogeocoenoses. (russ). *Woprossy Geografii (USSR)* 16.
Sukachew, V. N. 1953. *On the Exploration of the Vegetation of the Soviet Union. Proceedings of the 7th International Botanical Congress. Stockholm, 1950*. Uppsala, Sweden, pp. 659–60.
Sukachew, V. N. and N. Dylis. 1964. *Fundamentals of forest biogeocoenology*. Translated by M. MacLennan. Oliver & Boyd, Edinburgh, Scotland.
Tansley, A. G. 1935. The use and abuse of vegetational concepts and terms. *Ecology* 16:284–307.
Tansley, A. G. 1939. *The British isles and their vegetation*. Cambridge, United Kingdom.
Thienemann, A. 1941. *Leben und Umwelt*. Bios 12. Leipzig, Germany.
Tjallingii, S. P. and A. A. de Veer, eds. 1982. *Perspectives in landscape ecology, Proceedings International Congress Netherlands Society Landscape Ecology*, Veldhoven. PUDOC, Wageningen, The Netherlands.
Trautmann, W. 1966. *Erläuterungen zur Karte der potentiellen natürlichen Vegetation*

der bundesrepublik Deutschland, 1:200 000, Blatt 85 Minden. Schriften reihe für Vegetationskunde 1 Bonn-Bad Godesberg, Federal Republic of Germany.

Troll, C. 1939. Luftbildplan und ökologische Bodenforschung. *Zeitschrift der Gesellschaft für Erdkunde, Berlin*. Pp. 241–98.

Troll, C. 1950. *Die geographische Landschaft und ihre Erforschung*. Studium Generale 3. Heidelberg, German Democratic Republic, pp.163–81.

Troll, C. 1966. *Landscape ecology*. Publication S 4, ITC-UNESCO, Delft, The Netherlands.

Troll, C. 1968a. Landschaftsökologie. In R. Tüxen, ed., *Pflanzensoziologie und Landschaftsökologie. Berichte das Internalen Symposiums der Internationalen Vereinigung für Vegetationskunde*, Stolzenau/Weser 1963. The Hague, pp.1–21.

Troll, C., ed. 1968b. *Geo-ecology of the mountainous regions of the tropical Americas. Proceedings of the UNESCO Mexico Symposium 1966*. Colloquium. Geographicum 9, Bonn, Federal Republic of Germany.

Troll, C. 1970. Landschaftsökologie (Geoecology) und Biogeocoenologie: Eine terminologische Studie. *Revue Roumaine de Géologie, Géophysique et Géographie, Série de Geographíe (Bucarest)* 14:9–18.

Troll, C., ed. 1972a. *Geoecology of the high-mountain regions of Eurasia/ Landschaftsökologie der Hochgebirge Eurasiens*. Erdwissenschaftliche Forschung 4. Steiner, Wiesbaden, Federal Republic of Germany.

Troll, C. 1972b. Geoecology and the world-wide differentiation of high-mountain ecosystems. Landschaftsökologie der Hochgebirge Eurasiens. In C. Troll, ed., *Geoecology of the high-mountain regions of Eurasia*. Erdwissenschaftliche Forschung 4. Steiner, Wiesbaden, Federal Republic of Germany, pp.1–16.

Tüxen, R. 1957. Die heutige potentielle natürliche Vegetation als Gegenstand der Vegetationskartierung. *Berichte zur Deutschen Landeskunde* 19:200–19, 241–6.

Vink, A. P. A. 1975. *Land use in advancing agriculture*. Springer-Verlag, Berlin.

Vink, A. P. A. 1980. *Landschapsecologie en landgebruik*. Bohn, Schelkema & Holkema. Utrecht, The Netherlands.

von Humboldt, A. 1808. Ansichten der Natur mit wissenschaftlichen Erläuterungen. Cotta, Stuttgart, Germany.

Walter, H. 1976. *Die ökologischen systeme der kontinente (Biogeospäre)*. Fischer, Stuttgart, Federal Republic of Germany.

Warming, E. 1896. *Ökologische Pflanzengeographie*. Gebrüder Borutraegen, Berlin.

Weller, F. and R. Silbereisen, with the cooperation of K. -F. Schreiber and F. Winter. 1978. *Erläuterungen zur ökologischen Standorteignungskarte für den Erwerbsobstbau in Baden-Wüttemberg 1:250000*. Herausgeber: Ministerium für Ernährung Landwirtschaft und Umwelt Baden-Württemberg. Stuttgart, Federal Republic of Germany.

Woltereck, R. 1928. Über die Spezifität des Lebensraumes, der Nahrung und der Körperformen bei pelagischen Cladoceren und über "ökologische Gestalt-systeme." *Biologisches Zentralblatt* 28:521–51.

Zonneveld, I. S. 1975. Der zusammenhang zwischen dem horizontalgefüge der vegetation und dem edaphischen zustand in einem savannengebiet Nord-Nigerias. In *Vegetation und substrat. Berichte des Internationalen Symposiums der Internationalen Vereinigung für Vegetationskunde, Rinteln,* 1969. Jung, The Hague. Pp. 511–27.

Zonneveld, I. S. 1982. Land(scape) ecology, a science or a state of mind. In S. P. Tjallingii, and A. A. de Veer, eds., Perspectives in landscape ecology, *Proceedings of the international congress of the Netherlands society of landscape ecology*. PUDOC, Wageningen, The Netherlands, pp.9–15.

3. The Beginnings of Landscape Ecology in America

Richard T.T. Forman

With roots penetrating many disciplines and approaches to landscape study, and with ongoing development in Europe, landscape ecology has emerged recently and generated considerable interest in North America. The expansion of the field in the United States of America (USA) over six years: has been rapid; has attracted significant numbers of leading ecologists and scholars from allied disciplines; has involved many young researchers and practitioners; has attained a balance of focus between theory and application; has strengthened ecology, geography, landscape architecture, forestry, and other fields by creating recognized subfields; and has created an embryonic discipline with its own body of theory and application, and with people calling themselves landscape ecologists.

The objective of this chapter is to encapsulate the recent history of the field, with a focus on time, people, subjects, and trends. The approach is mainly detailed and statistical, rather than interpretive, as perhaps appropriate for a recent or current history. Such an overview is necessarily incomplete, but hopefully useful in gauging the beginnings or resurgence of a field of knowledge.

I will introduce in sequence the origins and international linkages, American landscape ecology meetings, role of the Ecological Society of America meetings, parallel events, role of landscape ecology in allied disciplines, and conclusion.

Origins and International Linkages

The concept of landscape ecology, as focusing on the spatial relationships, fluxes and changes in energy, materials and species across large land mosaics, apparently had its first major impact on five American participants (F. B. Golley, J. G. Fabos, R. Forman, S. A. Carlson, and D. M. Sharpe) at a 1981 meeting of the 600-member Netherlands Society of Landscape Ecology in Veldhoven (Tjallingii and de Veer, 1982). North American activity in the field, however, basically began with a 1983 National Science Foundation (NSF) funded Workshop on Landscape Ecology. Organized by P. G. Risser, J. R. Karr and R. Forman, 25 ecologists and scholars in several allied disciplines, including NSF directors G. W. Barrett and R. G. Woodmansee, met at Allerton Park, Illinois. Unusually diverse perspectives on the subject were expressed, but gradually initial guidelines in an enormous uncharted frontier developed (Risser et al., 1984).

An increase in Americans attending meetings in Europe followed establishment of the International Association for Landscape Ecology (IALE) in 1982. This organization was formed at the sixth international meeting on Landscape Ecological Research, hosted by M. Ruzicka in Piestany, Czechoslovakia (CSSR) (Ruzicka, 1982; Preobrazhensky, 1984). R. Forman was elected a Vice President and F. B. Golley the U.S.A. Representative for IALE.

Seven Americans participated in the 1984 International Seminar on Methodology of Ecological Research and Planning in Roskilde, Denmark, hosted by J. Brandt and P. Agger (Brandt and Agger, 1984). About ten attended the 1987 Landscape Ecology Symposium on Connectivity held in Münster, Federal Republic of Germany, hosted by K.-F. Schreiber (Schreiber, 1988). Four attended the seventh international meeting on Landscape Ecology and Planning in Czechoslovakia in 1988, at which G. Merriam was elected President and M. J. McDonnell Treasurer of IALE.

Thus, for Americans, the meetings in Western and Eastern Europe have provided an inspiration and impetus, an understanding of the roots of the field, and personal contact with both leaders of the field and younger scholars opening new areas of thought.

American Landscape Ecology Meetings

Nearly 100 participants attended the first North American meeting for landscape ecology, held in January 1986 at the University of Georgia and hosted by M. G. Turner and F. B. Golley (Turner, 1987). Plenary speakers were D. H. Knight, P. G. Risser, E. P. Odum, D. Morrison, F. H. Bormann, and R. Forman.

An American Section (Region) of IALE (IALE-US) was established, with D. M. Sharpe elected Chairman, M. G. Turner, Program Chairman, and W. H. Romme, J. I. Nassauer, P. G. Risser, and J. F. Franklin, Council Members.

A year later in March 1987 the second annual meeting, with nearly 200 participants attending, was held at the University of Virginia, hosted by H. H. Shugart and W. E. Odum. Plenary speakers were J. F. Franklin, E. H. Zube, Z. Naveh, H. H. Shugart, and F. B. Golley. J. A. Wiens was elected to the IALE-US Council. The third annual meeting, held in March 1988 at the University of New Mexico and hosted by B. T. Milne, attracted about 250 participants. Plenary speakers were C. Steinitz, H. R. Delcourt, F. Hall, F. Steiner, and P. G. Risser. G. W. Barrett was elected Chairman of IALE-US and V. Meentemeyer, Council Member. The fourth annual meeting, held in March 1989 at Colorado State University and hosted by I. C. Burke, attracted about 250 participants. Plenary speakers were S. J. McNaughton, J. K. Berry, R. A. Pielke, A. W. Spirn, and J. A. Wiens. Elected to the IALE-US Council were J. F. Thorne, T. R. Crow, B. T. Milne, and R. V. O'Neill.

Some of the regular contributers at these meetings are J. F. Ahern, C. D. Allen, T. H. Allen, J. C. Billing, R. Costanza, V. H. Dale, L. Fahrig, R. Forman, K. E. Freemark, R. H. Gardner, F. B. Golley, F. G. Hall, W. C. Johnson, C. A. Johnston, A. W. King, D. A. Kovacic, B. T. Milne, J. I. Nassauer, R. Park, D. M. Sharpe, G. R. Shaver, J. F. Thorne, M. G. Turner, D. Urban, and M. D. Walker.

In essence, the American landscape ecology meetings have provided a smaller and more interactive group than in the traditional large disciplinary meeting. With their ecumenical approach, the meeting have catalyzed a linkage within a unique set of disciplines, and have provided a ready forum for presenting polished, as well as in-progress, results in an emerging field.

Role of the Ecological Society of America Meetings

The annual meetings of the Ecological Society of America (ESA) provided the initial forum for the growth and interest in landscape ecology in the USA. Meetings at Indiana University (1981), Pennsylvania State University (1982), the University of North Dakota (1983), and Colorado State University (1984) contained only individual talks on the subject, e.g., by R. Forman and coauthors M. Godron, L. Goodrich, and B. T. Milne: "Toward a Landscape Ecology: Patches and Structural Components"; "The Peninsula Effect in Projections from Woods in an Agricultural Landscape"; "Corridors in a Landscape: Their Ecological Structure and Function"; and "Landscape Ecology on the Verge of Emergence."

The 1984 ESA meeting offered the first Contributed Papers session, Landscape Ecology and Biogeography, with a leadoff talk by J. A. Wiens, "A Conceptual Framework for Studies of Landscape Ecosystems." The 1985 meeting at the University of Minnesota offered the first Symposium, "Relations between Vegetation and Geomorphology in Terrestrial and Riverine Ecosystems" (C. R. Hupp and F. J. Swanson, organizers).

The concept of landscape ecology first became widely known among North American ecologists in 1986, when ESA met jointly with the International

Congress of Ecology (INTECOL) in Syracuse, New York. A plenary lecture by Z. Naveh and a symposium lecture by R. Forman both described the content, breadth, and boundaries of landscape ecology. The INTECOL meeting also included: one symposium, "Landscape Ecology" (S. M. ten Houte de Lange); two contributed paper sessions, "Regional Vegetation and Landforms" (G. W. Cox presiding) and "Nutrient Cycling and Landscape Ecology" (D. A. Schimel); and two poster sessions.

At the 1987 ESA meeting there were two symposia: "Heterogeneity: Interface between Populations, Communities and Landscapes" (J. Kolasa) and "Changing Patterns of Landscapes" (V. H. Dale and R. H. Gardner). And in 1988 four symposia were presented: "Satellite Remote Sensing for the Analysis of Landscape Properties (H. D. Grover and D. E. Wickland), "Spatial and Temporal Analysis Using Geographical Information Systems" (C. A. Johnston and L. Johnson), "Geomorphic Processes Underlying Landscape Heterogeneity and Vegetation Patterns in North American Deserts" (J. R. McAuliffe and G. L. Cunningham), and "Conservation of Natural Communities in Agricultural Landscapes" (C. Carroll), plus a contributed paper session (B. T. Milne).

In short, the ESA meetings provided the initial opportunity to pinpoint for the wider scientific community this new frontier with both theoretical and practical importance. That it was significantly different from existing work in, e.g., ecosystem science, biogeography, physical geography, and population ecology; that it represented a natural next step and an enhancement of the field of ecology; and that it transcended ecology, especially in often involving human roles, became clear. Diversification and a sifting process, of both subjects and people, began in these meetings.

Parallel Events

New textbooks or handbooks reviewing, synthesizing, and developing the field of landscape ecology appeared in 1984 (Naveh and Lieberman) and 1986 (Forman and Godron). Both texts were coauthored by an American and a scholar from abroad.

The journal, *Landscape Ecology,* was established in 1987 with the support of both IALE and INTECOL. F. B. Golley was appointed as Editor-in-Chief, with P. M. Golley as Editorial Assistant. The Editorial Board included S. Levin, H. H. Shugart, P. G. Risser, R. Forman, C. Steinitz, and E. H. Zube.

The first government agency to create an official unit for landscape ecology in America was the U.S. Forest Service in 1987, with establishment of the program, "Applying Principles of Landscape Ecology to Managing Temperate Forests," directed by T. R. Crow, Rhinelander, Wisconsin.

Next door in Canada, landscape ecology, perhaps building on P. Dansereau's prescience, has also emerged in the 1981 to 1988 period. Among the guiding lights have been G. Merriam, M. Phipps, C. Rubec, J. Thie and D. Moss.

Thus the texts provided timely overviews for people both in and out of the field, and the journal, by attracting leading scholars and high-quality articles from the outset, provided a focused ongoing source and outlet for advances in the field. The government unit underlined the long-term importance of the field in problem-solving of human, environmental, and resource issues.

Role in Allied Disciplines

Meanwhile landscape ecology has been pinpointed or featured at meetings of a range of allied disciplines, including the following:

July 1986: Plenary lecture on landscape ecology, Conference on Science in the National Parks, U.S. National Park Service, Fort Collins, Colorado (R. Forman) (Forman, 1987).

September 1986: Workshop on Array and Parallel Processing in Landscape Dynamics, Pingree Park, Colorado (D. A. Jameson).

April 1987: Two symposia on landscape ecology, Association of American Geographers, Annual Meeting, Portland, Oregon (V. Meentemeyer and R. E. Frenkel). Speakers were E. R. Hobbs, J. F. Franklin, R. A. Roundtree, R. Forman, R. G. Bailey, M. G. Turner, and D. M. Sharpe.

June 1987: Symposium "Ecosystem Faces and Interfaces: Implications of Edges for Animals," Society for Conservation Biology, Initial Meeting, Montana State University (L. D. Harris). Speakers were S. A. Temple, R. H. Yahner, D. M. Waller, R. Forman, and L. D. Harris.

October 1987: Symposium on landscape ecology, American Society of Landscape Architects, Annual Meeting, San Francisco, California. Speakers included G. W. Barrett.

January 1988: Workshop on Landscape History and Ecological Succession, Duke University (N. L. Christensen, J. Richards, and F. B. Golley).

February 1988: Symposium "Landscape Corridors: Structure and Function," American Association for the Advancement of Science, Annual Meeting, Boston, Massachusetts (J. I. Nassauer and G. W. Barrett). Speakers were G. Merriam, J. I. Nassauer, R. E. Chenoweth, R. Forman, and G. W. Barrett.

July 1988: Symposium "Landscape Ecology and Sustainable Development," International Federation of Landscape Architects, Meeting, Boston, Massachusetts (J. F. Ahern). Speakers were P. Jacobs, J. I. Nassauer, J. E. Rodiek, R. Forman, and J. F. Ahern (Proceedings of the IFLA, 1988).

October 1988: Two symposia, Society of American Foresters, National Convention, Rochester, New York. "Landscape Ecology, Biotic Diversity, and Professional Forestry" (R. L. Burgess) included speakers, F. W. Stearns, W. C. Johnson, W. Zipperer, R. Forman, J. F. Franklin, T. R. Crow, and H. Salwasser. "Is Forest Fragmentation a Wildlife and Fish Habitat Issue in the Northeast?" (R. M. DeGraaf) included speakers, D. S. Wilcove, D. E. Capen, H. Brocke, R. W. Hollingsworth, R. Brenneman, R. Forman, and K. E. Freemark (Proceedings of the SAF, 1988; Forman, 1988).

October 1988: Symposium "Landscape Ecology: Policy Implications and Effects," Applied Geographers Conference (J. I. Nassauer and E. R. Hobbs). Speakers were E. R. Hobbs, R. E. Chenoweth, P. J. Gersmehl, J. I. Nassauer, and D. M. Sharpe.

November 1988: Conference on "Predicting Across Scales: Theory Development and Testing," Oak Ridge National Laboratory, Tennessee (R. H. Gardner, M. G. Turner, and V. H. Dale).

December 1988: Workshop on "Landscape Boundaries: Consequences for Biotic Diversity and Ecological Flows," International Council of Scientific Unions, Paris, France (F. Di Castri and A. J. Hansen). Speakers included H. R. Delcourt, P. A. Delcourt, R. Forman, R. H. Gardner, J. R. Gosz, G. Grant, C. A. Johnston, R. P. Neilson, R. L. Peters, G. C. Ray, P. G. Risser, D. Urban, D. Weinstein, and J. A. Wiens.

These symposia, conferences, workshops, and lectures have made landscape ecology broadly known among environmentally related academics and practitioners. The meetings have contributed to institutionalizing the field within a number of disciplines, and have pinpointed leaders or spokespersons in each discipline. The extension of thought within each discipline, as well as the hybrid vigor provided within landscape ecology, are both noteworthy.

Conclusion

A list of key terms might best summarize the core of landscape ecology work in the USA today: heterogeneity, landscape fragmentation, spatial-temporal scale, hierarchy and pattern analysis, indices and fractals, natural and human disturbance, spatial models, patch shapes and configuration, boundaries and edges, corridors and networks, connectivity, riparian zones, material fluxes, changing landscape structure, landforms and geomorphic process, regional mosaic patterns, application in planning and design, linkages with visual quality, application in conservation, application in suburban development, application in wildlife management, application in forestry, and remote sensing and geographic information systems.

Landscape ecology in America is addressing new questions, as well as providing new answers to familiar questions. Both theoretical and applied dimensions are growing. Impressive intellectual resources and a distinct disciplinary mix have joined in a short period to open up this frontier. Published literature and some research programs show a monotonic acceleration in emphasis on landscape-level patterns and processes.

Yet real challenges remain. The theoretical foundation of landscape ecology needs to be strengthened and made more explicit. Key methods and equipment (e.g., geographic information systems, gaseous measurements, experimental alterations in landscape structure, scaling up from fine-scale studies, and spatially explicit models) should be developed and pinpointed. Examples of successful, and failed, applications should be regularly accumulated and published. Funding sources need to increasingly recognize the significance of and support for landscape ecology. The usual competition between a new discipline and traditional disciplines should be overcome with the ecumenism characteristic of this field. Courses in landscape ecology should be taught in every college and university. Linkages with modern European and other approaches in landscape ecology need to be strengthened and vibrant.

Acknowledgments

I am pleased to thank Paul G. Risser, David M. Sharpe, Gary W. Barrett, and Frank B. Golley for useful comments on the manuscript, and Monica G. Turner and Gray Merriam for helpful information. Inevitable omissions of fact and commissions of error are mine, and I hope will be corrected in the future.

References

Brandt, J. and P. Agger, eds. 1984. *Proceedings of the first international seminar on methodology in landscape ecological research and planning*. 5 vols. Roskilde University Center, Denmark, pp. 118, 150, 153, 171, 235.

Forman, R. T. T. 1987. Emerging directions in landscape ecology and applications in natural resource management. In R. Herrmann and T. Bostedt-Craig, eds., *Proceedings of the conference on science in the national parks*. U.S. National Park Service and the George Wright Society: Fort Collins, Colorado, pp. 59–88.

Forman, R. T. T. and M. Godron. 1986. *Landscape ecology*. Wiley, New York.

Forman, R. T. T. 1988. Landscape ecology plans for managing forests. In *Proceedings of the Society of American Foresters, 1988 National Convention*. Society of American Foresters, Bethesda, Maryland. pp. 131–5.

Hardt, R. A. and R. T. T. Forman. In press. Boundary form effects on woody colonization of reclaimed surface mines. *Ecology*.

Naveh, Z. and A. S. Lieberman. 1984. *Landscape ecology: Theory and application*. Springer-Verlag, New York.

Preobrazhensky, V. S. 1984. International symposium on landscape ecology. *Soviet Geography* 6:453–63.

Proceedings of the International Federation of Landscape Architects, 1988 World Congress. In press. American Society of Landscape Architects, Bethesda, Maryland.

Proceedings of the Society of American Foresters 1988 National Convention. 1988. Society of American Foresters, Bethesda, Maryland.

Risser, P. G., Karr, J. R., Forman, R. T. T. 1984. *Landscape ecology. Directions and approaches*. Illinois Natural History Survey Special Publication Number 2. Champaign, Illinois.

Ruzicka, M., ed. 1982. *Proceedings of the sixth international symposium on problems in landscape ecological research*. Institute for Experimental Biology and Ecology, Bratislava, Czechoslovakia.

Schreiber, K.-F., ed. 1988. *Connectivity in landscape ecology. Münstersche Geographische Arbeiten 29, Münster,* Federal Republic of Germany.

Tjallingii, S. P. and A. A. de Veer, eds. 1982. *Perspectives in landscape ecology, Proceedings of the international congress of the Netherlands society for·landscape ecology* PUDOC, Wageningen, The Netherlands.

Turner, M. G., ed. 1987. *Landscape heterogeneity and disturbance*. Springer-Verlag, New York.

Part II Energy, Nutrient, and Species Fluxes in a Mosaic

Landscape level processes, both natural and human, play key roles in forming the spatial structure that we see, use and depend on. In turn, landscape structure exerts a primary control over the processes or fluxes. Thus, the juxtaposition of sources and sinks, the connectivity of landscape elements, and the variable strengths of flows caused by wind, water, animals and people are key function-structure linkages. Part II presents examples of the movements of water and nutrients between landscape units, and the movement of plants, invertebrates and vertebrates through networks. Indeed, functional processes that link spatial units are the basic reason a landscape system differs from the sum of its parts, a holistic tenet of landscape ecology.

A mosaic model of farmland, generally containing a clear patch, corridor and matrix structure, is a simple and useful object for study. Though an agricultural area is produced by landscape fragmentation, the basic model mimics virtually any landscape mosaic, and is used here to gain numerous ecological insights into structure and movements.

Thus the ecological characteristics of a patch are strongly determined by interactions with the nearby surroundings. For example, forest interior bird species in an agricultural landscape were significantly correlated with the amount of wooded area within 1 km of a woods, and also with the density of hedgerows in the surrounding matrix. The population dynamics of small mammals in a patch was strongly related to whether the patch had a single functional connecting corridor or not, to how many other patches were connected in the corridor system, and to the quality of the corridors, especially the presence of narrow, poor-quality linkages. The mosaic model appears to offer more promise of understanding the ecological characteristics of a patch than traditional forest ecology or equilibrium biogeography.

A corridor network system is normally just as prominent as the scattered patches, and connectivity is a key to its understanding. For example, in moist hilly terrain ditches and banks in a hedgerow system address the apparent paradox of both too much and too little drainage in fields. Locomotion driven objects usually move in more circuitous directions based on animal behavior, and the removal or addition of specific links in the network system may alter the plant and animal communities, as well as material fluxes. A cluster of interacting adjacent sites may be tied together by the life history requirements of deer or monkey, for example, or alternatively may be linked as a catena in a topographic sequence.

Structure and processes change, often abruptly, at several key spatial and temporal scales involved in understanding landscapes. A hierarchy of spatial scales introduced is useful in studying the exchange of matter, and links the cell-to-organism hierarchy of biology to the aggregate-to-pedon hierarchy of soil science. The two merge at the site level (the local ecosystem, ecotope, or landscape element) composed of organisms and pedons, followed progressively by scales from the site cluster and landscape to higher levels where the atmosphere is the principal carrier of matter.

Finally, examining the ecology of populations with the mosaic model highlights the key role of spatial heterogeneity. A metapopulation that functions as a demographic unit may have subpopulations in several landscape elements. Here individual animals may move much greater distances than the same species within a single habitat. In landscapes with high connectivity, frequent species extinctions in patches are of minor significance, because of rapid recolonization. A model that predicts avian colonization (based on distance and matrix quality) and avian persistence (based on patch size and amount of wooded area surrounding it) is used innovatively for landscape planning, to evaluate the spatial arrangement and viability of proposed new woods in an agricultural/urban region.

4. Landscape Pattern and Its Effects on Energy and Nutrient Distribution

Paul G. Risser

Landscape ecology is the study of natural and human-influenced processes that operate within heterogeneous geographical areas of the dimension of several to many square kilometers. Thus, in a hierarchical structure landscape processes are recognized between ecosystems and regions. In this context, ecosystems are defined as relatively homogeneous because the internal processes are driven by characteristics of that ecosystem (e.g., a deciduous forest or a grassland). Regions, on the other hand, are very broad in dimension, and as such, involve so much complexity and spatial heterogeneity that simple process studies are not practical. Therefore, in a hierarchical sense, landscape ecology focuses on that crucial level where natural and human-influenced processes are consequences of the heterogeneous landscape composed of contrasting ecosystems, and at the spatial and temporal scale where these processes can be analyzed.

Many of the precursor concepts for landscape ecology have been developing for decades (Brandt and Agger, 1984; Leopold, 1949; Rowe, 1961; Watt, 1947), but within the past five years, the field has crystallized rapidly (Forman and Godron, 1981; 1986; Naveh, 1982; Naveh and Lieberman, 1984; Risser et al., 1984; Risser, 1987). This developmental sequence for the field of landscape ecology has followed the development of the necessary antecedent steps. Specifically, ecosystem proccesses were studied within natural ecosystems with the consequent understanding of processes at various spatial scales up to the watershed level (Lindeman, 1942; Bormann and Likens, 1979). These processes were primarily the movement of nutrients and materials, and the description of patterns of primary productivity. The configurations of patches of vegetation were also recognized as important, especially in determining the population dynamics of animal populations (Burgess and Sharpe, 1981; Forman and Baudry, 1984; Harris, 1984; Romme, 1982; Senft et al., 1987; Szaro and Jakle, 1984; Wegner and Merriam, 1979). Also, the concepts of hierarchy in ecological systems were enunciated (Allen and Starr, 1982; Allen et al., 1984; O'Neill et al., 1986) as were methods of quantifying spatial patterns (Kessell, 1979; Levin, 1978; Mandelbrot, 1983). Finally, there has been an increasing recognition that prudent management of the world's natural resources requires that human-defined parts of the landscape must be managed in the context of the interactions with the more natural parts of the landscape (Risser et al., 1984; Romme, 1982; Krummel et al., 1986).

The purpose of this paper is to build on these beginning steps, and to demonstrate that landscape ecology is the next logical expansion to understanding ecological processes of the biosphere. There is, of course, much still to be learned about specific ecosystems and the ecological processes within these systems. However, the flows of energy, materials, and organisms between and among ecosystems is now an obvious extension of our ecological investigations. While this paper will specifically focus on the landscape pattern and distribution of nutrients and energy, it should be obvious that these patterns also have direct influences on the patterns and distribution of organisms.

Pathways and Flow

Forman and Godron (1986) identify five vectors for the flow of energy, nutrients, and most species between and among ecosystems (hereafter referred to as landscape units): wind, water, flying animals, ground animals, and people. The forces that drive these flows at the landscape level are identified in broad terms to include diffusion, mass flow, and locomotion. Recognizing these basic vectors and forces is a useful beginning, but the challenge of landscape ecology is to understand the characteristics of the landscape which determine the flow of energy and materials in specific instances, and then to understand the generality of these instances.

Study of the causal mechanisms for energy and material flows has caused a reinforcement of several important ideas in ecology. First, these landscape flows vary over short and long temporal patterns, so long-term studies are frequently necessary. A related second point is the importance of geological processes and the short- and long-term interactions of geological and ecological processes. Third, landscape ecology clearly emphasizes the necessity of combining both natural and human-derived processes of energy and material flows. Finally, the couplet of structure and function becomes fundamental for describing and understanding pathways and flows at the landscape level of ecological investigation.

Geologic processes and landforms play major roles in regulating ecosystem structure and function by controlling the flow of energy and materials through landscapes (Carter, 1986; Decamps, 1984; Kelsey, 1982; Swanson et al., 1988). Climatic patterns, in conjunction with landform, also are determinants of ecosystems, as are human activities (Delcourt et al., 1983). Against this coarse scale of general geology, hydrology, climatology, and humans, the pathways of materials and energy in landscape ecology will necessarily be described by more mechanistic terms such as: degree and intensity of connectivity of landscape units, the juxtaposition and patterns of recipient and donor units (sink and sources), and the variable and relative strengths of flows caused by wind, water, animals, and humans. In the following paragraphs, these processes will be described, and general conclusions will be made about these patterns and redistributions of materials and energy across the landscape.

Patterns and Distribution of Energy

The description of vegetation patterns across the landscape, long a focus of ecological study, has been important in understanding successional sequences and the relationships between vegetation patterns and environmental conditions (Kessell, 1979; Reiners and Lang, 1979; Whittaker and Niering, 1965). More recently, attention has been paid to many aspects of the fine-scale patterns of heterogeneity (Figure 1) within ecosystems (Aarssen and Turkington, 1985; Arp and Helmut, 1984; Archer, 1984; Crozier and Boerner, 1984; Gibson and Greig-Smith, 1986; Loneragan and del Moral, 1984; Sterling et al., 1984;

Figure 1. Heterogeneous landscape in the state of Oklahoma (USA) where woody vegetation affects rate of water and material flows from the watershed.

Turkington et al., 1985; Welden, 1985; Whittaker et al., 1979). From the perspective of landscape ecology, the important attributes of these within-ecosystem studies is how ecosystem characteristics might affect the flow of energy and nutrients between landscape units (Senft et al., 1987). For example, Belsky (1986) studied a Tanzanian grassland mosaic of *Andropogon greenwayi* and *Chloris pycnothrix,* and found that the persistence of the mosaic itself depended upon the grazing activities of wild herbivores. In this Serengeti plain a predictable landscape-level pattern of grazing movement of herbivores depends upon rainfall distribution and vegetation type. Thus, the existence and persistence of this particular mosaic depends not only on the grazing and associated effects on the soil and vegetation in individual landscape units, but also on the grazing patterns throughout the Serengeti landscape mosaic.

Although the previous example demonstrated the importance of landscape-level grazing patterns on a particular grassland type, the general interaction of landscape-level processes on the energy and nutrient characteristics of specific ecosystems can be demonstrated in other landscapes. In riverine wetlands of the southeastern United States, the dominant floodplain forests are typically distributed along topographic and hydrologic gradients, and many wetland forest species have specific environmental requirements for seed germination and seedling establishment (Sharitz and Lee, 1985a; 1985b). The establishment, development, and maintenance of these floodplain forests depends upon the coincident availability of viable seeds and low water levels in order for germination and seedling establishment to occur. Primary productivity and organic matter processing by these wetland communities in turn depend upon species composition and the associated ecosystem processes (Minshall et al., 1983; Minshall, 1988). Thus, landscape-level processes, such as surface runoff patterns (Schlesinger and Jones, 1984) or the hydrological regime of the riverine wetlands (Naiman et al., 1986), determine not only the structural and functional characteristics of the ecosystems themselves, but also the rate at which carbon is fixed and processed across the landscape.

Just as landscape-level processes (e.g., grazing and flooding) can affect landscape units, there are reciprocal feedbacks where characteristics of landscape influence landscape-level processes. For example, McArthur and Marzolf (1986) examined the rate of dissolved organic carbon disappearance from a stream that flowed through a tallgrass prairie and then through a riparian forest. Uptake rates of microbial dissolved organic carbon were greater in the forested portion, and the grassland and forested portions appeared to have different microbial populations. Thus, in addition to carbon production varying across the landscape, terrestrial landscape heterogeneity influences the aquatic processing of organic carbon at the landscape level. In a more conjectural example, Schimel et al. (1985a) compared the soil-organic-matter dynamics of paired rangeland and cropland toposequences in the Northern Great Plains (USA). Prolonged agricultural practices caused soil-organic-matter losses, but these losses were different among the three soils present: sandstone, shale, and siltstone toposequences of montmorillonitic, typic Haploborolls, and Argiborolls. The sandstone

Table 1. Proportional loss of organic carbon from three soil toposequences in the state of North Dakota (USA) after 44 years of cultivation. Based on the current top 100 cm of soil. Adapted from Schimel et al. (1985a), and based in part on the studies of Aguilar (1984) and Kelly (1984). Toposequence goes from the crest of the slope (summit) to the base of the slope (footslope).

Toposequence position	Loss of organic carbon (%)		
	Sandstone	Stiltstone	Shale
Summit	54	45	45
Shoulder	40	49	53
Upper backslope	56	39	24[a]
Lower backslope	18[a]	49	24[a]
Footslope	34	37	24
Average (weighted by the relative area of each soil type)	34	46	35

[a]Identified by Kelly (1984) as soil deposited as erosional material rather than in situ formation.

soils were resistant to organic matter loss because the organic matter was distributed throughout a deeper soil profile. Thus, these soils maintained relatively stable organic matter concentrations despite erosion. The shale soils were resistant to erosional effects because the clayey soil retained the organic matter (Table 1). However, the siltstone sequence was unstable because much of the organic matter was near the soil surface and the soil had a low resistance to erosion. Thus, a landscape characterized by a number of soil types and topographic positions will exhibit a heterogeneous pattern of both organic matter production and organic loss to erosional processes.

Patterns and Distribution of Nutrients

Observations and manipulations of watersheds, especially experimental forest watersheds, have provided an understanding of nutrient dynamics within these ecosystems, as well as of the loss of soil and nutrients following disturbance and manipulation (Bormann and Likens, 1979; Bowden and Bormann, 1986; Cooper et al., 1986; Correll, 1981; Gilliam et al., 1986; McArthur et al., 1985; Osborn and Simanton, 1986; Pastor and Post, 1986; Woodmansee, 1978). Other watershed studies have examined correlations of land use patterns and watershed outputs (Correll et al., 1984; Lowrance et al., 1984b; 1985) or between atmospheric inputs and watershed outputs (Armentano and Menges, 1986; Correll and Ford, 1982). Of particular importance to landscape ecology are studies which measure the flow of nutrients from one ecosystem to another within the landscape (Correll, 1981; Hutchinson and Viets, 1969; Lowrance et al., 1984a; 1984b; 1985; Peterjohn and Correll, 1984; Schimel et al., 1985b; Schlosser and Karr, 1981a; 1981b). In fact, these are the types of studies that permit landscape ecology to move from a descriptive approach emphasizing structure to an analytical approach which emphasizes the structure and function of a landscape.

The need to understand landscape processes is exemplified by several studies conducted by D. L. Correll and associates on the Rhode River watershed in the state of Maryland, eastern USA. Early attempts to relate surface soil and subsoil composition to nutrient discharge for three watersheds of different land use were unsuccessful. It appeared that nutrient transformations or selective pathways were occurring as nutrients moved across the landscape, encountering various ecosystem types or landscape units. Thus Peterjohn and Correll (1984) investigated the nitrogen, phosphorous, and carbon transformations as water moved through a cropland and riparian watershed via surface runoff and shallow groundwater. As might be expected, nitrogen retention was much greater in the riparian forest (89 percent) than in the cropland (8 percent); similar values for phosphorous were 80 percent and 41 percent, respectively. Of more interest was the differentiation of nitrogen and phosphorous pathways of transport across the landscape. Groundwater was the predominant pathway of total nitrogen flux between the cropland and riparian forest, and the major pathway for loss of nitrogen from the forest. On the other hand, surface runoff was the dominant pathway for the transfer of phosphorus from the cropland and from the riparian forest. Furthermore, there was a significant decrease in the nitrate groundwater concentrations in the riparian forest, and this decrease was much greater than could be explained by uptake during the incremental growth of the forest.

This study (Peterjohn and Correll, 1984) and related ones (Lowrance et al., 1984a; 1985; Schlosser and Karr, 1981a; 1981b) identify several important ideas for landscape ecology. For example, the configuration of landscape units, such as riparian forests and croplands, plays a major role in determining the discharge characteristics of the landscape. In the above example, the riparian forest removed 45 kilograms per hectare of nitrate-nitrogen, and release of this amount would have doubled the total amount lost during the study year (March 1981–March 1982). Second, various landscape units are selectively effective at trapping and transforming nutrients, so simple watershed-discharge measurements do not capture the transformation characteristics of the landscape. Third, important nutrients such as nitrogen and phosphorus may be transferred by different hydrological pathways across the landscape, and these pathways play a part in determining the fate and consequences of nutrients within the landscape. Thus, careful management and placement of landscape units should allow the construction of a landscape that will not only retain nutrients, but also control their availabilities for plant uptake as well as vulnerability to loss by leaching or runoff.

An example of more subtle landscape flows involves nitrogen on a grazed shortgrass steppe (Schimel et al., 1986). The pattern of soil and forage properties within such a grassland influences cattle behavior and urine deposit (Senft et al., 1985; 1987). Within the landscape studied, higher slopes had a more coarsely textured soil than did lower slopes. Losses of nitrogen were greater from the coarse upland sites (0.016 $gN/m^2/yr$) than from the more finely textured lowland

Figure 2. Tallgrass prairie demonstrating subtle differences in vegetation pattern. These patterns of soil and plants represent different temporal dynamics of sources and sinks for organic matter, water, and nutrients.

sites (negligible), although the mechanisms of ammonia retention were not ascertained. However, even in this relatively uniform landscape, the assumption that nitrogen volatilization is the same in lowland sites as in upland sites would significantly overestimate the volatile loss of nitrogen from the landscape. Other landscape differences have also been found for N_2O flux and the processes of mineralization, nitrification, and denitrification (Schimel et al., 1985b). Figure 2 illustrates a small tallgrass-prairie-vegetation pattern in which the foreground is much more moist than the surrounding slope. These small patches influence ecosystem-level processes such as accumulation of organic matter and sources and sinks for nitrogen.

Conclusion

Landscape ecology is now moving past the early, descriptive stages, which focused on the structure of landscapes, to analysis of the processes in landscape systems. Of particular importance are the processes by which energy and nutrients move between and among ecosystems within the landscape. Studies to date have demonstrated that flows across the landscape can affect the behavior of landscape units (e.g., flooding events influence vegetation growth and organic matter processing). Reciprocal and feedback actions also occur, where characteristics of the landscape units affect the processes of the landscape level (e.g., movement of nutrients or maintenance of animal populations). In a hierarchical sense, these interactions suggest the utility, if not the necessity, of understanding ecological processes at the level of the landscape.

Most current studies of energy and nutrient landscape flows are empirical, but these data will lead to future generalities, advances that will result from both conceptual and technical developments. In 1963, Margalef advanced the notion that when considering the flows between adjacent ecosystems or elements, the younger may operate as a source of energy or material, and the more mature as a sink or recipient. Too few studies exist to adequately test the hypothesis, but the idea is attractive and consistent with portions of our knowledge about successional processes. Furthermore, the idea can be tested by field observations in which the definition of source and sink depend upon the entity under study.

Another possible extension of current thought into the landscape perspective involves the Universal Soil Loss Equation:

$$E = f(R, K, L, S, C)$$

where E is the amount of soil loss from a field or part of the landscape, R is a function of rainfall intensity, K is the erodibility factor of the soil, L is the length of the slope, S is the angle of the slope, and C refers to the vegetation cover (Jenny, 1980). Obviously this simple equation does not account for landscape heterogeneity, so amplification must entail mathematical incorporation not only of these coefficients, but also of a measure of the juxtaposition or configuration of the landscape units. A strength of this approach lies in the wealth of empirical data accumulated for use in the soil loss equation.

While there are measurements of the role of riparian vegetation in retarding the movement of dissolved and particulate matter from the terrestrial environment into the discharge stream (Peterjohn and Correll, 1984; Schlosser and Karr, 1981a, 1981b), little attention has been paid to the pattern of forested patches elsewhere in the landscape. Many of the investigative approaches used in describing the relationship between landscape structure and animal populations (Burgess and Sharpe, 1981; Senft et al., 1987; Wegner and Merriam, 1979) may eventually be applicable to descriptions of the landscape flows of organic matter and nutrients. It will also be necessary to relate measures of spatial pattern (e.g., grain, connectivity, strength of boundaries) to empirical observations of energy and nutrient flows. At first these relationships will undoubtedly be multiple regressions, but as mechanisms become more obvious, simpler measures should be possible (Schimel et al., 1988). In some instances, remote sensing techniques capable of making measurements at the landscape level may facilitate the identification of consistent relationships (Scott et al., 1987).

References

Aarssen, L. W. and R. Turkington. 1985. Vegetation dynamics and neighbour associations in pasture-community evolution. *Journal of Ecology* 74:585–603.

Aguilar, R. 1984. Parent material-topographic-management controls on organic and inorganic nutrients in semiarid soils. Ph.D. Dissertation. Colorado State University, Fort Collins, Colorado.

Allen, T. F. H., and T. B. Starr. 1982. *Hierarchy: Perspectives for ecological complexity*. University of Chicago Press, Chicago.

Allen, T. F. H., R. V. O'Neill, and T. W. Hoekstra. 1984. *Interlevel relations in ecological research and management: Some working principles from hierarchy theory*. USDA Forest Service General Technical Report RM-110. Rocky Mountain Forest and Range Experiment Station, Fort Collins, Colorado.

Archer, S. 1984. The distribution of photosynthetic pathway types on a mixed-grass prairie hillside. *American Midland Naturalist* 111:138–42.

Armentano, T. V and E. S. Menges. 1986. Patterns of change in the carbon balance of the organic soil-wetlands of the temperate zone. *Journal of Ecology* 74:755–74.

Arp, P. A. and H. H. Helmut. 1984. The forest floor: Lateral variability as revealed by systematic sampling. *Canadian Journal of Soil Science* 64:423–37.

Belsky, A. J. 1986. Population and community processes in a mosaic grassland in the Serengeti, Tanzania. *Journal of Ecology* 74:841–56.

Bormann, F. H. and G. E. Likens. 1979. *Patterns and process in a forested ecosystem*. Springer-Verlag, New York.

Bowden, W. B. and F. H. Bormann. 1986. Transport and loss of nitrous oxide in soil water after forest clear-cutting. *Science* 233:867–9.

Brandt, J. and P. Agger, eds. 1984. *Proceedings of the first international seminar on methodology in landscape ecological research and planning*. 5 vol. Roskilde University Center, Denmark.

Burgess, R. L. and D. M. Sharpe, eds. 1981. *Forest-island dynamics in man-dominated landscapes*. Springer-Verlag, New York.

Carter, V. 1986. An overview of the hydrologic concerns related to wetlands in the United States. *Canadian Journal of Botany* 64:364–74.

Cooper, J. R., J. W. Gilliam, and T. C. Jacobs. 1986. Riparian areas as a control of nonpoint pollutants. In D. L. Correll, ed., *Watershed Research Perspectives*. Smithsonian Institution Press, Washington, DC, pp. 166–92.

Correll, D. L. 1981. Nutrient mass balances for the watershed, headwaters, intertidal zone, and basin of the Rhode River Estuary. *Limnology and Oceanography* 26:1142–9.

Correll, D. L. and D. Ford. 1982. Comparison of precipitation and land runoff as sources of estuarine nitrogen. *Estuarine, Coastal and Shelf Science* 15:45–56.

Correll, D. L., N. M. Goff, and W. T. Peterjohn. 1984. Ion balances between precipitation inputs and Rhode River watershed discharges. In O. P. Bricker, ed., *Geological Aspects of Acid Deposition*. Butterworth, New York, pp. 77–111.

Crozier, C. R., and R. E. Boerner. 1984. Correlations of understory herb distribution patterns with microhabitats under different tree species. *Oecologia* 62:337–43.

Decamps, H. 1984. Towards a landscape ecology of river valleys. In J. H. Cooley and F. B. Golley, eds., *Trends in Ecological Research for the 1980s*. Plenum Press, New York, pp. 163–78.

Delcourt, H. R., P. A. Delcourt, and T. Webb III. 1983. Dynamic plant ecology: The spectrum of vegetational change in space and time. *Quaternary Science Reviews* 1:153–75.

Forman, R. T. T. and M. Godron. 1981. Patches and structural components for a landscape ecology. *BioScience* 31:733–40.

Forman, R. T. T. and J. Baudry. 1984. Hedgerows and hedgerow networks in landscape ecology. *Environmental Management* 8:495–510.

Forman, R. T. T. and M. Godron. 1986. *Landscape ecology*. Wiley, New York.

Gibson, D. J. and P. Greig-Smith. 1986. Community pattern analysis: A method for quantifying community mosaic structure. *Vegetatio* 66:41–47.

Gilliam, J. W., R. W. Skaggs, and C. W. Doty. 1986. Controlled agricultural drainage: An alternative to riparian vegetation. In D. L. Correll, ed., *Watershed research perspectives*. Smithsonian Institution Press, Washington, DC., pp. 225–43.

Harris, L. 1984. *The fragmented forest: Island biogeography theory and the preservation of biotic diversity*. University of Chicago Press, Chicago.

Hutchinson, G. L. and F. G. Viets, Jr., 1969. Nitrogen enrichment of surface water by absorption of ammonia volatilized from cattle feedlots. *Science* 166:514–15.

Jenny, H. 1980. *The soil resource: Origin and behavior*. Springer-Verlag, New York.

Kelly, E. F. 1984. Long-term erosional effects on cropland vs. rangeland in semiarid ecosystems. M.S. Thesis. Colorado State University, Fort Collins, Colorado.

Kelsey, H. M. 1982. *Hillslope and sediment movement in a forested headwater basin, Van Duzen River, north coastal California*. In F. J. Swanson, R. J. Janda, T. Dune, and D. N. Swanston, eds., USDA Forest Service Technical Report PNW-141, Corvallis, Oregon, pp. 86–96.

Kessell, S. R. 1979. *Gradient modeling*. Springer-Verlag, New York.

Krummel, J. R., R. V. O'Neill, and J. B. Mankin. 1986. Regional environmental simulation of African cattle herding societies. *Human Ecology* 14:117–30.

Leopold, A. 1949. *A Sand County almanac*. Ballentine, New York.

Levin, S. A. 1978. Pattern formation in ecological communities. In J. H. Steele, ed., *Spatial pattern in plankton communities*. Plenum, New York, pp. 433–65.

Lindeman, R. L. 1942. The trophic-dynamic aspect of ecology. *Ecology* 23:399–418.

Loneragan, W. A., and R. del Moral. 1984. The influence of microrelief on community structure of subalpine meadows. *Bulletin of Torrey Botanical Club* 111:209–16.

Lowrance, R., R. Todd, J. Fail, Jr., O. Hendrickson, Jr., R. Leonard, and L. Asmussen. 1984a. Riparian forests as nutrient filters in agricultural watersheds. *BioScience* 34:374–7.

Lowrance, R., R. L. Todd, and L. E. Asmussen. 1984b. Nutrient cycling in an agricultural watershed: I. Phreatic movement. *Journal of Environmental Quality* 13:22–7.

Lowrance, R. R., R. A. Leonard, L. E. Asmussen, and R. L. Todd. 1985. Nutrient budgets for agricultural watersheds in the southeastern coastal plain. *Ecology* 66:287–96.

Mandelbrot, B. 1983. *The fractal geometry of nature*. Freeman, New York.

Margalef, R. 1963. On certain unifying principles in ecology. *American Naturalist* 97:357–74.

McArthur, J. V., M. E. Gurtz, C. M. Tate, and F. S. Gillam. 1985. The interaction of biological and hydrological phenomena that mediate the qualities of water draining native tallgrass prairie on the Konza Prairie Research Natural Area. *Perspectives of nonpoint source pollution: Proceedings of a national conference*. Report 44015-85-001, Office of Water Regulations and Standards, U. S. Environmental Protection Agency, Washington, DC., pp. 478–82.

McArthur, J. V., and G. R. Marzolf. 1986. Interactions of the bacterial assemblages of a prairie stream with dissolved organic carbon from riparian vegetation. *Hydrobiologia* 134:193–9.

Minshall, G. W., R. C. Peterson, K. W. Cummins, T. L. Bott, J. R. Sedell, C. E. Cushing, and R. L. Vannote. 1983. Interbiome comparison of stream ecosystem dynamics. *Ecological Monographs* 53:1–25.

Minsall, G. W. 1988. Stream ecosystem theory: A global perspective. *Journal of the North American Benthological Society* 7:263–88.

Naiman, R. J., J. M. Melillo, and J. E. Hobbie. 1986. Ecosystem alteration of boreal forest stream by beaver *(Castor canadensis)*. *Ecology* 67:1254–69.

Naveh, Z. 1982. Landscape ecology as an emerging branch of ecosystem science. *Advances in Ecological Research* 12:189–237.

Naveh, Z. and A. S. Lieberman. 1984. *Landscape ecology: Theory and application*. Springer-Verlag, New York.

O'Neill, R. V., D. L. DeAngelis, J. B. Waide, and T. F. H. Allen. 1986. *A hierarchical concept of ecosystems*. Princeton University Press, NJ.
Osborn, H. B. and J. R. Simanton. 1986. Gully migration on a southwest rangeland watershed. *Journal of Range Management* 39:558–61.
Pastor, J. and W. M. Post. 1986. Influence of climate, soil moisture, and succession on forest carbon and nitrogen cycles. *Biogeochemistry* 2:3–28.
Peterjohn, W. T. and D. L. Correll. 1984. Nutrient dynamics in an agricultural watershed: Observations on the role of a riparian forest. *Ecology* 65:1466–75.
Reiners, W. A. and G. E. Lang. 1979. Vegetation patterns and processes in the balsam fir zone, White Mountains, New Hampshire. *Ecology* 60:403–17.
Risser, P. G. 1987. Landscape ecology: State of the art. In M. G. Turner, ed., *Landscape heterogeneity and disturbance,* Springer-Verlag, New York, pp. 3–14.
Risser, P. G., J. R. Karr, and R. T. T. Forman. 1984. *Landscape ecology—directions and approaches*. Illinois Natural History Survey Special Publication Number 2, Champaign, Illinois.
Romme, W. H. 1982. Fire and landscape diversity in subalpine forests of Yellowstone National Park. *Ecological Monographs* 52:199–221.
Rowe, J. S. 1961. The level-of-integration concept and ecology. *Ecology* 42:420–7.
Schimel, D. S., D. C. Coleman, and K. A. Horton. 1985a. Soil organic matter dynamics in paired rangeland and cropland toposequences in North Dakota. *Geoderma* 36:201–14.
Schimel, D., M. A. Stillwell, and R. G. Woodmansee. 1985b. Biogeochemistry of C, N, and P in a soil catena of the shortgrass steppe. *Ecology* 66:276–82.
Schimel, D. S., W. J. Parton, F. J. Adamsen, R. G. Woodmansee, R. L. Senft, and M. A. Stillwell. 1986. The role of cattle in the volatile loss of nitrogen from a shortgrass steppe. *Biogeochemistry* 2:39–52.
Schimel, D. S., Simkins, T. Rosswall, A. R. Mosier, and W. J. Parton. 1988. Scale and the measurement of nitrogen-gas fluxes from terrestrial ecosystems. In T. Rosswall, R. G. Woodmansee and P. G. Risser, eds., *Scales and global change. Spatial and temporal variability in biospheric and geospheric processes*. Wiley, New York, pp. 179–94.
Schlesinger, W. H. and C. S. Jones. 1984. The comparative importance of overland runoff and mean annual rainfall to shrub communities of the Mojave desert. *Botanical Gazette* 145:116–24.
Schlosser, I. J. and J. R. Karr. 1981a. Water quality in agricultural watersheds: Impact of riparian vegetation during base flow. *Water Resources Bulletin* 17:233–40.
Schlosser, I. J. and J. R. Karr. 1981b. Riparian vegetation and channel morphology impact on spatial patterns of water quality in agricultural watersheds. *Environmental Management* 5:233–43.
Scott, J. M., B. Csuti, J. D. Jacobi and J. E. Estes. 1987. Species richness. *BioScience* 37:782–8.
Senft, R. L., L. R. Rittenhouse, and R. G. Woodmansee. 1985. Factors influencing selection of resting sites by cattle on shortgrass steppe. *Journal of Range Management* 38:295–9.
Senft, R. L., M. B. Coughenour, D. W. Bailey, L. R. Rittenhouse, O. E. Sala and D. M. Swift. 1987. Large herbivore foraging and ecological hierarchies. *BioScience* 37:789–99.
Sharitz, R. R. and L. C. Lee. 1985a. Limits on regeneration processes in south eastern riverine wetlands. In *Riparian ecosystems and their management: Reconciling conflicting uses*. USDA Forest Service General Technical Report RM-120. Rocky Mountain Forest and Range Experiment Station, Fort Collins, Colorado, pp. 139–60.
Sharitz, R. R. and L. C. Lee. 1985b. Recovery processes in southeastern riverine wetlands. In *Riparian ecosystems and their management: Reconciling conflicting uses*.

USDA Forest Service General Technical Report RM-120. Rocky Mountain Forest and Range Experiment Station, Fort Collins, Colorado, pp. 449–501.

Sterling, A., B. Peco, M. A. Casado, E. F. Galiano, and F. D. Pineda. 1984. Influence of microtopography on floristic variation in the ecological succession in grassland. *Oikos* 42:334–42.

Swanson, F. J., T. K. Kratz, N. Caine, and R. G. Woodmansee. 1988. Landform effects on ecosystem patterns and process. *BioScience* 38:92–8.

Szaro, R. C. and M. D. Jakle. 1985. Avian use of a desert riparian island and its adjacent scrub habitat. *Condor* 87:511–19.

Tilman, G. D. 1984. Plant dominance along an experimental nutrient gradient. *Ecology* 65:1445–53.

Turkington, R., J. L. Harper, P. de Jong, and L. W. Aarssen. 1985. A reanalysis of interspecific association in an old pasture. *Journal of Ecology* 73:123–32.

Verry, E. S. and D. R. Timmons. 1982. Waterborne nutrient flow through an upland-peatland watershed in Minnesota. *Ecology* 63:1456–7.

Watt, A. S. 1947. Pattern and process in the plant community. *Journal of Ecology* 35:1–22.

Wegner, J. F. and G. Merriam. 1979. Movements by birds and small mammals between a wood and adjoining farmland habitats. *Journal of Applied Ecology* 16:349–58.

Welden, C. 1985. Structural pattern in alpine tundra vegetation. *American Journal of Botany* 72:120–34.

Whittaker, R. H. and W. A. Niering. 1965. Vegetation of the Santa Catalina Mountains, Arizona: A gradient analysis of the south slope. *Ecology* 46:429–52.

Whittaker, R. H., L. E. Gilbert, and J. H. Connell. 1979. Analysis of two-phase pattern in a mesquite grassland, Texas. *Journal of Ecology* 67:935–52.

Woodmansee, R. G. 1978. Additions and losses of nitrogen in grassland ecosystems. *BioScience* 28:448–53.

5. Biogeochemical Cycles and Ecological Hierarchies

R. G. Woodmansee

Biogeochemical cycling (also called nutrient cycling and elemental cycling) refers to the generally accepted concept in ecology that essential elements are transformed, transported, and reused by organisms in the environment. However, many problems beset investigations of biogeochemistry. Among the problems are technological and methodological limitations, lack of uniform theory that can guide understanding, an inherent complexity of nature, and spatial and meteorological variability. The first two problems are amenable to solution, but the latter two have causes beyond experimenter control and commonly are approached through modeling or correlative studies.

My intent in this chapter is to examine the problems that concern spatial variability. Specifically, I will attempt to: (1) identify some levels of ecological organization that can serve as models when dealing with concepts of biogeochemical cycling; (2) place those levels within a model hierarchy based on the transfer of energy or matter; and (3) discuss important processes and the vectors of transport of energy and matter within each level of the hierarchy.

Theoretical Construct

The structure of knowledge may be encapsulated as follows: "Reality, in the modern conception, appears as a tremendous hierarchical order of organized entities, leading in superposition of many levels, from physical and chemical to biological and sociological systems. Unity of science is granted, not by Utopian reduction of all sciences to physics and chemistry, but by the structural uniformities of the different levels of reality" (von Bertalanffy, 1950).

Our understanding of ecological systems thus is not predicated on the presumption that we can understand all basic chemical and physical laws that direct biological functioning; rather our goal is to understand the behavior of systems within appropriate levels of organization. A system, as defined by Forrester (1968), is merely a grouping of parts that operate together for a common goal or purpose. Forrester's concept of goal or purpose is especially appropriate in industrial systems, with which he was most familiar. In natural systems, the terms, goal, and purpose might be thought of as interactions, maintenance, and persistence. He went on to suggest that any level of organization within a hierarchy can be represented as a system. The precise definition or model of a system, however, will depend on the objectives of any given study of a well-defined problem, because no natural system can be described in complete detail. Additional insight can be found in Feibleman (1954), DeWit (1970), Weiss (1971), Allen and Starr (1982), and O'Neill et al. (1986), who suggested that the understanding of a system at any conceived level of organization requires knowledge of the levels immediately above and below. Thus, by integrating some of the basic tenets from systems science with concepts from biology and soil science, we can develop schemes to better understand the cycling of nutrients in ecological systems.

It is important to remember, however, that nature is not simply a grand hierarchy of systems at various levels of organization; rather, *nature simply is*. We, as humans, can derive great intellectual benefit by using models such as levels of organization and hierarchies as conceptual tools for synthesis and integration.

MacMahon et al. (1978) viewed the organism as the central system from which various biological relationships are constructed. Within one series of relationships, the organism is depicted as being composed of organ systems, which in turn are themselves composed of organs, each of which is composed of tissues, and so on through a hierarchical ordering, the bottom of which is a subatomic particle. In another series of relationships, the organism is seen as the fundamental level of organization in a phylogenetic sequence that ranges from organisms to species to genera to families on up to plant or animal kingdoms. The organism is also seen in a community-oriented complex that views organisms as being the components of demes which are themselves the components of populations, which are components of communities. Of particular interest to this discussion is the fourth set of biological relationships dealt with by MacMahon et al. (1978). In that hierarchical development, the organism is seen as the fundamental level of organization, which is at a lower level than an ecosystem, which is a component of the biosphere. Within this last set of relationships, matter and energy exchanges are the basis for distinguishing the levels within the hierarchy.

Another scheme for an ecological hierarchy, one that started with subatomic particles and ultimately ended in the universe, was presented by Rowe (1961). Within this scheme, cells made up organs, which made up organisms, which made up what Row called "single organism habitat ecosystems." These systems could then be grouped into "local ecosystems," which made up regional ecosystems, which were, in turn, components of the ecosphere and ultimately the universe, the highest level in the organization.

Integrating the MacMahon and Rowe concepts, I have developed a hierarchical ordering of ecological units that is based on energy and matter exchange. In this paper, I will limit the discussion to the exchange of matter (e.g., carbon, nitrogen, phosphorus, and sulfur).

Levels of an Ecological Hierarchy

The theoretical construct presented herein is based on the assumption that all levels within the ecological hierarchy are models that must be defined in space and time. When matter is exchanged between components within one level in the hierarchy, that exchange must take place through some vector or medium of transport. The levels and components are abstractions that represent real entities. The modeled matter exchange also represents real exchanges and can be quantified in terms of matter moved per unit area per unit time.

To develop this argument further, I will borrow concepts from the science of pedology to help clarify the definitions of ecological systems. Integration of the concepts of MacMahon et al. (1978), Rowe (1961), and Anderson et al. (1983) produces a hierarchical diagram, with the focal system being a site (Figure 1). However, no level within the hierarchy should be considered fundamental, because the hierarchy is simply a useful abstraction or model that serves to aid our ability to integrate information in order to solve problems. The diagrammatic model (Figure 1) depicts sites as being composed of organisms, which are in turn composed of cells, and so on. Clearly, many important ecological functions occur at the suborganism level, but I will restrict this discussion to those levels above the organism.

The Organism

Organisms, such as large or small animals, large or small plants, and microorganisms, are the living biotic components of site-scale ecosystems (Figure 1). Though exceptions exist, the principal assumption made at this level of organization is that the organisms are easily definable. They have epidermal layers or cell walls that define the limits of their bodies. Living organisms are responsible for most chemical syntheses, transformations, and degradations in ecological systems (Table 1). Higher plants, fungi, and mobile animals acquire and circulate large amounts of nutrients within their bodies and deposit those materials at

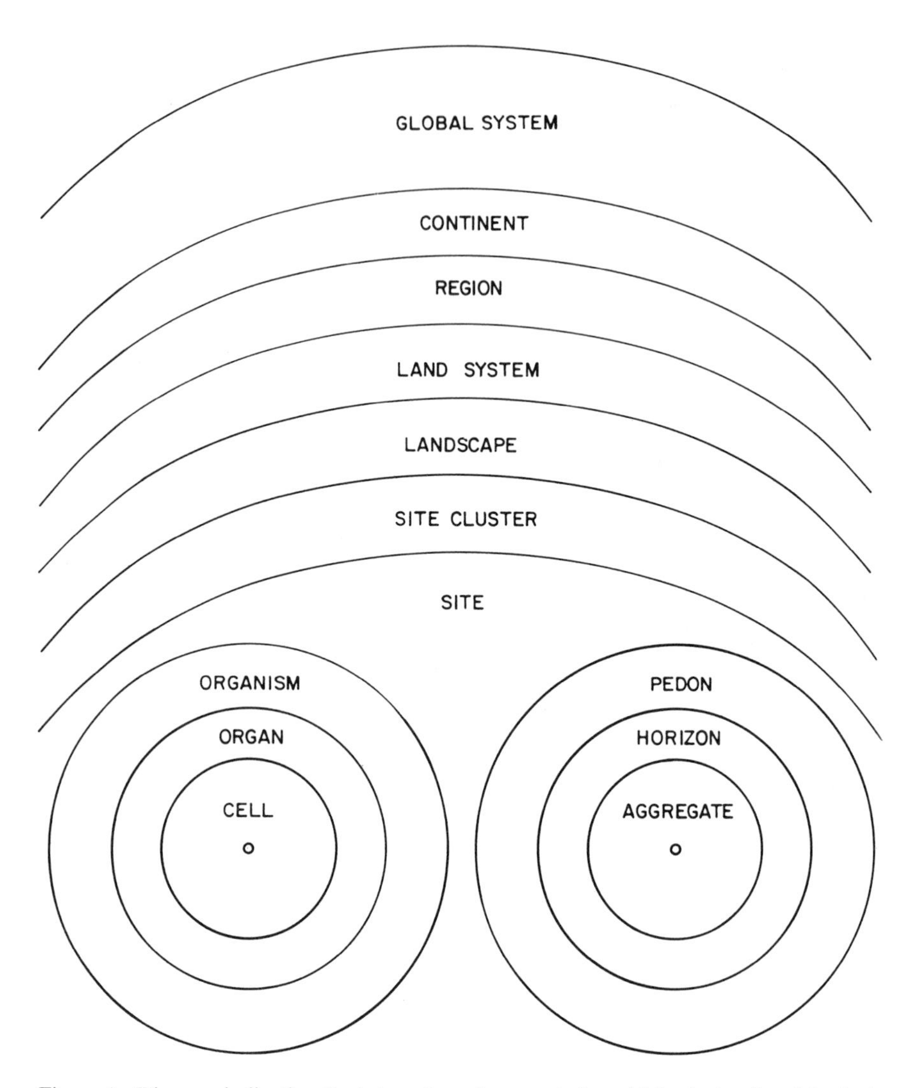

Figure 1. Diagram indicating the integration of concepts from biological and pedological hierarchies. Within each level of organization, energy or matter is the currency of exchange between components.

various locations, depending on their areal extent or ability to move. Plants, for example, have long been recognized for their role in taking up nutrients deep within the solum and translocating them to the upper horizons of the soil. The results of this process mean that most of those nutrients are kept in active circulation and are not lost from the rooting zone.

The parallel pedologic level of organization to the organism is the pedon (Anderson et al., 1983). A *pedon* is the smallest volume that can be considered a soil, and has a minimum area of 1 m^2 and maximum of 10 m^2.

The Site

The next level in the ecological hierarchy is the site. Many other names, such as land unit, ecotope, biotope, and landscape element (Forman and Godron, 1986) could be used. However, when considering a terrestrial portion of the earth's surface, a biotic community existing in a soil polypedon can be considered a fundamental unit, the site.

A *polypedon* is a grouping of contiguous pedons with similar characteristics. A polypedon is bounded on all sides by either nonsoil or pedons of significantly different characteristics (Soil Survey Staff, 1975). The boundary between polypedons may be sharp, as with a vertical scarp, or may be gradual and extend to a distance of kilometers. Generally, the boundary will extend over a few meters. This is the unit from which individual soil samples are taken and in which process-level measurements are made.

An acceptable, though abstract, definition of an ecosytem is that of Odum (1971): "Living organisms and their non-living (abiotic) environment are inseparably interrelated and interact upon each other. Any unit that includes all of the organisms in a given area interacting with the physical environment so that a flow of energy leads to a clearly defined trophic structure, biotic diversity, and material cycles (i.e., exchange of materials between living and nonliving parts) within the system is an ecological system or ecosystem." In the hierarchy proposed here, I consider the local ecosystem to be equivalent to a site, that is, a unit composed of a biotic community growing in association with a soil polypedon (and with associated parent material and atmosphere). Clearly, the soil and biota interact and a trophic structure is present in a site (e.g., Jenny, 1941; Coleman et al., 1978).

The other component of Odum's definition is that all organisms in a given area must be considered. Since soil and plants are fixed in space, if not in time, and are relatively easy to measure, it is convenient to use this relationship to define the boundaries of a site.

Within the concept of a polypedon, there is certainly variability within soil properties, but any parameter would have a definable distribution. For example, an *A* horizon should be the same depth, given a certain amount of describable variability throughout the polypedon. The soil could be given a series name, or in some cases, the name of a phase of a soil series—it is not a mapping unit that can include more than one soil type.

Numerous different biotic communities, developed under varying historical conditions including grazing, fire, or agricultural activities, may exist on a single soil type (Figure 2). Each plant community–soil complex represents a different site. Conversely, many situations exist in which one plant community (e.g., a corn or wheat field) may exist over several different soils (Figure 3). In this case soil type boundaries differentiate the sites. The soil characteristics differentiating sites here are based on functional differences (i.e., the water-holding capacity of the surface horizon or of the entire solum, or the capacity of the soil to mineralize nutrients), which significantly influence the rates of processes

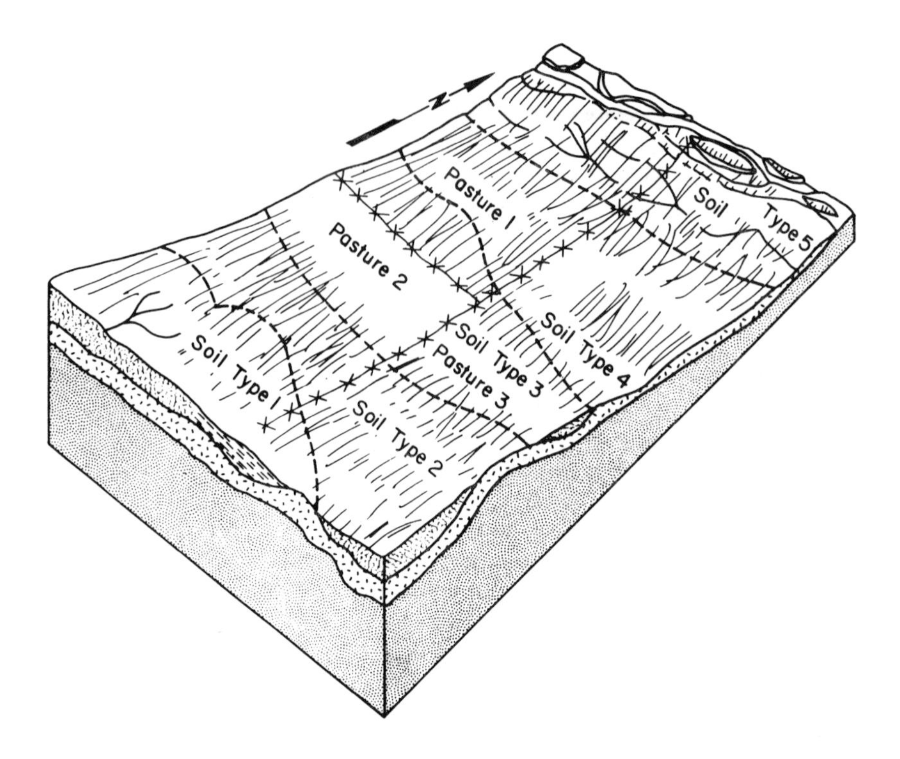

Figure 2. Three pastures superimposed on a catena of five soil types. Land-management often leads to drastic differences in vegetational composition on opposite sides of fences, here indicated by lines with crosses. In this figure, we assume each soil within each pasture has a unique associated biota; thus each soil/plant community combination represents a unique site or local ecosystem.

(Table 1) and thus, plant community development. Given functional differences in soils, we assume faunal and microbial communities would likewise vary, as would such responses as productivity and succession.

Important differences in input and output dynamics (Table 1) are observed between sites (either several biotic communities associated with one soil, or one vegetation type associated with several soils). For example, the capacity of a site to intercept and retain nutrients from the atmosphere as gases or dry deposition, or its capacity to retain debris or sediment from water, may vary greatly depending on both plant community structure and soil characteristics (Gorham et al., 1979). Losses of nutrients also depend on biotic and abiotic soil factors.

Several agents of transport of nutrients are important for the distribution and redistribution of nutrients within, into, and out of a site (Table 1). Nutrients are transported through water, atmosphere, and living animals and plants at this level of organization. Both horizontal and vertical transport are important.

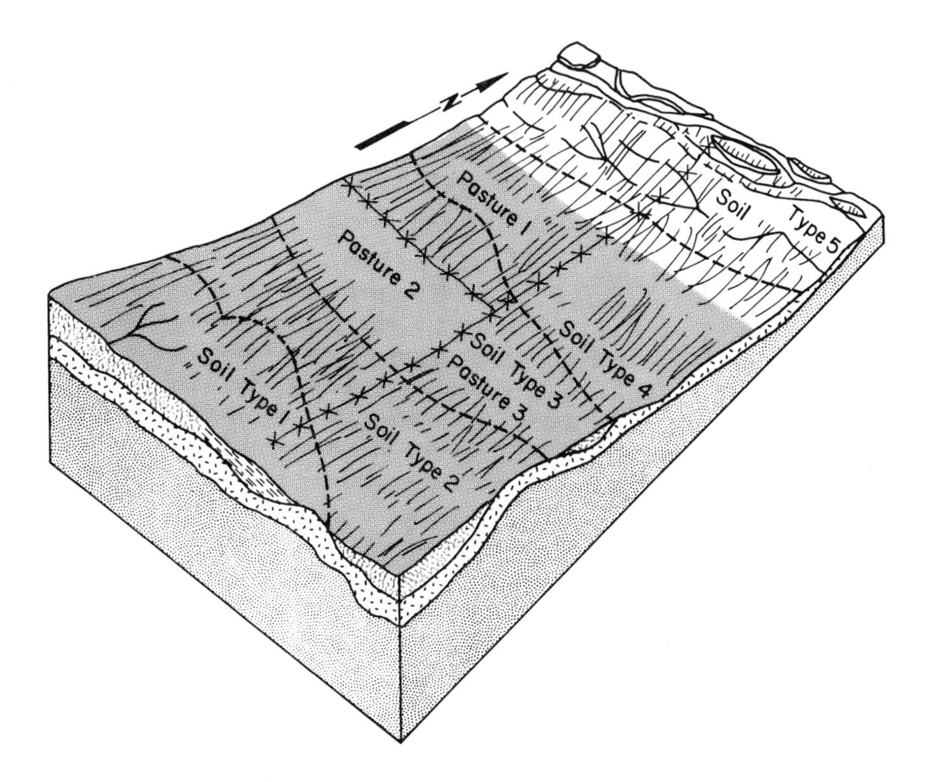

Figure 3. A single vegetation type (crop) superimposed on a soil catena. In the crop area, indicated by shading, each soil type represents a unique site or local ecosystem.

The Site Cluster

I have chosen to use the phrase *site cluster* to represent the level of organization in the ecological hierarchy whose components are sites. The concept of a site cluster is important for two reasons. First, it describes a series of sites connected by a significant exchange of matter. Second, it includes the concept of a catena, traditionally used by soil scientists to describe a connected series of soils on a slope (Figure 2). The connective element in a soil catena is typically the downslope movement of water that follows distinct flowpaths. Soil development and productivity are generally enhanced at the bottom slope position, due to downward transport of nutrients through leaching and/or erosion, plus deposition and increased water availability. The soil catena conveys an inherent concept of size, or length, characteristic of an ecological *flowpath* (a series of sites related to one another via a transported substance or object). However, a flowpath is not simply a linear downslope concept. Rather, it is a two-dimensional area concept where short-range atmospheric and animal transport of material also takes place among the cluster of neighboring sites. Thus, I use the term *flowweb* to convey this broadened meaning.

Table 1. Examples of processes and media of transport that are indicative of various levels of organization in ecological hierarchies.

Level of organization	Process	Agent of transport
Organism	Internal	
	Synthesis of chemical compounds	—
	Transformations of chemical compounds	—
	Death of parts	—
	Translocation	Aqueous solution in mycorrhizae, xylem, phloem, circulatory systems, renal systems
	Inputs	
	Absorption of gases	Atmosphere
	Uptake by plants and microorganisms	Aqueous solution
	Ingestion	Animals, mechanical movement
	Outputs	
	Excretion of wastes	Aqueous solution or organic debris
	Exudation and secretion	Aqueous solution
	Volatilization of gases	Atmosphere
	Respiration	Atmosphere
Site	Internal	
	Mineralization	—
	Immobilization	—
	Diffusion (gas)	Atmosphere
	Diffusion (liquid)	Soil solution
	Mass flow	Soil and surface solution
	Soil organic matter transformations	—
	Chemical equilibrium	Soil Solution
	Sloughing of parts (exudation and secretion)	Atmosphere or water, driven by wind or gravity
	Foraging/excretion	Animals
	Inputs	
	Absorption of gases by soils and plants	Atmosphere
	Wet deposition (ions, aerosols, particles)	Atmospheric water
	Dry deposition (ions, aerosols, particles)	Atmosphere by wind and gravity
	Debris deposition	Water or atmosphere by gravity or wind
	Body matter and excreta	Animals
	Interflow deposition	Soil solution and gravity

Level of organization	Process	Agent of transport
	Weathering (mineral soil formation)	Soil solution
	Fertilization	Humans and mechanization
	Outputs	
	Erosion	Atmosphere by wind, and water by gravity
	Leaching	Soil solution and gravity
	Interflow	Soil solution and gravity
	Volatilization of gases (conversion and export)	Atmosphere by diffusion or wind
	Grazing (of plant export)	Animals
	Cropping (of plant and animal export)	Humans and mechanization
Site Cluster	Internal	
	Debris transport	Atmosphere by gravity or wind
	Debris transport	Water
	Bodies and excreta	Animals
	Interflow	Soil solution
	Short-range gas, aerosol, and particle transport	Atmosphere
	Inputs	
	Fertilization to sites	Humans by mechanization
	Bodies and excreta to sites	Animals
	Interflow	Soil solution
	Outputs	
	Stream discharge of sediment and debris (net of sites leaching and interflow)	Surface water
	Stream discharge—subsurface (net of sites leaching and interflow)	Soil solution
	Mid-range air transport of gases, aerosols, and particles	Atmosphere by wind and gravity
	Grazing (sum of plant transport from sites)	Animals
	Crop removal from sites	Humans by mechanization
Landscape	Internal	
	Gases, ions, aerosols, particles, and debris transport	Atmosphere by wind and gravity
	Debris and sediment transport	Drainage water
	Bodies and excreta transport among catenas	Animals

Table 1. *Continued*

Level of organization	Process	Agent of transport
	Inputs	
	Gas, ion, aerosol, and particle deposition to sites	Atmosphere by diffusion, wind, and gravity
	Sediment and debris transport to bottom members of site clusters	Surface drainage water
	Fertilization to site clusters	Humans by mechanization
	Outputs	
	Stream discharge from catenas	Surface water
	Gas, ion, aerosol, and particles from site clusters	Atmosphere by diffusion and wind
	Crop and animal export	Humans by mechanization
Higher Levels	Internal	
	Transport of gases, ions, aerosols, and particulates	Atmosphere by diffusion, wind, and gravity
	Inputs	
	Transport of gases, ions, aerosols, and particulates	Atmosphere by wind and gravity
	Fertilization	Humans by mechanization
	Outputs	
	Transport of gases, ions, aerosols, and particulates	—
	Sediment transport	Surface water
	Plant and animal export	Humans by mechanization

Spectacular examples of wind-transported materials exist, such as pictures from the American dust-bowl era showing soil that has drifted to the top of fence lines. The travel distance of that material was often quite short, and the effects on the site being buried were marked. Similarly, in northeastern Colorado most of the undisturbed soils (i.e., native range) have an A11 horizon of about 2 cm of wind-reworked material that is presumed to move among adjacent sites. The formation of snowdrifts on the lee side of hills is also a short-distance, wind-transport phenomenon, since the half-life of a snowflake is rather short once it is in the air. In semiarid environments, the formation of snowdrifts and the increase in water where they occur leads to substantial differences in vegetation.

Kevin O'Connor (personal communication) described a vivid example of animal transport of matter in New Zealand. At a research site in the hill-farm region, sheep spend much of their time grazing in the lowlands next to a railroad track. When trains pass, the beasts become frightened and race up-slope to the ridge top, whereupon they defecate and urinate. This phenomenon has occurred for a sufficiently long period of time to allow a substantial buildup of nutrients on the ridge top.

Gains or losses of nutrients from a flowweb are, in fact, a summation of the net gains or losses realized by the individual sites in a site cluster (Table 1). For example, Swanson et al. (1980) suggested that upland forest sites may lose nutrients, but associated lowland sites probably immobilize nutrients as water moves through them on the soil surface or as interflow. Thus, the net loss from the flowweb is not reflective of any single site. The same phenomenon probably occurs in many grasslands.

The Landscape

A landscape is a hierarchical level whose component parts are site clusters. Processes and agents of material transport common to this level of organization are shown in Table 1, though the flux rates between site clusters are essentially unknown and represent an ecological frontier.

Matter can be directly exchanged between water- or wind-formed site clusters by animals that forage in one cluster (or in sites within that cluster), and defecate, urinate, or die in another, a typical phenomenon that occurs in rangeland (Senft et al., 1986; Schimel et al., 1986). Direct exchange of matter may also take place by wind transport of soil particles and organic debris.

According to this concept, some watersheds (*sensu* Bormann and Likens, 1967) are physiographically delineated landscapes. Of course, most watersheds or basins are within a landscape (or even a site cluster or site), and some large watersheds include several landscapes. Upstream site clusters in watersheds yield water and dissolved or suspended nutrients to downstream bottomland clusters.

Higher Levels of Organization

Scholars from diverse disciplines have described many higher levels of organization including land systems, regions, ecoregions, climatic zones, soil zones, vegetation zones, biomes, continents, and the globe or biosphere. The boundaries of these may be distinct or indistinct, but each area contains many landscapes (Figure 4). From the biogeochemical perspective, however, certain patterns are common to all of these areas (Table 1). Water that drains from the land is essentially lost from terrestrial ecosystems. Therefore, the atmosphere is the principal carrier of matter among landscapes. Gases, ions, aerosols, and small particles are carried along the path of wind or air-mass movement (Figure 4), where they are subject to various depositional phenomena.

An example of the importance of atmospheric transport of "nutrients" at this level of organization is the current acid rain problem (C.E.Q., 1980).

The next higher practical level of organization is the globe itself. Nutrients derived from various regions, including grasslands, may be admixed in the

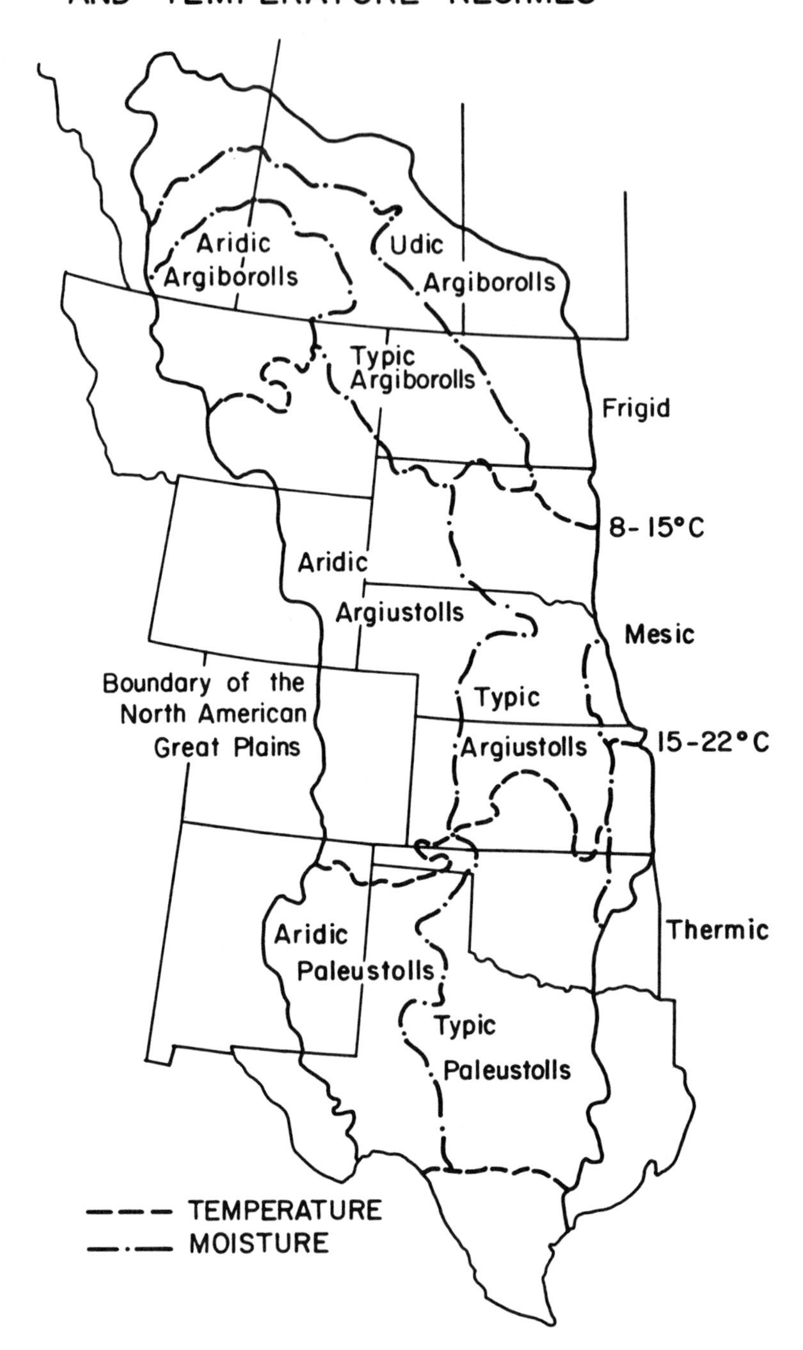

Figure 4. Map of the Great Plains of North America showing major soil zones that reflect major climatic regions. The potential vegetation in these regions was closely correlated to both climate and soils. (Map courtesy of R. D. Heil and P. C. Deutsch.)

atmosphere or ocean. Those nutrients then are dependent on general global circulation patterns for redistribution.

Discussion and Conclusion

An ecological hierarchy composed of organisms, sites, site clusters, landscapes, and higher-level systems is presented. Examples of biogeochemical processes and transport mechanisms at each level of organization are used to illustrate the unique properties of the systems discussed.

This hierarchy of ecological systems presented is, I hope, useful in building a conceptual framework for organization of many related or seemingly unrelated facts and observations about nutrient transfers in the environment within which we live. It organizes information and is based on tangible entities or objects that can be described in time and space and can be measured, albeit sometimes with difficulty. This scheme allows the placement of processes at meaningful levels within the hierarchy. For example, photosynthesis and nutrient uptake are directly associated with organisms and only indirectly with sites or landscapes. Likewise, erosion and leaching are site-scale rather than broad-scale processes. Though scientists may wish to consolidate information about carbon fixation into such generalizations as canopy, pasture, or stand photosynthesis, it is the enzyme systems in cells of leaves or stems that allow individual plants, not pastures, to fix CO_2 into carbohydrates. Processes at one level of organization must not be inappropriately applied to another level (e.g., transmutations) (O'Neill, 1979).

For components to form systems there must be some mode of interaction among the components. Furthermore, the interaction must be expressed as the transfer of energy or matter (or genes or information) between components. Thus, for example, "the shortgrass prairie ecosystem of North America" is a form of classification rather than a system with interacting subtypes. That is, the southern and northern shortgrass prairies are quite different, and are no more related by the direct exchange of matter than the mixed-grass prairie and the tallgrass prairie.

Finally, another advantage of a hierarchical approach is that the various levels of organization in the hierarchy lend themselves to mapping and aggregation according to the USDA/SCS Soil Classification System (Soil Survey Staff, 1975). Furthermore, the general environmental factors that influence system development at the various levels of organization can be characterized (Anderson et al., 1983). Thus, if we wish to estimate the amount of N_2O evolved from the Great Plains of North America, for example, we can aggregate our best information from individual sites to make an improved estimate compared to those currently available. We know from the work of Parton et al. (1988) that lowland sites in flowpaths produce several times more N_2O than do upland sites. By aggregating information about the extent of ecosystems according to the hierarchical concepts herein and those of Anderson et al. (1983), we can vastly improve our understanding of higher-order ecological systems.

Acknowledgments

This paper reports on work supported in part by National Science Foundation Grants DEB79-06009, "Additions, Losses and Transformations of Nitrogen in Grassland Ecosystems," and DEB79-11988, "Organic Matter and Nutrient Cycling in Semiarid Agroecosystems." I thank R. D. Heil for many conceptual ideas developed in this paper and Richard Forman for his critical review and helpful comments.

References

Allen, T. F. H. and T. B. Starr. 1982. Hierarchy: Perspectives for ecological complexity. University of Chicago Press, IL.

Anderson, D. W., R. D. Heil, C. V. Cole, and P. C. Deutsch. 1983. Identification and characterization of agricultural ecosystems at different integrative levels. R. Lowrance, R. Todd, L. Asmussen, and R. Leonard, eds., *Nutrient cycling in agricultural ecosystems*. Special Publication 23. University of Georgia, College of Agriculture Experiment Station, Athens, Georgia.

Bormann, F. H. and G. E. Likens. 1967. Nutrient cycling. *Science* 155:424–9.

C.E.Q. 1980. *The federal acid rain assessment plan*. Draft prepared by Acid Rain Coordination Committee, Council of Environmental Quality, Washington, DC.

Coleman, D. C., C. V. Cole, R. V. Anderson, M. Blaha, M. K. Campion, M. Clarholm, E. T. Elliott, H. W. Hunt, B. Shaefer, and J. Sinclair. 1978. An analysis of rhizosphere-saprophage interactions in terrestrial ecosystems. In U. Lohm and T. Persson, eds., *Soil organisms as components of ecosystems*. Ecology Bulletin (Stockholm) 25:299–309.

DeWit, C. T. 1970. Dynamic concepts in biology. In I. Sellik, ed., *Prediction and measurement of photosynthetic productivity*. PUDOC, Wageningen, The Netherlands, pp. 17–23

Feibleman, J. K. 1954. Theory of integrative levels. *British Journal of Philosophy and Science* 5:59–66.

Forman, R. T. T. and M. Godron. 1986. Landscape ecology. Wiley, New York.

Forrester, J. W. 1968. *Principles of systems*. Wright-Allen, Cambridge, MA.

Gorham, E., P. M. Vitousek, and W. A. Reiners. 1979. The regulation of chemical budgets over the course of terrestrial ecosystem succession. *Annual Review of Ecology and Systematics* 10:53–84.

Jenny, H. 1941. *Factors of soil formation: A system of quantitative pedology*. McGraw-Hill, New York.

MacMahon, J. A., D. L. Phillips, J. V. Robinson, and D. J. Schimpf. 1978. Levels of biological organization: An organism-centered approach. *BioScience* 28:700–4.

Odum, E. P. 1971. *Fundamentals of ecology*, 3rd ed. Saunders, Philadelphia.

O'Neill, R. V. 1979. Transmutations across hierarchical levels. In G. S. Innis and R. V. O'Neill, eds., *Systems analysis of ecosystems*. Statistical Ecology Series, International Cooperative, Fairland, Maryland, pp. 59–78.

O'Neill, R. V., D. L. DeAngelis, J. B. Waide, and T. F. H. Allen. 1986. A hierarchical concept of ecosystems. Princeton University Press, New Jersey.

Parton, W. J., A. R. Mosier, and D. S. Schimel. 1988. Rates and pathways of nitrous oxide production in a shortgrass steppe. *Biogeochemistry* 6:45–55.

Rowe, J. S. 1961. The level-of-integration concept and ecology. *Ecology* 42:420–7.

Schimel, D. S., W. J. Parton, F. J. Adamsen, R. G. Woodmansee, R. L. Senft, and M. A. Stillwell. 1986. The role of cattle in the volatile loss of nitrogen from a shortgrass steppe. *Biogeochemistry* 2:39–52.

Senft, R. L., L. R. Rittenhouse, and R. G. Woodmansee. 1986. Predicting patterns of cattle behavior on shortgrass prairies. *Proceedings of the American Society of Animal Science (Western Section)*. Vol. 31.

Soil Survey Staff. 1975. *Soil taxonomy*. USDA Soil Conservation Service Agricultural Handbook No. 436, Washington, D.C.

Swanson, F. J., R. L. Fredriksen, and F. M. McCorison. In press. Material transfer in a western Oregon watershed. In R. L. Edmonds, ed., *The natural behavior and response to stress of a western coniferous forest*. Dowden, Hutchinson, and Ross, Stroudsburg, Pennsylvania.

von Bertalanffy, L. 1950. An outline of general system theory. *British Journal of Philosophy and Science* 1:134–65.

Weiss, P. A. (ed.). 1971. *Hierarchically organized systems in theory and practice*. Hafner, New York.

6. Woods as Habitat Patches for Birds: Application in Landscape Planning in the Netherlands

W. Bert Harms and Paul Opdam

The western part of the Netherlands is one of the most densely populated areas of the world. A series of cities forms a ring of urban development surrounded by an open agricultural landscape (the *Randstad,* Figure 1). This urban-agricultural area separates a narrow coastal forest belt to the west from an agricultural landscape with intermingled woodlots and medium-sized forests on Pleistocene deposits to the east. This chapter explores how ecological prerequisites play key roles in the planning of new forest areas. Spatial relations between the existing and the planned forests will be particularly important in this analysis.

In a recent plan for the development of a green belt around the Randstad cities, new forests were projected to fulfill the increasing demands for timber production and recreation facilities. We developed a method to account for the ecological aspects of this plan, including both the localization and the management of these forests. It is an attempt to predict which species and which natural communities are to be expected, given a particular size and configuration of the forest patches, and a particular habitat quality. As an example, we describe this method for forest birds, a relatively well-studied group of animals.

For most species of temperate forest birds, habitat requirements can be identified at least in qualitative terms (Wiens and Rotenberry, 1981; Opdam and Schotman, 1986). It is becoming clear, however, that habitat quality is not the only predictor for the presence of forest-bird species. Species-area relationships have been known for several decades, but it is only recently that effects of area per se have been separated from the effects of habitat heterogeneity, which is

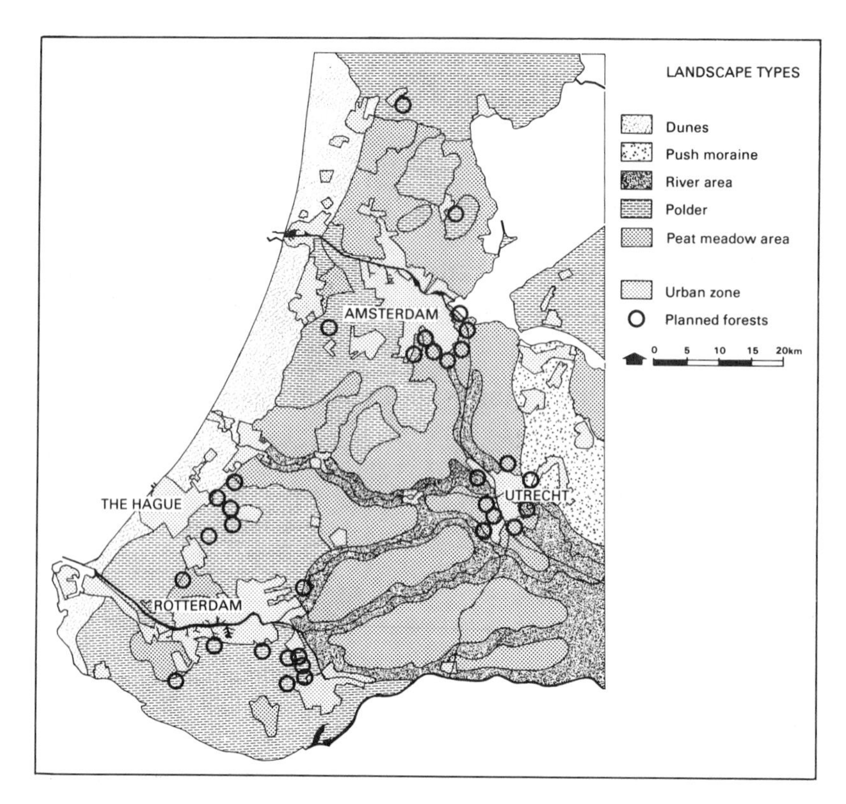

Figure 1. The Randstad area, the densely populated west-central portion of The Netherlands, and the planned forests of the Randstad Green Structure Plan.

commonly correlated with area (Freemark and Merriam, 1986). The area-per-se effect is associated with an area-dependent extinction rate, due to demographic stochasticity in the fluctuation in numbers of small populations (see, e.g., Goodman, 1987). A theory of metapopulation dynamics is emerging (Opdam, 1987; 1988; Merriam, 1988), based on the interaction between local extinction rates in habitat patches and recolonization rates. The role of isolation in the distribution of forest birds in woodlots has been clearly demonstrated. Usually, the amount of, or distance to, forest patches in the surrounding landscape is found to correlate with species number, or with the probability of occurrence of particular species (Howe, 1984; Lynch and Whigham, 1984; Opdam et al., 1985; Askins et al., 1987; Ford 1987; Blake and Karr 1987; van Dorp and Opdam 1987). Van Dorp and Opdam (1987) have also assessed the role of corridors.

Here, we use two Dutch studies (Opdam et al., 1985; Van Dorp and Opdam 1987) as a basis for a simulation model to predict the chance that forest-dwelling bird species will both colonize and persist in newly planted forests in the open landscape of the Randstad area (Harms and Knaapen, 1988).

Objectives

The aims of this chapter are to:

1. summarize the patterns of forest-interior bird species, as well as all bird species, as related to woods size and the isolation effects of matrix structure, including hedgerow density
2. analyze the generality of the results for application to the Randstad area
3. describe a method to predict colonization and persistence in a specific future forest, as a tool for landscape planning

In this paper we use bird communities as an example, and restrict ourselves to one level of spatial scale (the distribution of forest localities over the entire Randstad area) (Figure 2) and to the main landscape-ecological aspects of the problem. Space precludes a discussion about habitat aspects; here we assume that all forests will eventually develop into mature mixed deciduous forests, suitable habitat for any species of the forest-bird community. Also, the configuration of forest patches at a chosen site and the question of whether or not to connect them by corridors are not treated here (see Harms and Knaapen, 1988).

Major Trends in the Dutch Landscape

In prehistoric times, most of The Netherlands outside the coastal and peatland marshes was covered by mixed deciduous woodland. Due to extensive cutting and grazing, the landscape changed completely. In the beginning of the nineteenth century only 500 km^2 of woodland (1 percent of the nation's area) was left. Extensive areas of marshland, dry heathland, inland sand dunes, and moors encompassed most of the nation's semi-natural areas. In the second half of the nineteenth century, however, extensive reforestation started in the Pleistocene (eastern) part of The Netherlands, mainly aimed at future timber production. Heathland and moors gradually changed into monotonous pine plantations poor in plant species and in structural diversity. At present, 8 percent of the country is forested (Figure 2) and we are at the start of a second wave of afforestation (Anonymous, 1984), now mainly projected on clay and peat soils in the Holocene (western) part of the country (Figure 3).

The majority of the Dutch landscape is used for agricultural purposes (Figure 4). Agricultural production is very high, due to a high input of energy, cattle food (imported from developing countries), manure, and biocides. Being an important economic force, agricultural organizations have succeeded in intensifying land use to an extraordinarily high level, causing an enormous decrease of plant and animal species on semi-natural grassland and arable land.

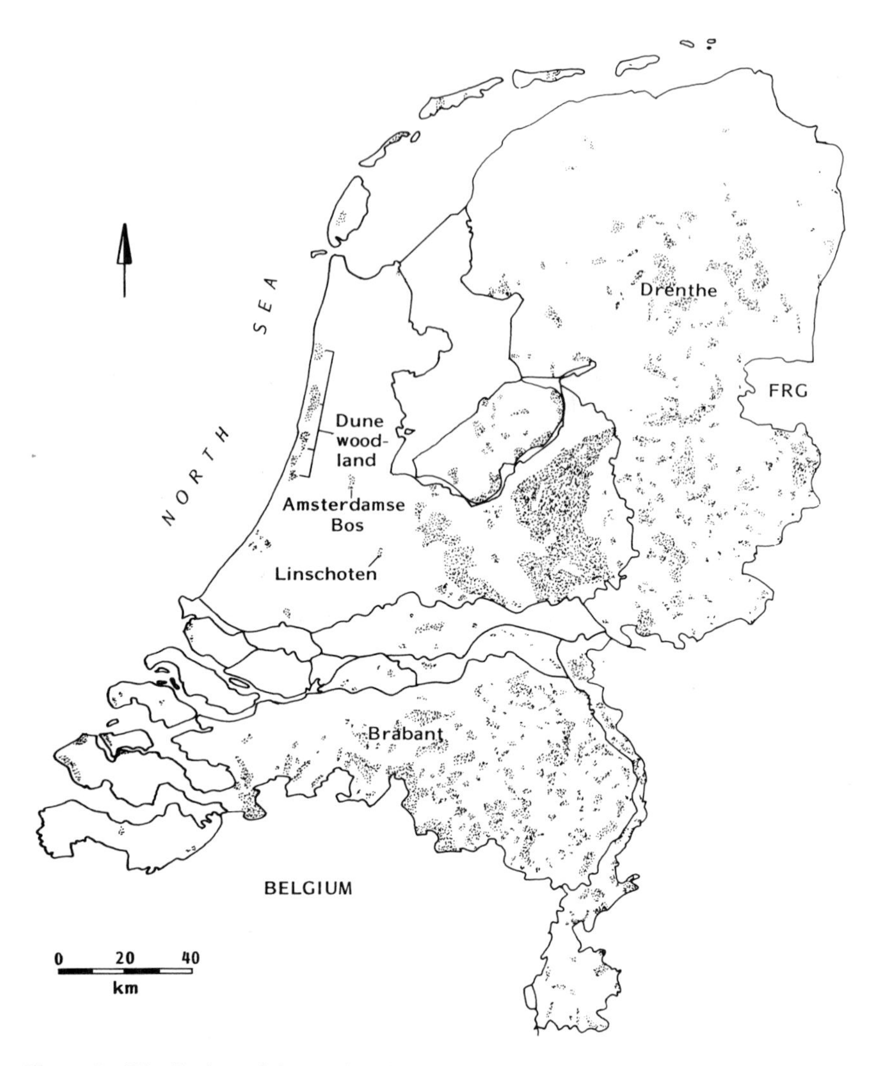

Figure 2. Distribution of forests in The Netherlands. (From Anonymous [1984].)

Small landscape elements, like woodlots and hedgerows have decreased in size or been removed to make the size of agricultural parcels more convenient to modern farming machines. This has altered the rural landscape in the eastern half of The Netherlands from a fine-grained, closed landscape to a coarse-grained open one (Harms et al., 1987). Plans are now being developed to reconstruct at least part of this characteristic network of woodlots and hedgerows and make these landscapes more suitable as a habitat for plant and animal populations. Ecological data are urgently needed as a basis for predicting the effects of these landscape changes, and to develop ecological criteria for landscape reconstruction.

Figure 3. Planted woods near town (Westland). Note glasshouses on left.

Forest Birds Relative to Woods and the Surrounding Mosaic

The impact of area and isolation on forest-bird communities was assessed in agricultural landscapes. Woods in this type of landscape are invariably in the order of magnitude of 1 to 20 ha (Figure 2), with the exception of a few large estates and some afforested former heath areas. These patches now form an archipelago of habitats situated in a matrix of cultivated land and sometimes interconnected by corridors. Quantification of the relationship between landscape structure and the distribution and persistence of species and populations offers a necessary ecological basis for landscape planning and land consolidation. Obviously, the presence of a species in a woodlot is primarily determined by habitat quality. In this study, however, we focused on size and isolation of woodlots and tried to minimize habitat variation between woodlots. The effect of woodlot size is supposed to be related to population size and thereby to extinction rate (Jones and Diamond, 1976). Thus, in these woodlots, small populations may

Figure 4. Intensive agriculture area with canals, homesteads, tiny woods, and a high density of cows. On former peat meadows in the western part of The Netherlands (Zoetermeer).

become frequently extinct, due to stochastic effects or genetic deterioration (Soulé, 1983). Persistence of a species in a fragmented landscape then depends on frequent recolonization from adjacent populations. Dispersal between patches is essential to survival. Dispersal frequency is assumed to be a function of interpatch distance and the density of corridors and stepping stones. It may also depend on the number and size of source populations. Since bird species differ in body size and territory size, species vary as to the mean population size in a patch of a given area. At the same time, we may expect species-specific dispersal abilities. Thus, species are expected to show different distribution patterns in a dissected landscape.

The effects of size and distance to nearest wood have been described for forest birds (e.g., Lynch and Whigham, 1984; Opdam et al., 1985), but the impact of corridors on isolation effects and the role of regional abundance is still unknown. We attempted to include all these variables of landscape structure to assess their relative importance to birds. For a detailed report see van Dorp and Opdam (1987).

Methods

Twenty-two regions of roughly homogeneous landscape structure were selected in the eastern, central, and southern parts of The Netherlands (Figure 2).

Interpatch distance, density of woodlots, amount of woods surrounding a woodlot, and density of connecting landscape elements varied considerably between, but not within, regions. Each region included 5 to 18 mature deciduous woodlots ranging from 0.1 to 35 ha in size. The total sample included 235 woodlots. All woodlots were selected for relative uniformity in tree layer, age class, and dominant tree species. Of course, complete homogeneity was not attainable, and the remaining variation in habitat quality was measured for twelve habitat variables.

Data on the presence of bird species were collected during four short standardized visits. The time spent in a woodlot was held constant at 15 minutes per visit for woodlots up to 5 ha. In larger woodlots, more time was needed to scan the whole area, up to approximately 30 to 45 minutes per visit. The presence of a species was derived from the combined four censuses on the basis of species-specific criteria described by Hustings et al. (1985). The criteria were chosen such that all species were surveyed at a minimal accuracy level of 90 percent. We derived the following parameters for birds: (a) number of forest species, that is, species restricted to forest habitat (these birds occur in sampled woods as well as other unsampled woody habitats of the area), but excluding birds nesting in the patch and feeding in the surrounding landscape; (b) number of forest-interior species, that is, species restricted to mature forest (the selected woodlots are the only patches of suitable habitat available in the area); (c) presence or absence of species.

The relationship between bird parameters and habitat and landscape variables was analyzed by multiple regression and multiple logistic regression methods.

Results

Both the number of forest species and the number of forest-interior species were correlated with woodlot size. As shown earlier (Opdam et al., 1985), individual species differed considerably in their response to increasing patch size. In the 235 woodlots, 32 birds were found to be forest species, of which 15 birds were considered as forest interior species (Table 1). Five of the 32 forest species—chaffinch *(Fringilla coelebs)*, blackbird *(Turdus merula)*, great tit *(Parus major)*, blue tit *(Parus caeruleus)* and chiffchaff *(Phylloscopus collybita)*—were in all woodlot sizes, though sometimes absent from the smallest woodlots (< 1 ha). The 27 remaining species (84 percent of the total) were all found to be size-dependent (i.e., were present only in woodlots down to a minimum size). Progressively smaller woodlots contained fewer and fewer forest-bird species. The wood warbler *(Phylloscopus sibilatrix)* was most affected, occurring only in woodlots exceeding 10 ha in size (0.4 probability of occurrence).

At the species level, the probability of occurrence in a particular woodlot size class differed regionally (Figure 5). To clarify this pattern, we separated eastern woodlots from central and southern ones (Figure 2), the former area having considerably more mature deciduous woodland than the other areas. The distance between these regions is roughly 100 to 150 km, but even over this short distance obvious geographical trends in species frequency (and presumably also in species

Table 1. Occurrence of bird species in deciduous woodlands in different regions of The Netherlands. Column 1: YES = evidence of regular occurrence; yes = evidence of occasional occurrence; no = no evidence; ? = evidence equivocal. Other columns: + = present; – = absent; ⊗ = habitat unsuitable.

Species		Woodlots in east/center/ southeast	Woodlots in west (Randstad)			Woodlots in south (Brabant)
			Linschoten	Amster-damse bos	Dune woodland	
Nuthatch	*Sitta europaea*	YES	–	⊗	+	–
Marsh tit	*Parus palustris*	YES	–	–	+	–
Hawfinch	*C. coccothraustes*	yes	–	–	–	–
Green woodpecker	*Picus viridus*	YES	⊗	⊗	+	+
Pied flycatcher	*Ficedula hypoleuca*	yes	–	–	–	–
Wood warbler	*Phylloscopus sibilatrix*	?	–	–	+	+
Black woodpecker	*Dryocopus martius*	?	–	⊗	–	+
Tawny owl	*Strix aluco*	?	+	+	+	–
Lesser spotted woodpecker	*Dendrocopos minor*	?	+	+	+	+
Great spotted woodpecker	*Dendrocopos major*	Yes	+	+	+	+
Tree creeper	*Certhia brachydactyla*	yes	+	+	+	+
Long-tailed tit	*Aegithalos caudatus*	yes	+	+	+	+
Golden oriole	*Oriolus oriolus*	yes	+	+	+	+
Blue tit	*Parus caeruleus*	no	+	+	+	+
Chaffinch	*Fringilla coelebs*	no	+	+	+	+

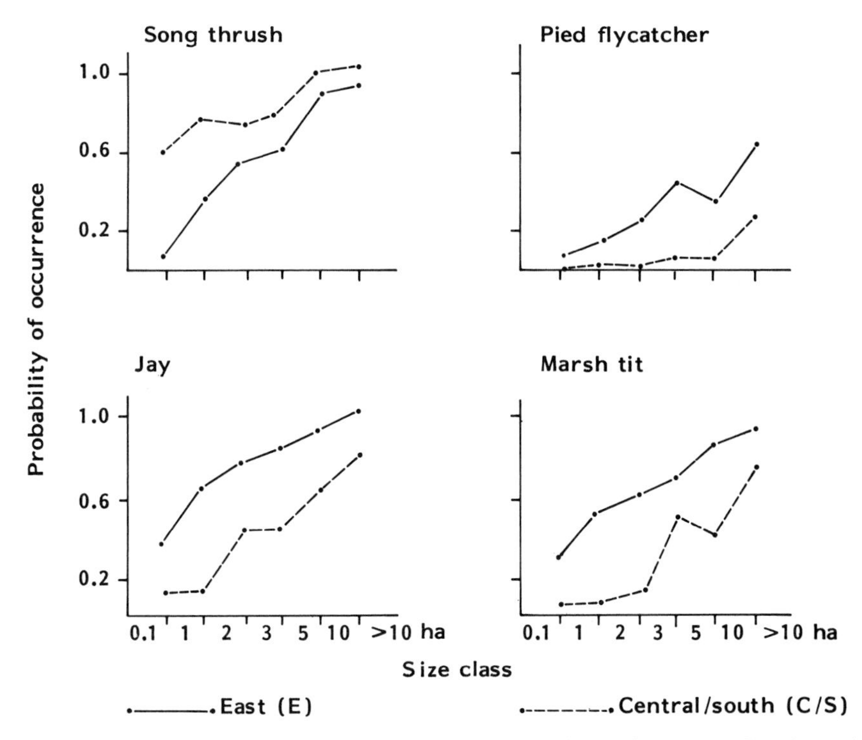

Figure 5. Probability of occurrence of four bird species in relation to woodlot size and regional difference. Song thrush *(Turdus philomelos)* and jay *(Garrulus glandarius)* are forest species; pied flycatcher *(Ficedula hypoleuca)* and marsh tit *(Parus palustris)* are forest-interior species. (From van Dorp and Opdam [1987].)

density) were apparent. Nine species were significantly more frequent at eastern sites, and seven species showed higher values in central/southern woodlots.

A similar regional difference was found in the effect of isolation and connectivity. After fitting a regression model for area and habitat variation, the residual variation was correlated with isolation factors. For forest-interior species, several isolation variables correlated significantly with this residual (which may be interpreted as a measure of species richness), but only after accounting for the regional differences. Of these isolation variables, the density of hedgerows explained the largest part of the variation (Figure 6). The trends in the two geographical regions are similar, but there is a constant difference in species richness, in favour of the eastern woodlots.

Van Dorp and Opdam (1987) found that 11 of the 32 forest bird species (34 percent) show isolation effects. Seven of these were ranked as forest-interior species. Green woodpecker *(Picus viridis),* marsh tit *(Parus palustris)* and nuthatch *(Sitta europaea)* showed the most distinct effects (i.e., were most dependent on connecting hedgerows).

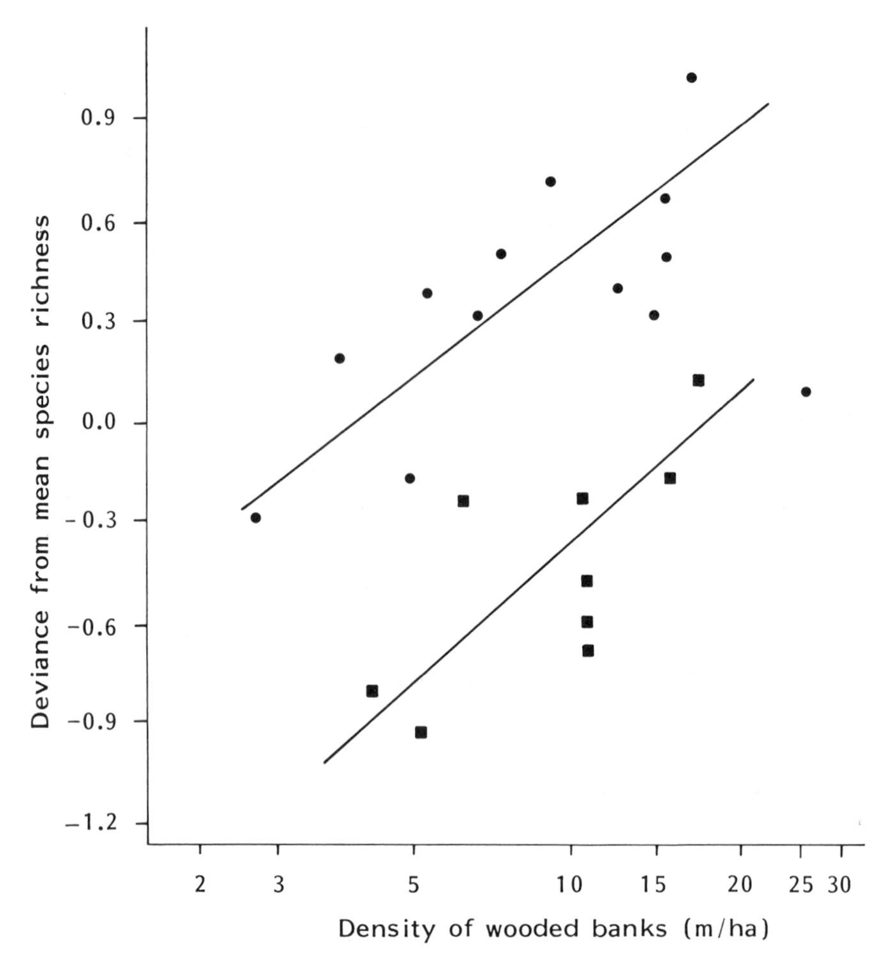

Figure 6. Relationship between density of wooded banks (hedgerows) and species richness. Hedgerow density is recorded as total length (m) of hedgerows per hectare. Along the vertical axis, the deviance from the mean species number for a region is plotted (after correcting for woodlot area and habitat effects). (From van Dorp and Opdam [1987]. ● = eastern regions; ■ = central and southern regions.)

Discussion

In summary, area accounted for 58 percent of the variation in species number for both forest species and forest-interior species. Habitat explained only 2 percent of the variation, though this does not imply that habitat is not an important factor. It is merely the result of the way we selected samples, trying to minimize habitat variation. For forest species the variation unexplained by area and habitat was not related to any isolation factor. However, for forest-interior species 82 percent of the residual variance was explained by a combination of regional difference, total wooded area within 1 km, and density of wooded banks. This result supports the prediction that these woodlots are habitat islands for forest-interior birds, but less so for other forest birds.

We conclude that these effects on distribution patterns are caused by extinctions being most frequent in the smallest and most remote woodlots, and recolonization being more frequent the less isolated the patch is. Admittedly, these conclusions are based on a one-year pattern analysis. To test the validity of this reasoning, we carried out a three-year survey of 68 of the woodlots. The results indicate a considerable yearly turnover in species composition. Van Noorden (1986) found a distinct negative correlation between extinction probability and woodlot size, whereas the amount of wood in the surrounding landscape had a similar, though much weaker effect. The frequency of recolonization was demonstrated to correlate positively with the number of surrounding woodlots. Thus, these results support our conclusions about distribution dynamics of forest bird species in a dissected landscape.

Another source of evidence is a study to evaluate isolation effects on territorial density (van Noorden, 1986).* This study confirmed the positive relation between population density in woodlots and frequency of occurrence suggested by Hanski (1982) and Brown (1984) (Figure 7). After accounting for variation in sample plot area, van Noorden found the total density of forest-interior birds to be positively correlated with a combination of area of wood in the surrounding landscape and presence of wooded banks or lanes (46 percent of the variation explained). Presumably due to the relatively small sample size (40 woodlots), only five species were found to correlate with isolation factors, four of which were also found to be affected by isolation in the one-year study described above.

We therefore conclude that the results of the one-year study of an extensive area are sufficiently supported to permit application in landscape planning. Although the validity of the regression model has not yet been tested with an independent data set, the model may be used to predict the expected effects of landscape changes, such as a decrease in hedgerow density (Figure 8).

Generalization of the Bird and Landscape Structure Results

The results presented in the preceding section are based on a study at the landscape scale. For forest-interior species, isolation effects were demonstrated to occur over distances ranging from a few to several tens of kilometers. Also, at a broader level of scale, differences were found within species as to the probability of occurrence in woodlots of the same size and in the same landscape type, but in different geographical areas.

*Assuming a roughly constant habitat quality, and hence a constant carrying capacity, he tested the prediction that if vacant territories would occur, they should be found in isolated woodlots more often than in woodlots surrounded by woods in the landscape or connected to a corridor network.

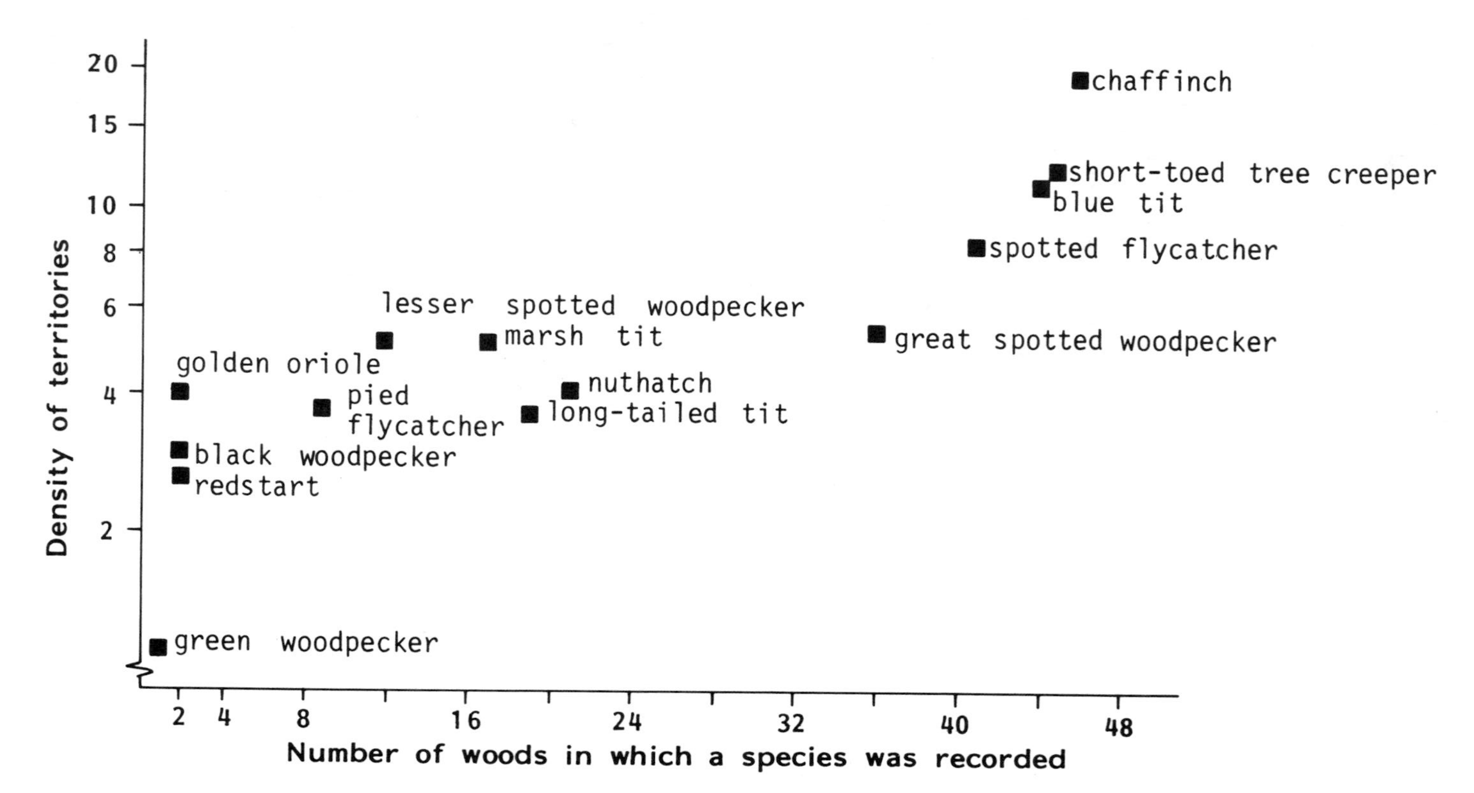

Figure 7. Relationship between density (pairs/10 ha) of bird territories in suitable habitat patches and frequency of species occurrence. Measured in 40 woodlots in the Gelderse Vallei, central part of The Netherlands (after van Noorden, [1986]).

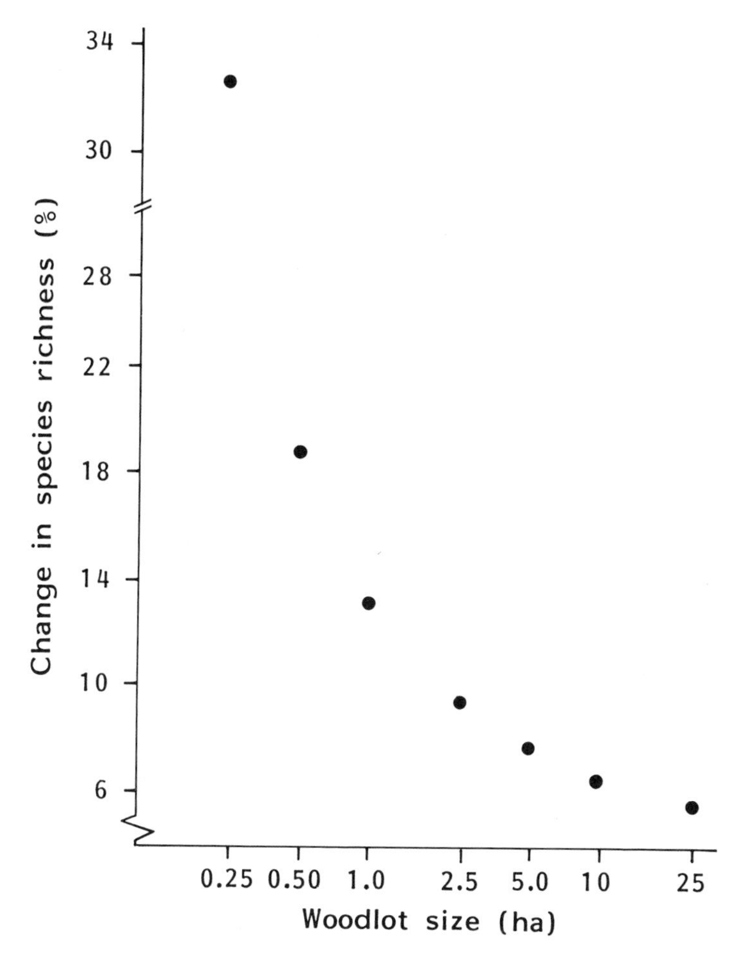

Figure 8. Predicted loss of bird species richness in woodlots after an imaginary decrease in surrounding hedgerow density from 60 to 12 m/ha. Based on a regression model by van Dorp and Opdam (1987).

Is it possible to generalize these results to predict colonization and survival in future forests planned for the Randstad area (Figure 1)? Only eight of the woodlots investigated are situated in the Randstad landscape, which is much more open (Figure 4), so that forest birds have to cross broader strips of treeless agricultural landscape than in the regions of The Netherlands where our study was carried out. It is uncertain whether or not forest birds would disperse into very open landscape without being able to reach a group of trees or a bush within their range of vision. Another difficulty arises from the regional abundance trend: Extrapolating this trend further west, beyond the range of forest on Pleistocene deposits, produces another source of uncertainty. Fortunately, certain sources of evidence may serve as a qualitative test of our results.

To evaluate the applicability of the woodlot bird studies to the Randstad area, we used the Dutch breeding bird atlas (Teixeira, 1979) and the regional atlas of the province of Southern Holland (Vogelwerkgroep Avifauna West-Nederland, 1981). In Table 1 the evidence for isolation effects found by Opdam et al. (1985), van Dorp and Opdam (1987), and van Noorden (1986) is summarized by species in the east/center/southeast of The Netherlands, and compared to the presence of these species in other regions. For the wood warbler and the black woodpecker *(Dryocopus martius)* we were unable to assess isolation effects, due to scarcity of these species in the woodlots. However, this scarcity by itself may be regarded as evidence for an isolation effect, since the species are not too uncommon in extensive forests. The evidence for the lesser spotted woodpecker *(Dendrocopos minor)* is equivocal.

Two forest areas, Linschoten and Amsterdamse bos, are situated in the Randstad (Figure 2), at 20 and 15 km from a large forested source area. The former area is an old estate (ca. 20 ha in size) and the latter (ca. 500 ha) was planted some 50 years ago. In both areas about half of the forest-interior bird species are absent (Table 1). For some of these absences, habitat unsuitability cannot be excluded as a possible cause. For example, the green woodpecker prefers sandy-soil woodland because of a dependence on wood ants as a food source, whereas the tree layer in the Amsterdamse bos may still be unsuitable as a habitat for black woodpecker and nuthatch. The other absences are probably attributable to isolation effects.

A strip of mainly deciduous mature forest which presumably has always been separated from the forest areas on Pleistocene deposits by a gap of 40 to 60 km is situated along the coastal dunes (see Figures 1 and 2). These isolated but extensive woods have been reached by nuthatch, marsh tit, and green woodpecker, which have succeeded in founding large populations. Hawfinch *(Coccothraustes coccothraustes)* and pied flycatcher *(Ficedula hypoleuca)* are rare and irregular breeding birds. The black woodpecker has recently founded a small population, notably in the part of the woodland belt where the distance to the source population is smallest.

A century ago, the province of Brabant (situated in the southern part, along the Belgian border) was covered by much heathland and peat moors, but forest was scarce. Since then, extensive afforestations have changed the aspect of the province from an open to a fairly wooded landscape (Figure 2). Estates with mature deciduous forest mainly occurred along the Belgian border, 70 to 100 km southeast of the main distribution cores of forest-interior birds. In these remote forest patches, nuthatch, marsh tit, hawfinch, pied flycatcher and tawny owl *(Strix aluco)* never succeeded in founding a persistent population (Table 1). The situation in the province of Drenthe (in the Northeast) was similar. Recently, presumably due to the planting of (mainly coniferous) forests, which function as a transition habitat, populations of these species are extending their ranges and will sooner or later spread over all the suitable forest patches in these provinces.

For at least half of the forest-interior species, these general comparisons (Table 1) support our conclusions about isolation-affected distribution patterns in a fragmented landscape. The absence of nuthatch and marsh tit from suitable habitat patches is especially interpreted as an indication of isolation effects. Depending on large trees for feeding, the nuthatch is typical for late successional stages of deciduous forest types. In the future forest of the Randstad, the presence of the nuthatch will be a key indicator of a species-rich forest-bird community, as well as a biogeographically favourable locality.

What size should a forest have to support a viable population of nuthatches? The species was found in over 90 percent of the woodlots exceeding 10 ha in size. However, because of the decreasing east to west trend in abundance (e.g., Figure 5), the minimal forest-island size in the Randstad will be larger (cf., the existing Linschoten estate of 20 ha, where the nuthatch is missing). We assume 50 ha to be the minimal habitat area to support a population with a low probability of going extinct. In a mature forest of that size, 10 to 15 pairs of nuthatches may have breeding and winter territories. A population of that size is not very likely to disappear by stochastic demographic processes alone. Yet some immigration remains necessary. Rare cases of local extinction will have to be compensated by recolonization. Also, immigration should prevent genetic deterioration, which may readily occur in populations smaller than about 50 pairs (Franklin, 1980). Except for the green woodpecker (which prefers sandy soils), and perhaps the black woodpecker, hawfinch, and pied flycatcher (with at present strong geographical trends, yet absent in the western dunes), all forest-interior species are expected to coexist in a 50-ha mature forest in the Randstad.

Application in Landscape Planning

The Planning Area: The Randstad of Holland

The area enclosed by the four largest cities of The Netherlands can be divided into six landscape types (Figure 1). A coastal zone of sand dunes to the west is covered by dune shrubs on the shore side and forests on the interior side. The forests are mainly deciduous wood, but in the northern part planted pine woods predominate. The eastern border of the Randstad is marked by a Pleistocene push moraine, a ridge up to 100 m above sea level. This ridge is covered by extensive deciduous and planted pine woods. Between these two woodland areas two very open landscape types are evident: (a) grasslands on peat-soil (Figure 4) and (b) polders, reclaimed from the sea or from peat-bog lakes, and used mainly as arable land. Both landscapes are almost free of forests and even single trees are scarce.

The Randstad area is crossed by several rivers (Figure 1). In this peatland, the clayey riverbanks have long been the favorite site of human settlers, and old estates surrounded by deciduous forest remain. The sixth landscape type is the urban zone represented by the four cities—Amsterdam, Rotterdam, The Hague, and Utrecht (Figure 1)—and by a number of smaller cities in between. Several green space areas (maximum 50 years old) are present in the urban landscapes

(Figure 3), and locally the tree density is relatively high due to road plantation and other kinds of landscaping.

An outline plan called the Randstad Green Structure (Anonymous, 1985) has been developed for the Randstad for urban afforestation. This plan entails the development of about 10,000 ha of forests and recreation areas within the next 15 years. The Randstad Green Structure Plan is based on the following considerations:

1. Green areas in and near the towns are threatened in many ways and have to be reconstructed and enlarged.
2. The recreational benefits of the green areas have to be improved.
3. Forests will have to be planted in light of the policy to increase the national timber production from 8 to 25 percent of the consumption (Anonymous, 1984).

The forests are mainly planned in or near the urban zone (Figure 1). From a landscape-ecological point of view these future forests could be considered as a favorable landscape-ecological framework for the dispersal and persistence of forest species. To evaluate this hypothesis we developed a spatial model to predict the probability of colonization and the probability of persistence of species in suitable habitat patches—i.e., mature deciduous woodlands of variable size in various locations in the Randstad (Harms, 1987).

Predicting Colonization

The chance of colonization of the new forests depends on several factors apart from habitat quality. Most important is the existence of a source population. Large areas with high densities of forest species are assumed to function as source areas for dispersing individuals. Based on the results discussed in the preceding sections, we assumed that for forest-interior species, the minimal size of a source area should be 50 ha.

However, species with downward population trends are not likely to expand their ranges and colonize remote patches. So we have to consider regional population trends too. These trends we derived from the Dutch Breeding Bird Monitoring Scheme (SOVON, unpubl. results). Taking into account these population trends, we used the distribution maps available for all breeding-bird species (Teixeira, 1979) to select the source areas from which potential colonizers may be expected. Other factors determining colonization probability are the distance of the new forest from the source area, the resistance of the landscape to the dispersing wood-dwelling birds, and the dispersal capacity of the species. We estimated the two latter variables from the results mentioned before. The dispersal resistance of the landscape is based largely on assumptions about the relative resistance of various landscape elements to forest-interior species. The lack or amount of woody vegetation is an important factor considered in this estimate (Figure 9).

With these data sets we were able to determine the input variables for a simple

Figure 9. Extensive glasshouses (greenhouses) and high landscape resistance to avian dispersal. In Westland.

spatial model. The model simulates dispersal from the population source to the new forests, taking into account the dispersal rate determined by the resistance values of the landscape. The simulation model and its underlying assumptions is discussed in detail by Scheffer and Knaapen (in preparation). The model is implemented in a Geographical Information System MAP2 (Tomlin, 1983; van den Berg et al., 1985). The results of the simulation are belts of relative accessibility to forest species centered around the source areas. By superimposing the locations of the planned forests on the accessibility zones, a relative probability of colonization can be given.

As mentioned before, the nuthatch *(Sitta europaea)* is a key indicator for a group of characteristic bird species of mature deciduous forests. The source areas of the nuthatch (Figure 10) are mainly located in the dune strip and on the push moraine ridge, although some old estates along small rivers may also be minor species sources. The resistance of the landscape to dispersal of forest-interior species is then plotted (Figure 11). The results of the dispersal simulation (Figure 12) are expressed as four zones of relative accessibility of the landscape, which is a simplification of the simulation results. We emphasize that accessibility is

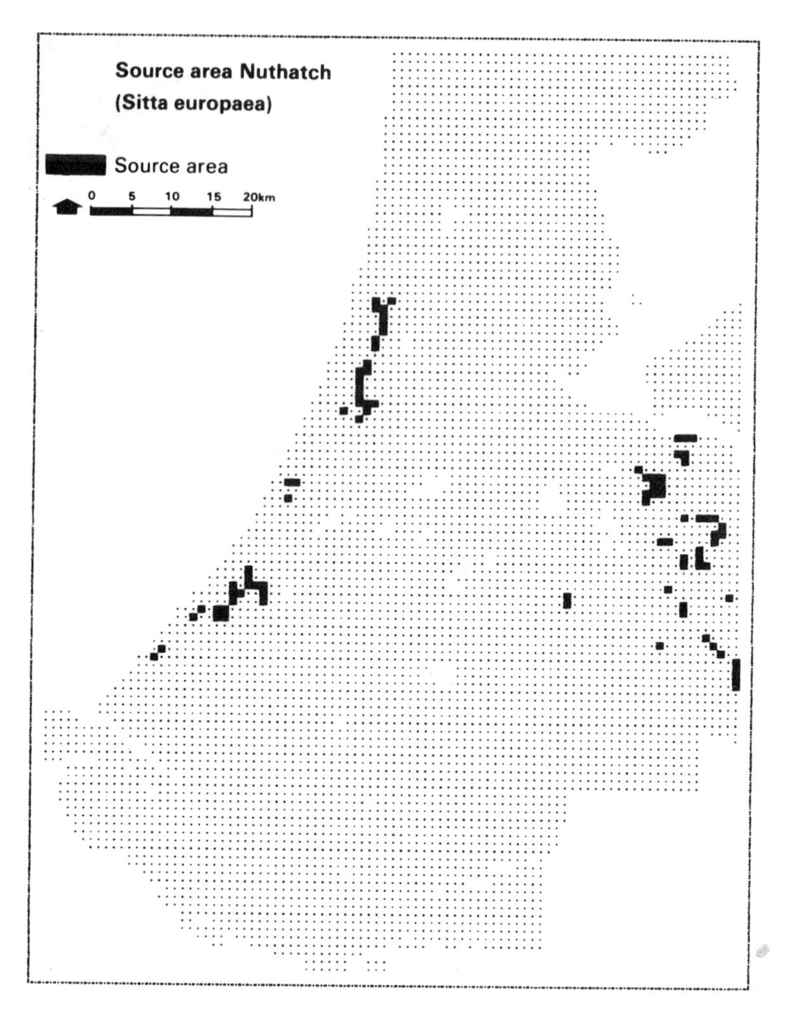

Figure 10. The source areas of the nuthatch *(Sitta europaea)*. Source areas are mainly located in the dune strip to the west, and on the push moraine ridge to the east (compare with Figure 1). Based on breeding-bird surveys (Teixeira, 1979).

expressed in a relative time scale and not in absolute time units. In Figure 13 these four zones of accessibility have been compared to the localities proposed in the outline plan of the Randstad Green Structure.

Predicting Persistence

Once a species has settled in a new forest, an important question that remains is whether or not the population will be large enough and viable enough to persist

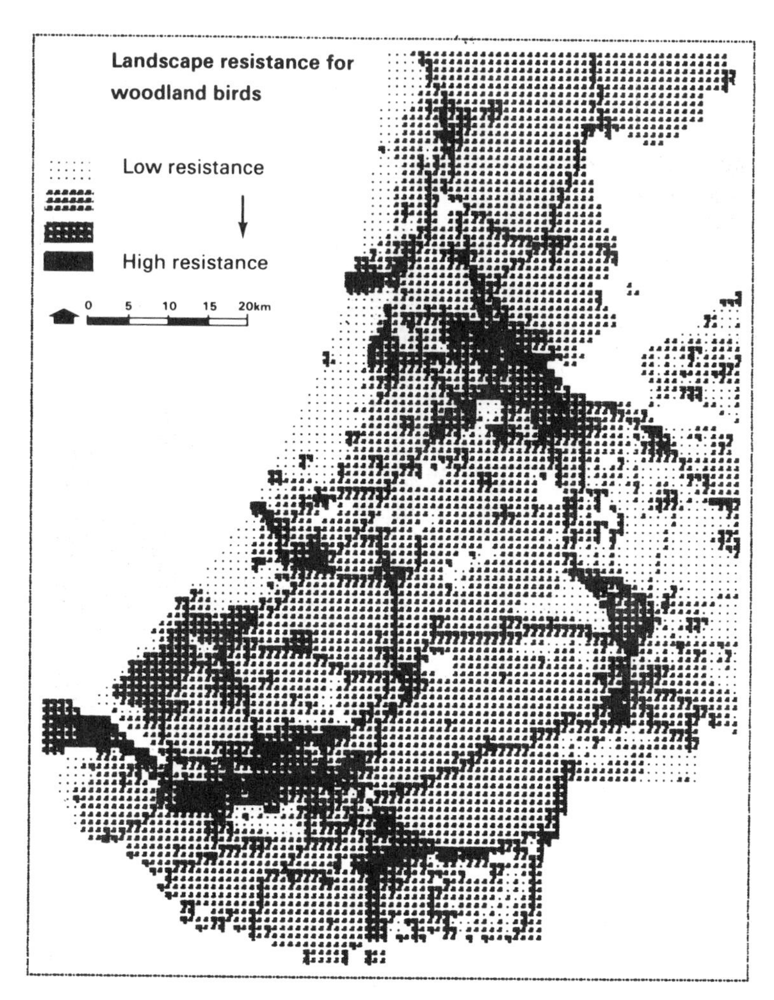

Figure 11. The resistance of the landscape for dispersal of forest-bird species. The dispersal resistance value is expressed here in four classes, but in the simulation model more classes were used. Cities, greenhouses (glasshouses), and busy traffic roads have the highest landscape resistance. Simulation model results.

for long. To estimate the probability of persistence we have to consider the following factors.

First, persistence is likely to depend on the area of the patch and on the population dynamics of the species. From the results presented earlier in this chapter we derived a woodlot size of 50 ha as the minimal area with a occurrence approaching 100 percent for nuthatch or other related species.

A second factor is the existence of suitable patches in the surroundings of the planned forests. As we have seen, the total area of woodlots and the distance to

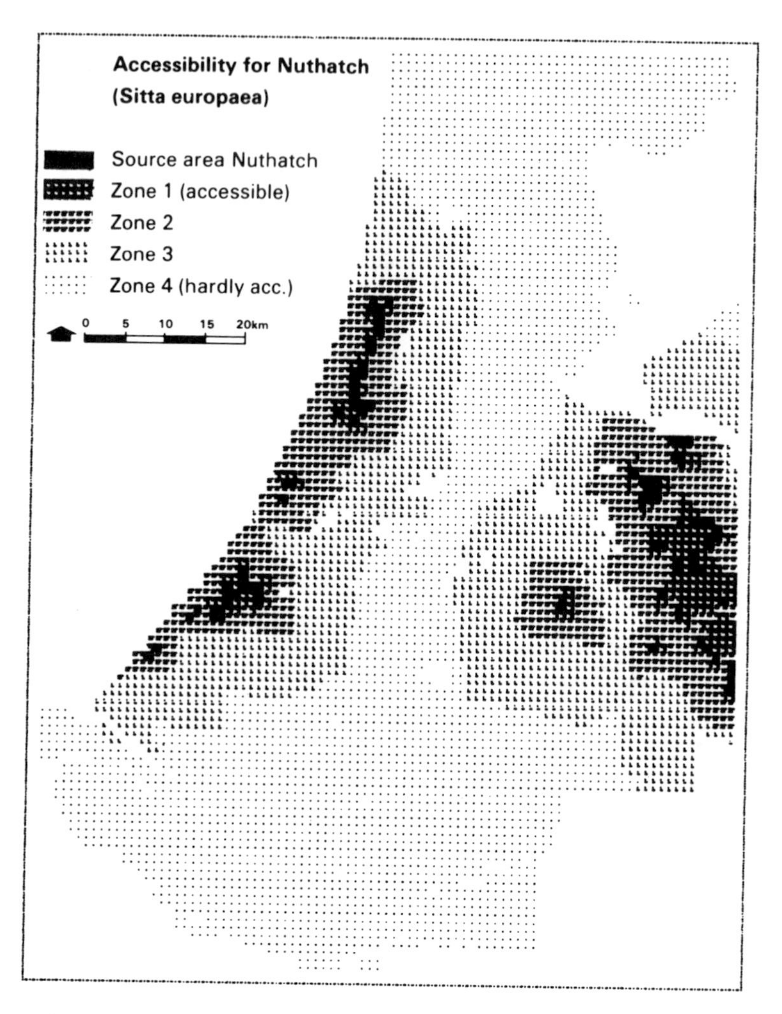

Figure 12. The relative accessibility for the nuthatch *(Sitta europaea)*. Note that the South and the North are hardly accessible due to distance and high landscape resistance. Simulation model results.

the nearest wood are positively correlated with the probability of occurrence of forest-interior species. However, this factor is only relevant to woods less than 50 ha. This factor was determined by scanning the gridcells of the geographical information system for suitable habitat within approximately 3 km of the planned forests. For new forest locations less than 50 ha in size the probability of persistence was taken proportional to the amount of nearby forest. For each planned forest the relative chance of persistence of an avian species is given in three classes (Figure 14).

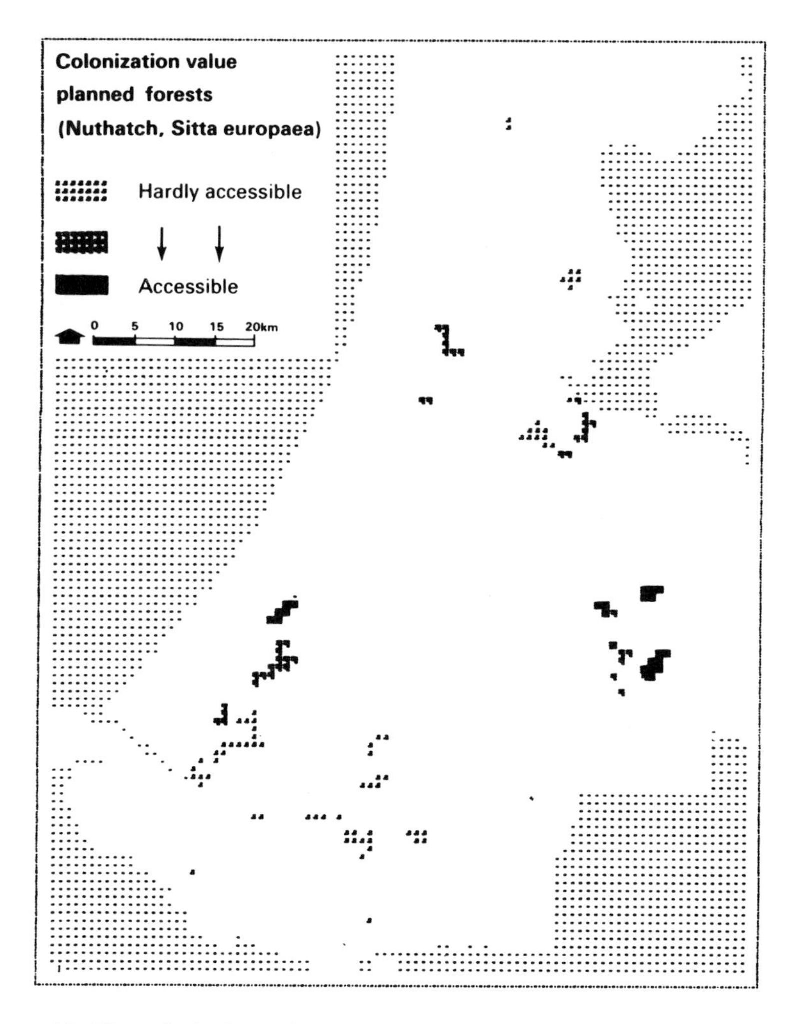

Figure 13. The colonization value of the planned forests for the nuthatch *(Sitta europaea)*. Locations near Rotterdam to the south (Figure 1) are hardly accessible. Simulation model results.

Results

A comparison of colonization probability versus persistence probability for the nuthatch (figures 13 and 14) provides several insights. In the neighborhood of Amsterdam and The Hague the values of these parameters are approximately similar. However, in the surroundings of Utrecht and Rotterdam the probabilities of colonization and persistence are inversely correlated. Near Rotterdam some locations show a poor colonization chance, due to the absence of a nearby source area and a high resistance of the landscape, but a good survival chance, because

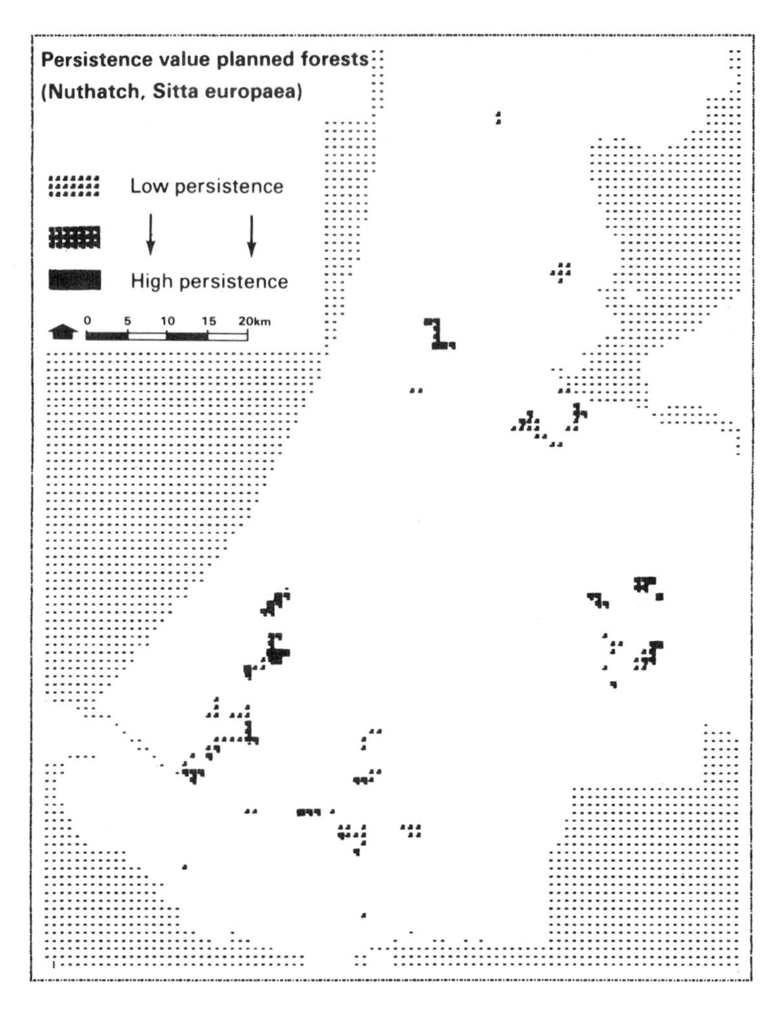

Figure 14. The persistence value of the planned forests for the nuthatch *(Sitta europaea)*. Note the high persistence value of some locations near Rotterdam to the south (Figure 1), even though they are hardly accessible. Simulation model results.

several small woodlots are grouped together. By contrast, some locations near Utrecht illustrate a better accessibility, but persistence is less probable, due to the lack of any suitable habitat in the direct surroundings of the planned forest. In this case, planners may decide to enlarge these planned forests to at least 50 ha. Near Rotterdam, these results may lead to reducing the resistance of the landscape between the new forests and the source areas, or redistributing the planned localities of new forests.

Discussion

The method presented here could be a useful tool for landscape planning and decision making, providing a relatively objective means for incorporating critical ecological information. We also note that application of this preliminary model is liable to a number of restrictions. For instance, in determining the source areas, we did not have population density information, which should soon be available. At the moment, the applicability of the model is mainly limited by the lack of data to estimate the resistance of the landscape to migrating organisms. For the case described in this chapter, that resistance was estimated partly on the basis of research data and partly on best professional judgement.

The validity of the model will be further tested in the future by comparing the results with the actual distribution pattern of the species concerned, by monitoring range expansions, and by measuring population turnover in landscapes with scattered woods. For the present, the method should be considered as an aid in choosing the best localities for urban afforestation, rather than a model to precisely account for the probability of species colonization and persistence.

Acknowledgments

The authors gratefully acknowledge their colleagues D. van Dorp, J. Knaapen, and M. Scheffer for contributions to the ideas expressed in this chapter and for constructive criticism of the manuscript.

References

Anonymous. 1984. *Meerjarenplan bosbouw*. Tweede kamer 1984–1985, 18630 numbers 1–2.

Anonymous. 1985. *Nota ruimtelijk kader randstadgroenstructuur*. Ministerie van Volkshuisvesting, Ruimtelijke Ordening en Milieubeheer en Ministerie van Landbouw and Visserij. Staatsuitgeverij, 's-Gravenhage.

Askins, R. A., M. J. Philbrick, and D. S. Sugeno. 1987. Relationship between the regional abundance of forest and the composition of forest bird communities. *Biological Conservation* 39:129–52.

Blake, J. G. and J. R. Karr. 1987. Breeding birds of isolated woodlots: area and habitat relationships. *Ecology* 68:1724–34.

Brown, J. H. 1984. On the relationship between abundance and distribution of species. *American Naturalist* 124:255–79.

Ford, H. A. 1987. Bird communities on habitat islands in England. *Bird Study* 34:205–18.

Franklin, I. R. 1980. Evolutionary change in small populations. In M. E. Soulé and B. A. Wilcox, eds., *Conservation biology*. Sinauer Associates, Sunderland, Massachusetts, pp. 135–49.

Freemark, K. E. and Merriam, H. G. 1986. Importance of area and habitat heterogeneity to bird assemblages in temperate forest fragments. *Biological Conservation* 36:115–41.

Goodman, D. 1987. The demography of change extinction. In M. E. Soulé, ed., *Viable populations for conservation*. Cambridge University Press, United Kingdom, pp. 11–34.

Hanski, I. 1982. Dynamics of regional distribution: The core and satellite species hypothesis. *Oikos* 38:210–21.

Harms, W. B., ed. 1987. *Ecologische infrastructuur en bosontwikkeling in de Randstad*. Rapport Rijksinstituut voor onderzoek in de bos- en landschapsbouw "De Dorschkamp," Number 484, Wageningen, The Netherlands.

Harms, W. B., A. H. F. Stortelder, and W. Vos. 1987. Effects of intensification of agriculture on nature and landscape in the Netherlands. In M. G. Wolman and F. G. A. Fournier, eds., *Land transformation in agriculture*. SCOPE 32. Wiley, New York, pp. 357–80.

Harms, W. B. and J. P. Knaapen. 1988. Landscape planning and ecological infrastructure: The Randstad study. In K.-F. Schreiber, ed., Connectivity in landscape ecology. *Münstersche Geographische Arbeiten* 29. Münster, Federal Republic of Germany, pp. 163–8.

Howe, R. W. 1984. Local dynamics of bird assemblages in Australia and North America. *Ecology* 65:1585–1601.

Hustings, M. F. H., R. G. M. Kwak, P. F. M. Opdam, and M. J. S. M. Reijnen. 1985. *Vogelinventarisatie: Achtergronden, richtlijnen en verslaglegging*. Natuurbeheer in Nederland, Deel 3. PUDOC, Wageningen, The Netherlands.

Jones, H. L. and J. M. Diamond. 1976. Short-time base studies on turnover in breeding bird populations on the California Channel Islands. *Condor* 78:526–49.

Lynch, J. F. and D. F. Whigham. 1984. Effects of forest fragmentation on breeding bird communities in Maryland, USA. *Biological Conservation* 28:287–324.

Merriam, G. 1988. Landscape dynamics in farmland. *TREE* 3:16–20.

Opdam, P. 1987. Metapopulatie, model van een populatie in een versnipperd landschap. *Landschap* 4:289–306.

Opdam, P. 1988. Populations in fragmented landscape, *In* K.-F. Schreiber, ed., *Connectivity in Landscape Ecology*. Proceedings of the 2nd international seminar of the international association for landscape ecology. Münstersche Geographische Arbeiten 29. Münster, Federal Republic of Germany, pp. 75–77.

Opdam, P., G. Rijsdijk, and F. Hustings. 1985. Bird communities in small woods in an agricultural landscape: Effects of area and isolation. *Biological Conservation* 34:333–52.

Opdam, P. and A. Schotman. 1986. De betekenis van structuur en beheer van bossen voor de vogelrijkdom. *Nederlands Bosbouwtijdschrift* 58:21–33.

Soulé, M. E. 1983. What do we really know about extinction? In C. M. Schonewald-Cox, S. M. Chambers, B. MacBruyde and W. Z. Thoms, eds., *Genetics and conservation*. Benjamin/Cummings, Menlo Park, California, pp. 111–24.

Soulé, M. E. and D. Simberloff. 1986. What do genetics and ecology tell us about the designing of nature reserves? *Biological conservation* 35:19–40.

Teixeira, R. M. 1979. *Atlas van de Nederlandse broedvogels*. Vereniging tot Behoud van Natuurmonumenten in Nederland, 's-Graveland.

Tomlin, C. D. 1983. *Digital cartographic modeling techniques in environmental planning*. Yale University Press, New Haven, Connecticut.

van den Berg, A., J. van Lith, and J. Roos. 1985. Toepassing van het computerprogramma MAP2 in het landschapsbouwkundig onderzoek. *Landschap* 2:278–93.

van Dorp, D. and P. Opdam. 1987. Effects of patch size, isolation and regional abundance on forest bird communities. *Landscape Ecology* 1:59–73.

van Noorden, B. 1986. *Dynamiek en dichtheid van bosvogels in geïsoleerde loofbosfrag-*

menten. Report, Research Institute for Nature Management, Leersum, The Netherlands.

Vogelwerkgroep Avifauna West Nederland. 1981. *Randstand en broedvogels*. Tilburg, The Netherlands.

Wiens, J. A. and J. T. Rotenberry. 1981. Censusing and the evaluation of avian habitat occupancy. *Studies in Avian Biology* 6:522–32.

7. Hedgerow Network Patterns and Processes in France

Françoise Burel and Jacques Baudry

Hedgerows are conspicuous in agricultural landscapes around the world, and when combined into networks, offer a remarkable richness in pattern and consequent function. In contrast with eastern North American hedgerows, which have mainly developed from seeds dropped by birds along fences, in western Europe hedgerows have primarily originated by human planting or as remnants of formerly forested land. These hedgerows in Europe have been constructed, managed, and removed by farmers over centuries, depending on economic, political, and technological changes in society (Meynier, 1970; Dufour, 1976; Rackham, 1986). Such differences in origin and maintenance play key, though incompletely known, roles in hedgerow functioning and change (e.g., Brandt and Agger, 1984; Forman and Godron, 1984; Baudry and Merriam, 1988).

The functions of hedgerows are diverse, including direct contributions to society such as property boundaries, fences, source of wood, and protection against wind. Equally critical roles include protection against erosion, effect on stream water quality and fish populations, travel lanes for wildlife, and maintenance of species diversity (Pollard et al., 1974; Les Bocages, 1976; Forman and Baudry, 1984).

Hedgerows and hedgerow networks are part of agricultural systems, and thus have had, and still have, many economic roles. Their persistence in the landscape is linked to the degree to which they are integrated in changing agricultural systems over time. Because hedgerows in France were considered useless or inconvenient by farmers using modern technology such as large machines, many have been cleared since the 1950s. Changes in landscape aesthetics, as well as negative effects on ecological processes of erosion, flooding, and biological communities, have led to strong public reactions. Therefore, landscape design, done by government agencies today, including land consolidation programs, must take into account these ecological functions of hedgerows (Lefeuvre, 1979; Baudry and Burel, 1984; Burel, 1984).

Differences in the structure of individual hedgerows, as well as in the overall spatial pattern of hedgerow networks, are striking in France. Some differences may significantly affect functioning, while others may not. The objective of this chapter is to consider this linkage between structure and function, particularly for differences in network structure. We will examine this linkage specifically for the colonization patterns of woody plants and carabid beetles, and qualitatively for the physical processes of water flux, erosion, and nutrient movement.

Most hedgerow research has focused on individual hedgerows rather than on networks (Les Bocages, 1976). In this study we will particularly relate the results to the theory of landscape ecology, which focuses on (1) the interactions among landscape elements (Forman, 1981; Risser et al., 1983), (2) the role of spatial configuration on biological processes (Merriam, 1984 and Chapter 8; Baudry 1984; 1985; in press; Baudry and Merriam, 1988; Burel, 1988; in press; Blandin and Lamotte, 1988; Amoros et al., 1988), and (3) the dynamics of landscapes (Berdoulay and Phipps, 1985; Décamps et al., 1988; Forman and Godron, 1986; Godron and Forman, 1983; Phipps, 1984; Phipps et al., 1986a,b; Romme and Knight, 1982; Turner, 1987). In addition, results will be presented with the hope of enhancing understanding among ecologists, agriculturalists, and planners.

Hedgerow Network Patterns in France

Historians and geographers have described different types of networks of hedgerows in France, as well as their component individual hedgerows (Bloch, 1931; Lebeau, 1979; Defour, 1976; Palierne, 1975). Individual hedgerows differ according to their three-dimensional structure, species composition, and interactions with adjacent fields. The key three-dimensional characteristics are: (a) one or more rows of trees; (b) presence or absence of a shrub layer; (c) raised on an earthen bank or at the field level; (d) presence or absence of a ditch; and (e) pruned or unpruned (Figure 1). Species composition particularly includes whether the trees are broad-leaved or needled. Also bird populations are especially useful as assays of hedgerow wildlife (Constant et al., 1976; Yahner, 1983; Arnold, 1983; Osborne, 1984; Lack, 1988).

At the landscape level we distinguish two main types of network patterns: (1) those primarily linked to topographic features; and (2) others that are geometric (with mainly straight lines) or follow sequential lines of forest clearance. The first network type usually controls water fluxes, and has an important role in protecting soils against erosion. The second reflects human colonization patterns, as during the Roman period in Europe, or recent centuries in North

Figure 1. Structure of a lane and two parallel hedgerows in Brittany, France.

America (Pitte, 1983). In every region of France containing hedgerows, both patterns coexist, though their relative importance varies widely.

Complete networks (i.e., without breaks in their structure) are rare and only persist for short interludes. Hedgerows change by addition or removal of trees, and networks change by addition or removal of hedgerows (Leonard and Cobham, 1977; Dufour, 1976; Fox, 1976; Agger and Brandt, 1988). Many parameters, such as length of hedgerow, connectedness, and heterogeneity of field size, change concurrently with additions and removal (Baudry and Burel, 1985; Braekevelt, 1988; Notteghem, 1986; Figure 2). As will be seen shortly, such changes additionally affect interactions between hedgerows.

Preliminary studies in Brittany, France indicate that changes are related to soil type. In the most fertile areas hedgerow clearance, leading to large fields and low network connectedness, is widespread. In wet meadow areas almost no change is evident, as hedgerows provide protection for cattle against sun, wind, and rain. In former moorland areas cultivation is undergoing abandonment, leading to a more connected network with attached nodes of natural vegetation. Thus, today, highly contrasting network structures are emerging, whereas they were quite homogeneous in the mid-nineteenth century. Such distinct hedgerow networks result from the dynamics of agricultural systems in a changing economic and technological society.

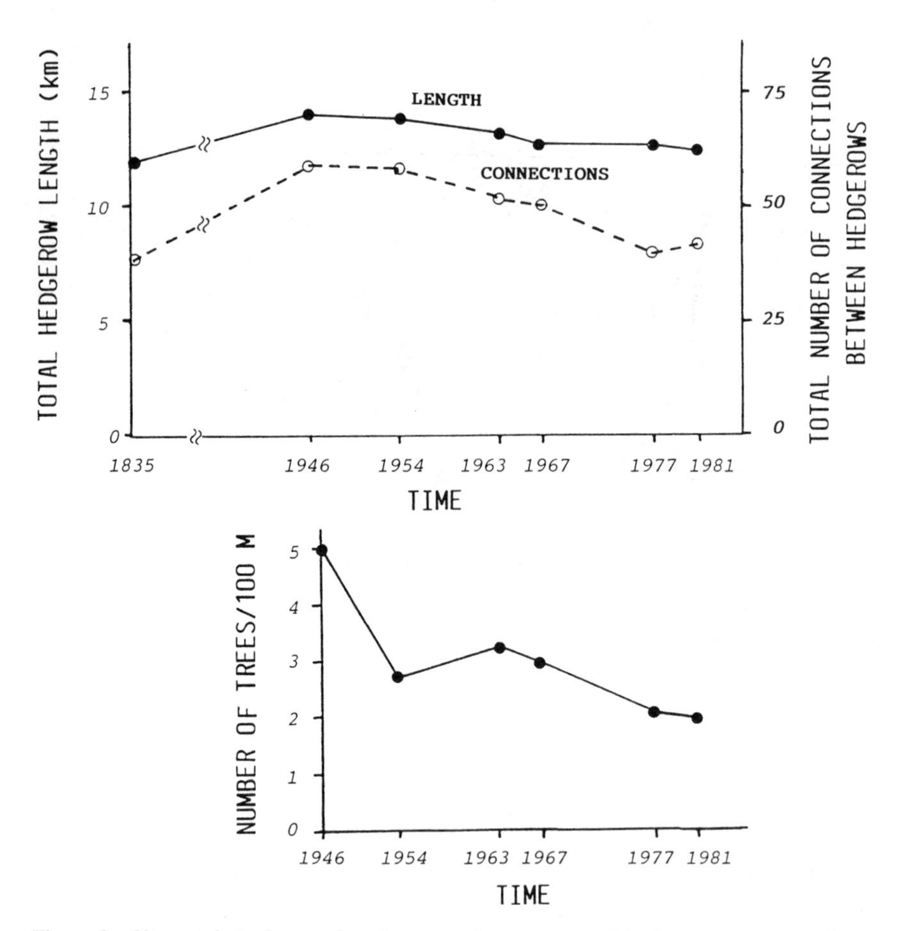

Figure 2. Changes in hedgerow length, network structure, and hedgerow trees over time. Based on a 200-ha area in central France. (After Notteghem, [1986].)

Biological Patterns and Processes

The first question posed is whether the species structure of a hedgerow is independent of or significantly affected by the adjacent connected hedgerows, by neighboring hedgerows, or by the network structure as a whole. For example, how similar are hedgerows connected to one another? Is understanding the interaction between a hedgerow and other landscape elements, especially adjacent fields, sufficient to understand the species structure of a landscape? The answers to these questions will determine whether a landscape-level approach, here involving network structure, is required.

To address this, we will examine the relationship between species composition in a hedgerow network and that in an adjacent forest. This is an indicator of the corridor function of hedgerows, that is, the degree to which species colonize along hedgerows from a forest, and hence produce different links in a network with predictable species composition. Woody plants and ground beetles are analyzed in hedgerow networks of Brittany in northwestern France.

Woody Plant Distribution in Two Brittany Networks

In the municipality of Malansac (southern Brittany), two adjacent networks, containing 76 and 77 hedgerows respectively, were selected in an area of 125 ha on a plateau (Figure 3). Elevation varies from 75 to 82 m, and the soils developed from a granitic bedrock are relatively homogeneous, mainly varying with apparently minor differences in water-table depth. Moorland with *Pinus spp*. still occupies the wettest parts. Species of trees and shrubs were sampled in each hedgerow, that is, in each length between hedgerow intersections, woods, or hedgerow ends (Baudry, 1984; 1985; Forman and Baudry, 1984).

Twenty-three woody plant species were recorded. Though the two networks are contiguous, their species composition differs significantly (Table 1). Hedgerows in network 1 contained more species (average 5.6, with a standard deviation of 0.38) than those in network 2 (4.4 ± 0.31). *Fraxinus excelsior* and *Evonymus europaeus* are only present in the first network, and *Sorbus torminalis* only in the second. *Ruscus aculeatus, Corylus avellana,* and *Prunus spinosa* are more frequent in network 1, and *Ulex europaeus* in network 2.

Three types of woody vegetation can be distinguished in the first hedgerow network and two types in the second network (Table 1). Both networks contain a moorland type (represented by *Ulex, Pinus,* and *Erica*) and a forest type *(Fagus, Ilex, Pyrus)* (Figure 3), though no forest is present within 1 km. In network 1, a "forest-edge" type where *Fraxinus, Sambucus,* and *Evonymus* are distinctive is present south of a village (Figure 4). Thus, in a relatively homogeneous physical environment we find a mosaic of vegetation types in the network. The hedgerow vegetation is neither homogeneous, regular, nor random in spatial distribution. A pattern of aggregation or species assemblages in the network is present at the landscape level. Time and space appear to be the key parameters explaining this result.

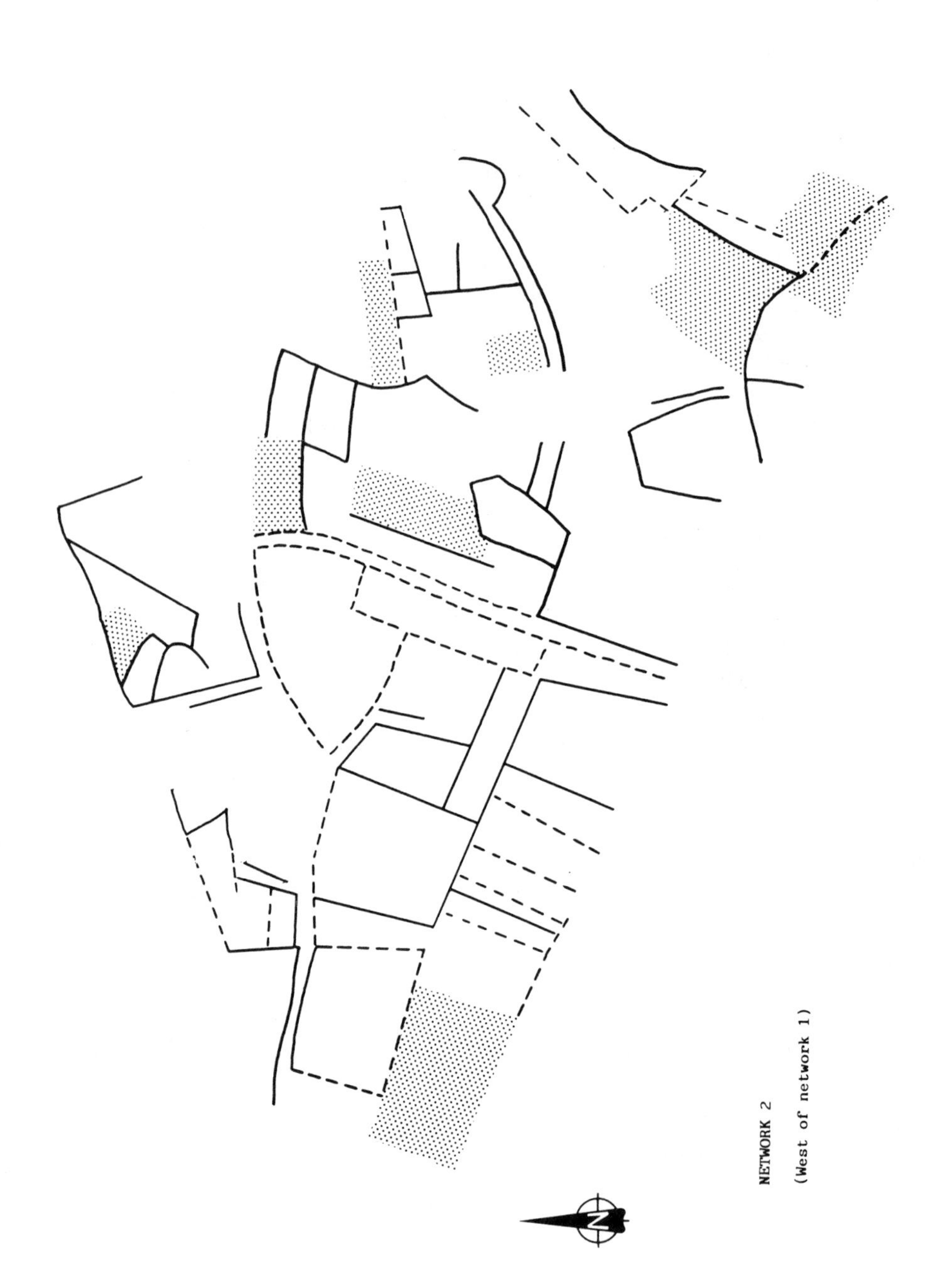
NETWORK 2
(West of network 1)
N

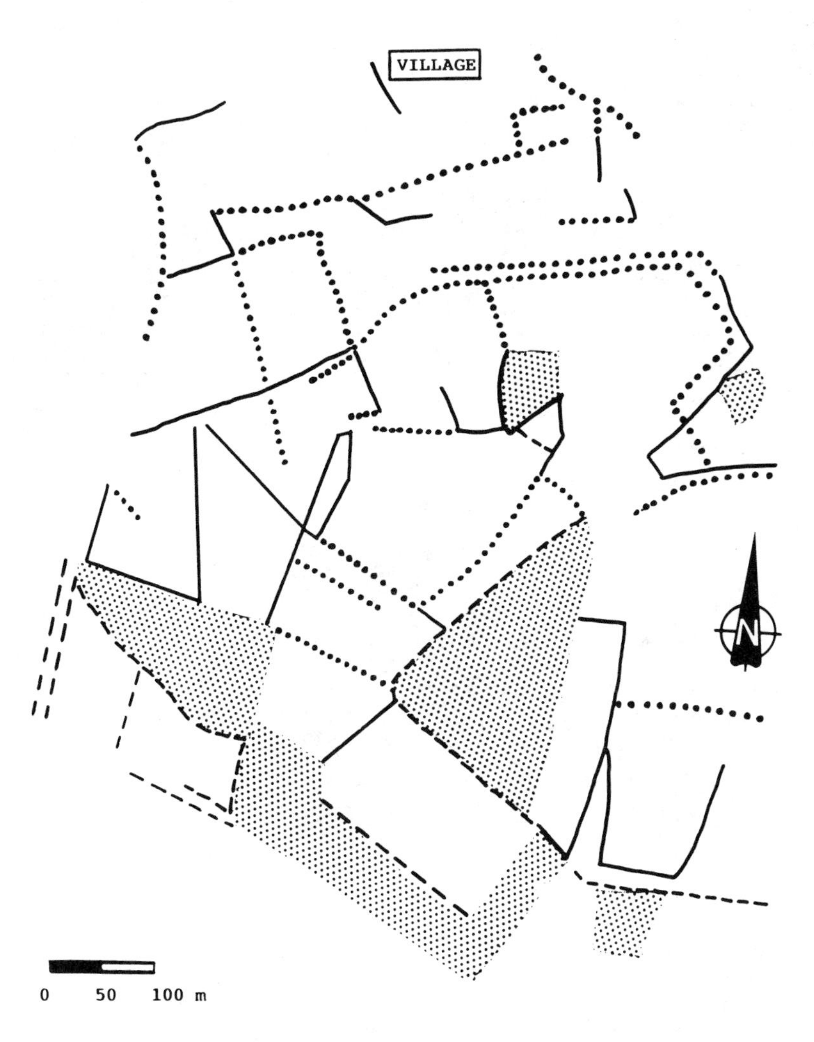

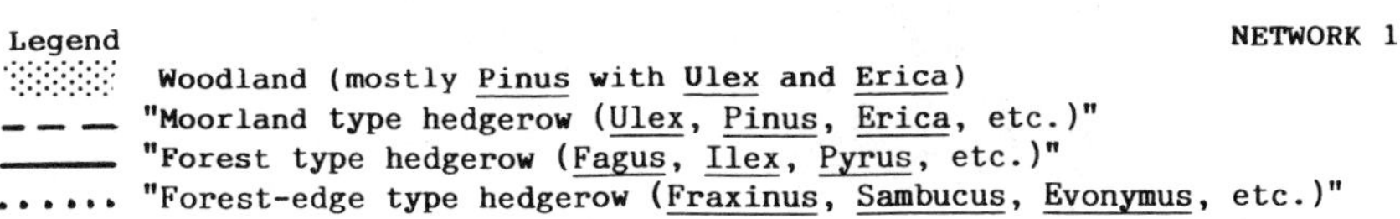

Figure 3. Woody vegetation types in two adjacent networks in Malansac, southern Brittany. Total area of both networks = 125 ha. Note that network 2 has been reduced in scale and turned 90° in orientation.

Table 1. Types of vegetation in two adjacent hedgerow networks in Malansac, Brittany. The three types of hedgerow—moorland, forest, and forest-edge—are differentiated based on the woody vegetation present, and are mapped in Figure 3. Network 1 = 76 and network 2 = 77 hedgerows. Numbers in lower portion indicate percent of hedgerow in which a species is present.

	Network 1			Network 2	
Hedgerow vegetation type	Moorland hedgerows	Forest hedgerows	Forest-edge hedgerows	Moorland hedgerows	Forest hedgerows
Number of hedgerows in the group	13	31	32	30	47
Mean species richness	4.8	4.9	5.7	4.3	4.9
Calluna vulgaris	+			07	–
Erica spp.	23			07	
Ulex europaeus	100	13		100	23
Sarothamnus scoparius	15	10	+	13	11
Rhamnus frangula				–	06
Betula spp.	15	06		17	11
Castanea sativa	38	43	34	50	49
Pinus spp.	61	–	–	17	08
Fagus sylvatica		35	–	–	25
Ilex aquifolium	46	97	84	43	91
Pyrus communis	–	42		–	17
Sorbus torminalis				–	08
Crataegus monogyna	+	29	37	–	23
Prunus spinosa		19	53	–	04
Sambucus nigra		–	37	–	–
Corylus avellana	31	71	56	07	17
Prunus avium	+	13	12	13	30
Fraxinus excelsior		–	22	–	–
Ulmus campestris		+	–	–	04
Ruscus aculeatus		32	100	–	04
Evonymus europaeus		–	22	–	–
Salix atrocinerea	38	55	19	77	49
Quercus pedunculata	92	97	97	97	100

The Effect of Time: New and Old Hedgerows

The bulk of the 125-ha area was occupied by moorland until the end of the nineteenth century, when most was cleared and new hedgerows were established. Because moorland is characteristic of infertile soils here, land clearing resulted primarily in pastures on small farms. Rather than being planted, these hedgerows—characterized by *Ulex, Pinus,* and *Erica*—reflect the original moorland vegetation type, and can be considered as remnant hedgerows (Figure 3).

In contrast, the hedgerow type characterized by forest-edge species, such as

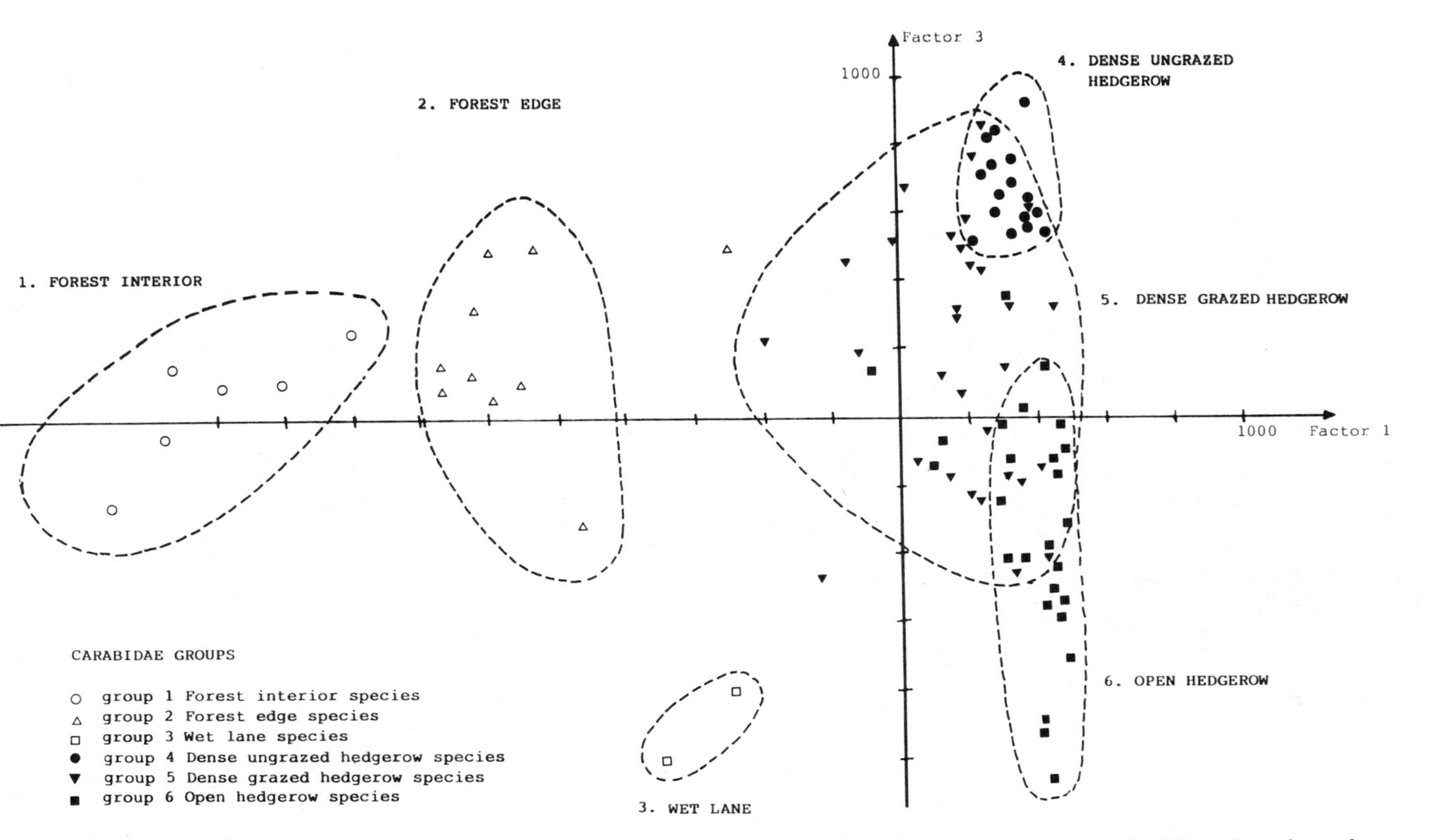

Figure 4. Forest and hedgerow ground beetle samples related to correspondence-analysis gradients. The horizontal axis differentiates forest from hedgerow, and the vertical axis differentiates dense ungrazed hedgerows from open hedgerows. See text.

Fraxinus, Sambucus, and *Evonymus,* is only found in an area of network 1 that was cultivated before 1850 (Figure 3). Thus, these are old hedgerows (although due to subsequent management activities, the individual stems present today are no older than those in other hedgerow types).

In addition, these early hedgerows with edge species may be related to proximity of a village. Indeed, a survey of a large area elsewhere in Brittany shows that the distribution of *Fraxinus excelsior* in networks is significantly and directly correlated with proximity to villages (Baudry, 1985). This species was frequently planted as fodder for cattle in late summer, and several of the forest-edge species in these hedgerows probably gain a competitive advantage from the abundance of nutrients from animal wastes near villages.

The Effect of Space: Landscape Structure

Is the species composition of a hedgerow related to the composition of an adjoining connected hedgerow? We compared case A, where a species is present in a hedgerow and is also in a connected hedgerow, with case B, where a species is present in a hedgerow but absent from any connected hedgerow. A test of proportion (Bertaud and Charles, 1980) for the 20 species found in more than 10 hedgerows showed that 70 percent of the species exhibited a significant fit with case A. The probability of finding a species in a hedgerow is higher if the species is present in a connected hedgerow. The reverse, however, was not true for any species: The presence of a species does not ensure colonization of other hedgerows.

For those species that show no evidence of colonization from a connected hedgerow, we assume that the physical conditions (mainly microclimate and the soil atop the earthen bank that results from ditch digging) are not suitable for growth (Figure 1). Overall, two types of processes—the dispersion of seeds between hedgerows by animals and wind, and the environmental conditions that constrain seed germination and seedling survival—interact concurrently to produce a biotic mosaic in the hedgerow network.

Carabid Distribution Near a Forest in an Agricultural Landscape

Ground beetles were sampled in a large old forest (Forêt de Liffré, near Rennes, Brittany) and adjacent hedgerows and fields to determine their spatial distribution relative to individual hedgerows and relative to landscape structure, as indicated by location in a hedgerow network. The agricultural land is a mixture of cultivation and little-grazed "natural" meadows on rather hydromorphic soils. The hedgerow network is in a continuous open area 800-m wide and extending 1 km out from the forest boundary (see Burel, in press, for details). All hedgerows have an earthen ridge or bank and in a few cases a pair of hedgerows borders a dirt lane or track for vehicles or livestock movement (Figure 1).

Pitfall traps (Drach et al., 1981; Greenslade, 1964; Thiele, 1977) were located in a network of connected hedgerows extending out from the edge of the forest. There were 25 sets of pitfall traps, with sets located at 50-m intervals. Each set consisted of three traps, one atop the earthen bank and one on each side of the bank at approximately mid-elevation. Two transects of eight traps each were set up inside the forest, perpendicular to the edge, and 300 m apart. One was in a beech–oak stand *(Fragus–Quercus)*, the other in a wet ironwood stand *(Carpinus betula)*.

A correspondence analysis (Benzecri, 1973; Benzecri and Benzecri, 1984) was performed on the matrix using 94 samples times 49 beetle species, followed by a cluster analysis of the samples using the three leading factors (Jambu and Lebeaux, 1978; Balent et al., 1988). In this manner, clusters or groups of species with similar environmental requirements were identified. Using these species groups for analysis, in lieu of individual samples, permits a clearer understanding of the environmental gradients present (Legendre and Legendre, 1984).

Six groups were recognized with characteristic species listed for each:

Group 1. Forest-Interior Species. Samples from the interior of the oak–beech forest: *Argutor oblongopunctatus, Abax parallelus, Platysma nigrum, Hadrocarabus purpurescens*.

Group 2. Forest-Edge Species. Samples from the edge of the oak–beech forest, a coppice stand, and the tops of banks of hedgerows connected to the forest: *Abax ater, Abax ovalis, Platysma nigrum, Agonum livens*.

Group 3. Wet-Lane Species. Two samples from the inward side of hedgerow banks along a lane with wet soil: *Agonum lugens, Philochtus sp., Anchus ruficornis*.

Group 4. Grazed-Dense-Hedgerow Species. Samples along a lane or from hedgerows with a dense vegetation cover: *Nebria salina, Abax ater, Metallina lampros, Platysma nigrita, Steropus madidus, Hadrocarabus purpurescens*.

Group 5. Ungrazed-Dense-Hedgerow Species. Mainly samples from ungrazed tops and sides of banks: *Chaetocarabus intricatus, Abax ater, Metallina lampros, Philochtus sp., Steropus madidus, Argutor strennus, Carabus auratus*.

Group 6. Open-Hedgerow Species. Samples on the side of earthen banks in poorly vegetated hedgerows (i.e., with discontinuous cover of shrubs and/or scattered trees): *Poecilus cupreus, Amara* sp., *Brachinus scolopeta, Percosia equestris*.

Based on factor analysis of the sample and the groupings of samples, the leading axes appear quite interpretable ecologically. The first axis, which accounts for the greatest amount of variability, is a gradient from old forest to hedgerow conditions (Figure 4). The second axis was based on a single "outlier" sample from a wet area, and therefore is not plotted. The third axis is a gradient of hedgerow vegetation density, from ungrazed dense hedgerows of woody and herbaceous cover to open hedgerows with only scattered woody cover (Figure 4).

The spatial patterns for these groups or species assemblages appear to be related to both environmental conditions at a location and to features of landscape spatial structure. Control by environmental conditions is well known (e.g.,

Pollard, 1968; Thiele, 1977) and is illustrated by species of group 6 growing primarily in open conditions. Also, carabid species composition (as well as woody plants) on the tops of earthen banks differed from that of the bank sides. The importance of lanes lined by two hedgerows in supporting a distinct species assemblage (group 3) is noteworthy.

However, landscape structure also plays a role in determining the beetle community in this landscape. With increasing distance from the forest some carabid species become less abundant *(Abax ater, Abax ovalis, Argutor oblongopunctatus)* (see Burel, in press, and Burel and Baudry, in press, for details). At the same time other species become more abundant (*Poecilus cupreus, Amara* sp.). At the group level, the pattern is similar to that observed for woody plants. When a group of carabids is present in a hedgerow, there is a significant probability that it is also present in a connected hedgerow. This is consistent with the observation that ground beetles do not move very far relative to the size of the hedgerow network (Rivard, 1965).

In short, the probability of the presence of certain carabid beetle species varies according to location of a hedgerow in a network. Since intervening agricultural fields are considered not to be suitable habitat, hedgerows appear to be corridors for movement, as well as for colonization of new elements such as attached old fields and planted hedgerows.

The Ecological Importance of Lanes

The carabid study suggested that lanes (Figure 1) bordered by two hedgerows serve as corridors for forest species in agricultural landscapes. This is consistent with Richards's (1928) interpretation of the distribution of mosses on earthen banks and forest environments. The shade provided by the double hedgerows of lanes probably plays a key role, as for instance it probably does for herbaceous species (Forman and Baudry, 1984; Forman and Godron, 1986).

If indeed lanes play a key role for biological richness in an agricultural landscape, we should examine their origin and woody-species composition. Are the hedgerows a remnant of former forest edges, or have they been planted and subsequently colonized by forest species?

Some indications may be gained from a survey of the woody species of 145 hedgerows in a 1000-ha municipality (Concoret) in Brittany. These hedgerows were dominated by the forest trees *Fagus, Ilex, Sorbus torminalis, Taxus baccata, Quercus sessiliflora,* and *Pyrus communis*. Hedgerows with three or more forest species were significantly more frequent along lanes than anywhere else in the landscape (22 out of 31, chi square = 12.77, $p < .05$). A preliminary conclusion is that many hedgerows along a lane are a remnant of former forest edges. Frequently the two hedgerows along a lane have quite different vegetation, one with few woody species (hypothesized to have been planted), and one with many woody species (hypothesized to be a remnant). Methods for assessing hedgerow origin and age need to be improved (Delelis-Dusollier, 1986).

Conclusion

The evaluation of both plant and animal species distribution in a hedgerow network has been useful to understand the effects of landscape structure. Distributions of individual species as well as groups of species indicate that both the type of hedgerow (ungrazed, open, lane, etc.), and its location in the network (near a forest, distant from forest, presence or absence of species in connecting hedgerow, etc.), are important controls. From the perspective of hierarchical spatial levels, one may hypothesize that species distributions result from habitat conditions at the individual hedgerow level and movement in the network at the landscape level. Movement is a function of both landscape structure and species-dispersal behavior (Peterken and Game, 1984; Burel and Baudry, in press).

Hedgerows of course are not isolated elements in a landscape; rather the hedgerow network is combined with fields and the soil mosaic to produce an integrated system. Within this structure, connectedness (spatial linkage among elements) is important because it allows connectivity (continuous fluxes of species) (Baudry and Merriam, 1988; Merriam, Chapter 8). The biological processes in a hedgerow-network landscape may be synthesized by integrating fluxes with habitat conditions.

Physical Processes

Knowledge of the influence of hedgerows and networks on physical processes is important from both ecological and agronomic perspectives. While the ecologist may focus on energy budgets, water fluxes, and nutrient cycling (and leaching), the agronomist is more concerned with the control of physical processes, at both field and landscape levels, to provide optimum microclimate, water supply, erosion control, and so on, for plant and animal growth.

Studies of windbreaks (hedgerows, shelterbelts, etc.) have provided detailed insight into influences on microclimate, though less so for water fluxes and erosion control (Loucks, 1983). Distinct seasonal differences in hydrological budgets in two watersheds in France one with and one without hedgerows (Merot, 1976), and hydrological differences between a field and landscape are recognized (Pihan, 1976; Gulinck, 1985). Here we will present some qualitative observations from the hedgerow-network landscapes we studied to consider how landscape structure may affect water flow and erosion. This will be useful in generating hypotheses as well as in landscape planning and design procedures (Baudry et al., 1988; Baudry, in press).

The frequency of hedgerows at the boundaries between soil units (Figure 5, hedgerow C) was noted by Carnet (1976). Thus, hedgerows associated with soil catenas are of particular interest because soil type boundaries are generally roughly parallel with slope contours. Such hedgerows, oriented across the slope, provide a barrier to both surface and subsurface water flow. Both the bank and the ditch of these hedgerows (Figure 1) are required to address the apparent

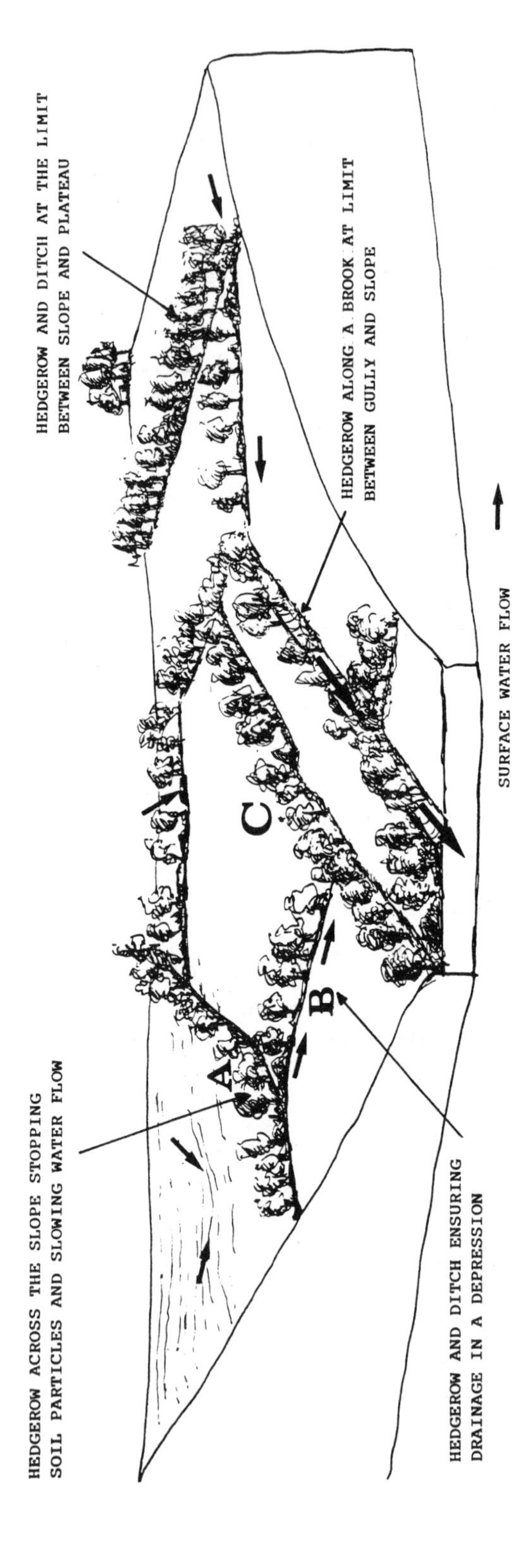

Figure 5. Major water fluxes in a hedgerow network constructed to fit with topography.

paradox of both too much and too little drainage in fields. A connected hedgerow network can be constructed so that the earthen banks inhibit direct downhill water flow, while the ditches provide an effective drainage system.

Erosion, especially sheet erosion, is a second major parameter controlled by hedgerows on slopes. Erosion increases with the speed of surface water flow (kinetic energy). Soils saturated with subsurface water also are more sensitive to erosion.

Hedgerow removal on a slope changes both the surface and subsurface water conditions to increase erosion. Sheet erosion by surface water accelerates without the blocking effect of the earthen bank. And without the ditch, subsurface water drains less rapidly, producing saturated soils that erode more readily.

Mineral nutrients transported by water flows are also controlled by the hedgerow structure. This may play a key role in determining water quality and fish populations in streams and lakes, by limiting the runoff of agricultural inputs, such as nitrogen, phosphorus, and pesticides, from fields (Risser, Chapter 4).

It appears that the more a hedgerow network fits topography, the more effective it is in controlling water flows, erosion, and nutrient runoff. For example, when a link in the network is removed, erosion may increase significantly. This is illustrated in Figure 5, where hedgerow A concentrates water at the lowest point of the upper field; if hedgerow B is removed, a gully will be formed in the field, which will be more severe than if hedgerow A were removed. Thus the particular link broken in the network is critical, another example of the effect of landscape structure.

The large landscape mosaic is a key level for planning. In a network-design procedure, a planner or farmer attempting to control physical processes has few alternatives involving the establishment, maintenance, or removal of hedgerows. Hedgerows may be located between or within soil units, and oriented across or up and down the slope. Hedgerows oriented up and down the slope aid in soil drainage, but often result in severe erosion and nutrient runoff. Hedgerows crossing the slope and located on the boundary between soil units (Figure 5, hedgerow C) are especially valuable in functional and long-term economic terms. Removal of such hedgerows blurs the boundary, especially between wet and mesic soils. Water fluxes are thus changed in both soil units, and farmers cannot plow them at the same time. The wettest parts are then abandoned, resulting in a possible gain for wildlife but a loss in crop production. Such cases of change in land use following change in landscape structure are common in France.

Discussion

Landscapes are heterogeneous, in part because of interactions among the component landscape elements, interactions that depend on spatial organization. In hedgerow-network landscapes the landscape elements such as pastures, cultivated fields, hedgerows, ditches, woodlots, and lanes, vary markedly in structure from one another, resulting in a high degree of landscape heterogeneity.

The patterns of element assemblages may be described in terms of contiguity, connectedness, shape, heterogeneity, etc. Baudry and Burel, 1982; 1985; O'Neill et al., 1988). Some of these patterns are closely related to topography and soil distribution, in which case they strongly influence water fluxes and erosion processes.

Strong biological differences are also evident within a type of landscape element. Thus species composition varies according to vegetation structure, former or present land use, and, as illustrated in this chapter, interactions among landscape elements (Merriam, 1984; Pollard et al., 1974; Les Bocages, 1976).

Such spatial patterns and functional processes can be expected to vary with level of scale, as well as with how different levels are functionally linked or integrated. Our understanding of the ecology of a landscape will be significantly enhanced by combining empirical studies at different levels of scale with the concepts of hierarchy theory (Allen and Starr, 1982; O'Neill et al., 1986).

Hedgerow network landscapes resulting from human activity, have many beneficial aspects for wildlife and resource conservation (Marshall, 1988). According to Middleton and Merriam (1983) an agricultural landscape with woodlots interconnected by hedgerows will contain 90 percent of the species found in a large forest. Although the missing 10 percent that depends on large forest patches are of key importance in planning and conservation, the 90 percent overlap illustrates the large degree of compatibility possible with agriculture and conservation.

Altieri et al. (1983) suggested that a sustainable agro-ecosystem should have some features of a mature ecosystem, an observation consistent with our studies in Brittany landscapes. Nevertheless, we must also consider how important the species of a mature forest are for, e.g., pest control (van Emden and Williams, 1974), without forgetting that hedgerows can be sources of pests (Thresh, 1981). That is, how do we design landscapes that not only have loops which maintain tight nutrient cycles, but also contain loops for biological control? At this point, landscape ecology and agro-ecology (Cox and Atkins, 1979; Lowrance et al., 1984; Gulinck, 1986; Paoletti and Stinner, in press) merge. The challenge remains to develop more ecological theory at the landscape level, the key level of human activity.

In some areas of France, hedgerows and ditches forming a continuous network are maintained for water control, whereas in other areas the network is maintained as windbreaks. Both types of networks have provided optimal conditions for the movement and dispersion of plants and animals using hedgerows as corridors. Hence, what was designed for agricultural systems turned out to be of major importance for the survival of many species in cultivated areas.

Finally, we should also consider how useful the knowledge of ecological processes at the landscape level is for design that focuses on the visual dimension of the landscape. Zube (1987) pointed out that there may be no significant relationship between ecological integrity and aesthetic constraints at this level, yet our preliminary research on the incorporation of aesthetic constraints in the design of an agricultural landscape suggests the opposite (Baudry et al., 1987). Changes in the visual dimension of the landscape resulting from hedgerow removal were viewed as degradation by most segments of society other than farmers.

Farmers know that hedgerows are useful for agriculture, but do not know which ones are important to nonfarmers from the visual perspective. Thus, farmers in these landscapes will typically remove any hedgerow except those on the boundaries between soil units, which are often parallel with slope contours. In effect, the farmers and landowners in charge inadvertently alter, and often downgrade, the visual dimension valued by the general population. We expect that landscape beauty, here a by-product of agricultural systems, can only be maintained if the specific landscape elements and their overall landscape structural pattern also have agro-ecological functions recognized by farmers.

Conclusion

Shifting from research on homogeneous units such as individual fields or hedgerows to research on landscapes as ecological systems allows recognition of different ecological processes, the emergence of a new theory, and the development of guidelines for planners and agriculturalists. The importance of landscape structure was inherent in earlier research on hedgerows (e.g., Constant et al., 1976; Lefeuvre et al., 1976; Cameron et al., 1980), but scholars at that time generally did not develop the necessary concepts and methods to evaluate the effects of spatial structure on biological patterns and processes.

Two areas of research in agricultural landscapes would be particularly informative:

1. The identification of key functional units within a landscape, where major ecological processes (e.g., population survival, tight nutrient cycles, etc.) are sustained or highlighted, is needed. For example, for animal and plant distribution, are such units the individual tesserae or landscape elements, the watershed, or clusters of sites as small mosaics within a landscape?
2. Analysis of overall landscape structures and their dynamics is equally significant. This spatial structure provides the framework for, and the interface between, landscape production/land use (Vink, 1983; Phipps, 1981; Berdoulay and Phipps, 1985; Phipps et al., 1986a,b) and ecological processes (e.g., this volume). How do agricultural systems dynamics and alterations in agricultural policy induce changes in the landscape structure? How do these changes affect ecological processes? What are the feedbacks on agricultural systems themselves? How can we design and manage landscapes for an optimum balance of multiple uses (Golley and Ryzskowski, 1989; Missonier and Ryskowski, 1986; Baudry, 1988)? Answers to these questions will require both development of theory at the landscape level and interactions between basic and applied ecology.

Acknowledgments

Data on plants were collected during an ecological survey done for the Direction Départementale de l'Agriculture du Morbihan. The Minstère de l'Environnement gave financial support for research on carabids (contract no. 85059 Ministère de l'Environnement) and erosion (contract no. 58/84). The latter was also supported by the Ministère de l'Agriculture. We thank Y. Le Flem for Fig. 1.

References

Agger, P. and J. Brandt. 1988. Dynamics of small biotopes in Danish agricultural landscapes. *Landscape Ecology* 1:227–40.

Allen, T. F. H. and T. B. Starr. 1982. *Hierarchy: Perspectives for ecological complexity*. University of Chicago Press, Illinois.

Altieri, M. A., D. K. Letourneau, and J. R. Davis. 1983. Developing sustainable agroecosystems. *BioScience* 33:45–9.

Amoros, C., J. P. Bravard, J. L. Reygrobellet, G. Pautou, and A. L. Roux. 1988. Les concepts d'hydrosystème et de secteur fonctionnel dans l'analyse des systèmes fluviaux à l'échelle des écocomplexes. *Bulletin d'Ecologie* 19:531–46.

Arnold, G. W. 1983. The influence of ditch and hedgerow structure, length of hedgerow, and area of woodland and garden on bird numbers on farmland. *Journal of Applied Ecology* 20:731–50.

Balent, G., M. Genard, and F. Lescourret. 1988. Analyse des patrons de répartition des oiseaux nicheurs en Midi-Pyrénées. *Acta Oecologica- Oecologi Generalis* 9:247–63.

Baudry, J. 1984. Effects of landscape structure on biological communities: The case of hedgerow network landscapes. In J. Brandt and P. Agger, eds., *Proceedings of the first international seminar on methodology in landscape ecological research and planning*. Roskilde University Center, Denmark, Vol. 1, pp. 55–65.

Baudry, J. 1985. *Utilisation des concepts de landscape ecology pour l'analyse de l'espace rural: Occupation du sol et bocage*. Thèse de Doctorat d'etat, Université de Rennes, France.

Baudry, J. 1988. Hedgerows and hedgerow networks as wildlife habitat in agricultural landscapes. In J. R. Park, *Environmental Management in Agriculture. European perspectives*. Belhaven, London and New York. pp. 111–124.

Baudry, J. In press. Interactions between agricultural and ecological systems at the landscape level. In M. Paoletti and B. Stinner, eds., *Agricultural Ecology and Environment*. Special issue of *Agriculture Ecosystem and Environment*.

Baudry, J., and F. Burel. 1982. La mesure de la diversité spatiale: Utilisation dans les évaluations d'impact. *Acta Oecologica Oecologia Applicata* 3:177–90.

Baudry, J. and F. Burel. 1984. Landscape project: Remembrement: Landscape consolidation in France. *Landscape Planning* 11:235–41.

Baudry, J. and F. Burel. 1985. Système écologique, espace et théorie de l'information. In V. Berdoulay and M. Phipps, eds., *Paysage et système*. Presses de l'Université d'Ottawa, Canada, pp. 87–102.

Baudry, J., F. Burel, and M. C. Trotel. 1987. *Assistance paysagère dans les opérations de remembrement*. CERESA, Ministère de l'Agriculture, Imprimerie Nationale, Paris.

Baudry, J. and G. Merriam. 1988. Connectivity and connectedness: Functional versus structural patterns in landscapes. In K.-F. Schreiber, ed. Connectivity in Landscape Ecology, Proceedings of the 2nd IALE seminar. *Münstersche Geographische Arbeiten* 29:23–8.

Baudry, J., M. C. Trotel, F. Burel, and A. Asselin. 1988. *L'érosion des terres dans le massif armoraicain*. CERESA, Ministères de l'Environnement et de l'Agriculture, Imprimerie Nationale, Paris.

Benzecri, J. P., ed. 1973. *L'analyse des données*. Tome 1: La taxinomie. Dunod, Paris.
Benzecri, J. P. and F. Benzecri. 1984. *L'analyse des données: analyse des correspondances*. Dunod, Paris.
Berdoulay, V. and M. Phipps. 1985. *Paysage et systéme*. Editions de l'Université d'Ottawa, Canada.
Bertaud, M., and B. Charles. 1980. *Initiation à la statistique et aux probabilités*. Les Presses de l'Université de Montréal, Editions Eyrolles, Paris.
Blandin, P. and M. Lamotte. 1988. Recherche d'une entité écologique correspondant à l'étude des paysages: La notion d'écocomplexe. *Bulletin d'Ecologie* 19:547–55.
Bloch, M. 1931. *Les caractères originaux de l'Histoire Rurale Française*. 1955, 2nd ed., Librairie Armand Colin, Paris.
Braekevelt, A. 1988. Evolution of the spatial structure of hedgerows in the Hautland (NW-Belgium). In K.-F. Schreiber, ed., Connectivity in landscape ecology, Proceedings of the second International seminar of IALE. *Münstersche Geographische Arbeiten* 29:153–61.
Brandt, J. and P. Agger, eds. 1984. *Proceedings of the first international seminar on methodology in landscape ecological research and planning*. 5 vol. Roskilde University Center, Denmark.
Burel, F. 1984. Use of landscape ecology for the management of rural hedgerow network areas in Western France. In J. Brandt, and P. Agger, eds., *Proceedings of the first international seminar on methodology in landscape ecological research and planning*. Roskilde University Center, Denmark, Vol. 2, pp. 73–81.
Burel, F. 1988. Biological patterns and structural patterns in agricultural landscapes. In Schreiber K.-F., ed., Connectivity in landscape ecology, *Münstersche Geographische Arbeiten* 29:107–110.
Burel, F. In press. Landscape structure effects on carabid beetles' spatial patterns in western France. *Landscape Ecology* 2:215–26.
Burel, F. and J. Baudry. In press. Hedgerows as habitats for forest species: Some implications for colonization of abandoned agricultural land. In B. Bunce, ed., *Dispersal in agricultural habitats*. Belhaven, London.
Cameron, R. A. D., Down, K., and Pannett, D. J. 1980. Historical and environmental influences on hedgerow snail faunas. *Biological Journal of the Linnean Society* 13:75–87.
Carnet, C. 1976. Role du bocage sur la distribution des sols et la circulation de l'eau dans les sols. In *Les Bocages: Histoire, ecologie, economie*. I.N.R.A., C.N.R.S., E.N.S.A. et Université de Rennes, France, pp. 159–162.
Constant, P., M. C. Eybert, and R. Maheo. 1976. Avifaune reproductrice du bocage de l'Ouest. In *Les Bocages: Histoire, ecologie, economie*. I.N.R.A., C.N.R.S., E.N.S.A. et Université de Rennes, France, pp. 327–32.
Cox, G. W. and M. D. Atkins. 1979. *Agricultural ecology*. Freeman, New York.
Décamps, H., M. Fortune, F. Gazelle and G. Pautou. 1988. Historical influence of man on the riparian dynamics of a fluvial landscape. *Landscape Ecology* 1:163–73.
Delelis-Dusollier, A. 1986. Histoire du paysage par l'analyse de la vegetation: L'exemple des haies. Actes du colloque: Du pollen au cadastre Lille. *Hommes et Terres du Nord* 2–3:110–15.
Drach, A., G. Benest and J. P. Cancela da Fonseca. 1981. Analyse comparative de différents types de pièges basée sur l'étude de deux peuplements de carabiques (Col. Carabidae) *Revue d'Ecologie et de Biologie du sol* 18:91–114.
Dufour, J. 1976. Un bocage tardif et éphémère: Le bocage de la Champagne de Coulie (Nord de la Champagne Mancelle). In *Les bocages: Histoire, ecologie, economie*. I.N.R.A., C.N.R.S., E.N.S.A. et Université de Rennes, France, pp. 49–54.
van Emden, H. F. and G. F. Williams. 1974. Insect stability and diversity in agroecosystems. *Annual Review of Entomology* 19:455–75.
Forman, R. T. T. 1982. Interaction among landscape elements: A core of landscape

ecology. In S. P. Tjallingii and A. A. de Veer, eds. *Perspectives in landscape ecology. Proceedings of the international congress of the Netherlands society of landscape ecology*. PUDOC, Wageningen, The Netherlands, pp. 57–64.
Forman, R. T. T. and J. Baudry. 1984. Hedgerows and hedgerow networks in landscape ecology. *Environmental Management* 8:495–510.
Forman, R. T. T. and M. Godron. 1986. *Landscape ecology*. Wiley, New York.
Forman, R. T. T. and M. Godron. 1984. Landscape ecology principles and landscape function. In J. Brandt and P. Agger, eds, *Proceedings of the first international seminar on Methodology in landscape ecological research and planning*. Roskilde University Center, Denmark, Vol. 5, pp. 4–15.
Fox, H. S. A. 1976. The functioning of bocage landscapes in Devon and Cornwall between 1500 and 1800. In *Les bocages: Histoire, ecologie, economie,* I.N.R.A., C.N.R.S., E.N.S.A. et Université de Rennes, France, pp. 55–61.
Godron, M. and R. T. T. Forman. 1983. Landscape modification and changing ecological characteristics. In H. A. Mooney and M. Godron, eds. *Disturbance and ecosystems,* Springer-Verlag, New York, pp. 12–28.
Golley, F. and L. Ryzskowski, In press. *Ecological consequences of changing agricultural policy and practices*. Special issue of *Ecology International Bulletin* 16:1–75.
Greenslade, P. J. M. 1964. Pitfall trapping as a method for studying populations of *Carabidae (Coleoptera)*. *Journal of Animal Ecology* 33:311–33.
Gulinck, H. 1985. Agriculture, conservation du sol et gestion paysagère. *Revue de l'agriculture*. 38:37–48.
Gulinck, H. 1986. Landscape ecological aspects of agro-ecosystems. *Agriculture, Ecosystems and Environment* 16:79–86.
Jambu, M. and M. O. Lebeaux. 1978. *Classification automatique pour l'analyse des données*. Dunod, Paris.
Lack, P. C. 1988. Hedge intersections and breeding bird distribution in farmland. *Bird Study* 35:133–6.
Lebeau, R. 1979. *Les grands types de structures agraires dans le monde*. Masson, Paris.
Lefeuvre, J. C. 1979., Les études scientifiques, un préalable indispensable à la restructuration foncière et à l'aménagement des zones bocagères. In J. C. Lefeuvre, G. Long, and G. Ricou, eds., *Les connaissances scientifiques écologiques, le développement*. Ecologie et Développement, Editions C.N.R.S., Paris.
Lefeuvre, J. C., J. Missionnier and Y. Robert. 1976. Caractérisation zoologique. Ecologie animale (des bocages), Rapport de synthèse. In *Les bocages: Histoire, ecologie, economie*. I.N.R.A., C.N.R.S., E.N.S.A. et Université de Rennes, France, pp. 315–26.
Legendre, L. and P. Legendre. 1984. *Ecologie numérique*. 2nd ed. Masson, Presses de l'Université du Québec, Canada.
Leonard, P. L. and R. O. Cobham. 1977. The farming landscape of England and Wales: A changing scene. *Landscape Planning* 4:205–36.
Les Bocages: Histoire, ecologie, economie. 1976. I.N.R.A., C.N.R.S., E.N.S.A. et Université de Rennes, France.
Loucks, W. L. 1983. *Windbreak bibliography*. Great Plains Agricultural Council Publication #113 Kansas, USA.
Lowrance, R., B. R. Stinner and G. J. House, eds. 1984. *Agricultural ecosystems*. Wiley, New York.
Marshall, E. J. P. 1988. The ecology and management of field margin floras in England. *Outlook on Agriculture* 17:178–82.
Merot, P. 1976. Hydrologie de deux bassins versants élémentaires granitiques, bocages et ouvert. In *Les bocages: Histoire, ecologie, economie,* I.N.R.A., C.N.R.S., E.N.S.A. et Université de Rennes, France, pp. 177–84.
Merriam, H. G. 1984. Connectivity: a fundamental characteristic of landscape pattern. In

J. Brandt, and P. Agger, eds. *Proceedings of the first international seminar on methodology in landscape ecological research and planning*. Roskilde University Center, Denmark, Vol. 1, pp. 5–15.

Meynier, A. 1970. *Les paysages agraires*. Armand Colin, Paris.

Middleton, J. and H. G. Merriam. 1983. Distribution of woodland species in farmland woods. *Journal of Applied Ecology* 20:625–44.

Missonier, J. and L. Ryszkowski. 1986. *Impacts de la structure des paysages agricoles sur la protection des cultures/impact of structure of agricultural landscape on crop protection*. Les colloques de l'I.N.R.A. 36, Paris.

Notteghem, P. 1986. *Incidences aux niveaux socio-économique et écologique du nouveau contexte agricole et énergétique sur la question du bocage*. Ministère de l'Environnement, SRETIE, Paris.

O'Neill, R. V., D. L. DeAngelis, J. B. Waide, and T. F. H. Allen. 1986. *A hierarchical concept of ecosystems*. Princeton University Press, New Jersey.

O'Neill, R. V., J. R. Krummel, R. H. Gardner, G. Sugihara, B. Jackson, D. L. DeAngelis, B. Milne, M. G. Turner, B. Zygmunt, S. W. Christensen, V. H. Dale, and R. L. Graham. 1988. Indices of landscape patterns. *Landscape Ecology* 1:153–62.

Osborne, P. 1984. Bird numbers and habitat characteristics in farmland hedgerows. *Journal of Applied Ecology* 21:63–82.

Palierne, J. M. 1975. *Les forêts et leur environnement dans les pays ligéro-atlantiques nord*. Thèse de doctorat d'etat, Université de Haute Bretagne, Rennes, France.

Paoletti, M. and B. Stinner, eds. In press. Agricultural ecology and environment. *Agriculture, Ecosystem and Environment*.

Peterken, G. F. and M. Game. 1984. Historical factors affecting the number and distribution of vascular plant species in the woodlands of central Lincolnshire. *Journal of Ecology* 72:155–82.

Phipps, M. 1981. Entropy and community pattern analysis. *Journal of Theoretical Biology* 93:253–73.

Phipps, M. 1984. Rural landscape dynamics: The illustration of some key concepts. In J. Brandt and P. Agger, eds., *Proceedings of the first international seminar on methodology in landscape ecological research and planning*. Roskilde University Center, Denmark, Vol. 1, pp. 47–54.

Phipps, M., J. Baudry, and F. Burel. 1986a. Ordre topoécologique dans un espace rural, les niches paysagiques. *C.R. Acad. Sc. Paris* T. 302, Série 3, 20:691–6.

Phipps, M., J. Baudry, and F. Burel. 1986b. Dynamique de l'organisation écologique d'un paysage rural: Modalités de la désorganisation dans une zone peri-urbaine. *Comptes Rendus de l'Académie des Sciences, Paris* T. 303, Série 3, 7:263–8.

Pihan, J. 1976. Bocage et érosion hydrique des sols en Bretagne. In *Les bocages: Histoire, ecologie, economie*. I.N.R.A., C.N.R.S., E.N.S.A. et Université de Rennes, France, pp. 185–192.

Pitte, J. R. 1983. *Histoire du paysage français*. 2 tomes, Tallandier, Paris.

Pollard, E. 1968. Hedges. III. The effect of removal of the bottom flora of a hawthorn hedgerow on the *Carabidae* of the hedge bottom. *Journal of Applied Ecology* 5:125–39.

Pollard, E., M. D. Hooper, and N. W. Moore. 1974. *Hedges*. W. Collins, London.

Rackham, O. 1986. *The history of the countryside*. J. M. Dent, London/Melbourne.

Richards, P. W. N. 1928. Ecological notes on the bryophites of Middlesex. *Journal of Ecology* 16:267–300.

Risser, P. G., J. R. Karr, and R. T. T. Forman. 1984. *Landscape ecology—directions and approaches*. Illinois Natural History Survey, Special Publication Number 2, Champaign, Illinois.

Rivard, I. 1965. Dispersal of ground beetles *(Coleoptera: Carabidae)* on soil surface. *Canadian Journal of Zoology* 43:465–73.

Romme, W. T. and D. H. Knight. 1982. Landscape diversity: The concept applied to Yellowstone Park. *BioScience* 32:664–70.

Thiele, H. U. 1977. *Carabid beetles in their environments*. Springer-Verlag, Berlin/Heidelberg/New York.

Thresh, J. M., ed. 1981. *Pests, Pathogens, and Vegetation*. Pitman, Boston/London/Melbourne.

Turner, M. G., ed. 1987. *Landscape heterogeneity and disturbance*. Springer-Verlag, New York.

Vink, A. P. A. 1983. *Landscape ecology and land use*. Longman, London and New York.

Yahner, R. H. 1983. Seasonal dynamics, habitat relationships, and management of avifauna in farmstead shelterbelts. *Journal of Wildlife Management* 47:85–104.

Zube, E. H. 1987. Perceived land use patterns and landscape values. *Landscape Ecology* 1:37–46.

8. Ecological Processes in the Time and Space of Farmland Mosaics

Gray Merriam

The research described here is illustrative, but not representative of landscape ecology today. The investigations presented depict processes at several time and space scales, and obtain parameter values for their elements. This permits parallel modeling of the dynamics of a species in a dynamic landscape (Merriam, 1988). My approach is specialized and focuses on the farmed landscape. Early studies concentrated on small forest areas as residual fragments of formerly nearly continuous, although probably very heterogeneous, forest (Wegner and Merriam, 1979; Middleton and Merriam, 1981). My colleagues and I also have given priority to wooded fencerows that provide a variable network through a matrix of cultivated land (Figure 1), and thus potentially interconnect woodlots (Middleton and Merriam, 1983; Merriam, 1984; Henderson et al., 1985). These fencerows arose primarily by neglect of uncultivated margins of fields after fences were built. Fencerow vegetation here rarely includes relict strips of forest.

The landscape-ecology approach to studying farm landscapes has provided significant gains in understanding the ecology of these landscape systems. It is now clear that forest-ecology principles are not sufficient to understand forest fragments in farmland. Rather than being simply fragments of forest, these are elements of a larger, integrated farm-landscape system. Properties of the landscape cannot be acceptably understood, either intellectually or for purposes of predicting effects of intervention, by reference to wooded fragments as isolated units. Even as isolated units they are not ecologically independent; they survive only because they are functionally connected into a landscape.

Figure 1. Representative portion of agricultural landscape studied, showing farmsteads, cultivated fields and grasslands, and a fencerow network. Radio-marked white-footed mice *(Peromyscus leucopus)* moved unexpectedly large distances along fencerows in this southeastern Ontario, Canada, study area.

At the same time these farm woodlots, brushy copses, riverine borders, wooded ravines, and fencerows have specific individual characteristics and properties—aesthetically, geographically, hydrologically, pedalogically, biologically, and in other ways. Each is an assembly of interrelated real parts, and for some processes, each can be viewed as an ecological whole. But study of wooded patches in isolation from the landscape, no matter how exact, will not answer questions such as: How do species populations in these patches survive? Such questions must be answered in functional terms. The answers will be dynamic, not static, and therefore are dependent on a time scale. We have to select the time scale from a continuum, and more than one may apply. Structural characteristics require a spatial scale, again chosen from a continuum. The same question asked at different scales can have opposing, but correct, answers. Good questions asked at ill-chosen scales can give misleading answers (Ulanowicz, 1989).

Results and Discussion

Recolonization and Functional Connectivity

We worked in farm woodlots with white-footed mice *(Peromyscus leucopus)* and chipmunks *(Tamias striatus)*. Often, there are only a few females of these species surviving in a woodlot at the end of winter (Middleton and Merriam, 1981; Henderson et al., 1985). Such small numbers can easily result in a local population extinction by many kinds of stochastic events. We have found 2 to 5 percent of over 20 isolated woodlots (i.e., without effective fencerow connections) have local extinctions of white-footed mice at the time of spring snowmelt (Merriam and Wegner, in preparation). But these local extinctions are easily and rapidly recolonized from other woods, from fencerow populations or from farmsteads where white-footed mice establish breeding refuge populations over winter (Merriam and Martin, unpublished). This ease of recolonization is one major distinction between these systems and those modeled by equilibrium island biogeography theory.

Recolonization across agricultural landscapes can be enhanced by fencerows, which produce corridors for movement across the isolating cropland matrix. I have used *connectivity* to refer to the qualities of a landscape that facilitate movements among habitat patches (Merriam, 1984).

An interconnected set of subpopulations that function together as one demographic unit is what Levins (1970) called a metapopulation. Connectivity was shown by modeling (Fahrig and Merriam, 1985), both to enhance recolonization of local extinctions within a metapopulation and to reduce frequency of local extinctions by enhancing growth in the subpopulations. Lefkovitch and Fahrig (1985) used the same model to isolate the elements of connectivity. The primary determinants were whether a patch was isolated (no functional corridor) or not, and how many other patches it was connected to (but *not* how many corridors per patch). Henein and Merriam (in press) showed that quality of corridors interacts with patch geometry as an additional element of connectivity.

Research in landscape ecology must use a variety of approaches to provide explanations at many scales of time and space. Even ingenious experimentation meets its limit somewhere in the geographical extent, amount of biomass, and functional complexity of landscapes. Useful experiments can be done to investigate causal mechanisms with rodents, arthropods, some birds, many herbaceous plants, some shrubs, and many aspects of crop plants and farm practice. Experimentation can also be useful in providing predictive power for prescriptive management where causal mechanisms are not a goal.

Fencerow Corridor Effects on Movement

We have tried experimentally to determine the features of fencerows that make them acceptable to mice as movement corridors (Morrison, 1986). We fabricated experimental fencerows from snow fence (or sand fence) and from camouflage netting (Figure 2). We tested three configurations by releasing white-footed mice into an array of experimental fencerows and a control, and recording which were selected as corridors for movement. Mice chose corridors that provided both overhead cover and cover at ground level. With working experimental fencerows it should be possible to manipulate landscapes experimentally, by inserting corridors where patches are isolated. This will be useful both in research and in landscape planning and management.

We are extending the study of preference for fencerow structure by releasing mice bearing radio transmitters into a previously sampled array of natural

Figure 2. An artificial fencerow of camouflage netting for corridor preference experiments with white-footed mice.

fencerow types. The mice are tracked continuously as they move through the landscape for a period of 48 hours (Figure 1). Preliminary results show that mice moving through a strange landscape move almost entirely in fencerows (Figure 3). They do not distinguish among several classes of well-vegetated fencerows that the investigators can readily distinguish, but early results show that mice do select wide fencerows as corridors over less wooded and narrower structures (Figure 4; Merriam and Lanoue, in press).

The spatial scale is qualitatively greater than the norm for this species when resident in woodlands (Stickel, 1968). Several bursts of movement by an animal over two nights can cover 10 to 20 ha of farmland (Table 1). Exposure while moving (such as risk of predation) is not proportional to distance moved. These mice move in bursts; an individual may stay under cover for hours and then move tens of meters in ten minutes.

The corridor function of fencerows may depend on more than just vegetation and "furniture" or the geometrical structure of cover (cf. M'Closkey, 1975). Behavioral reactions to other individuals of a population and to other species must be considered. For example, short-tailed shrews *(Blarina brevicauda)* are common in some fencerows in farmland. Individuals have a *strong* characteristic

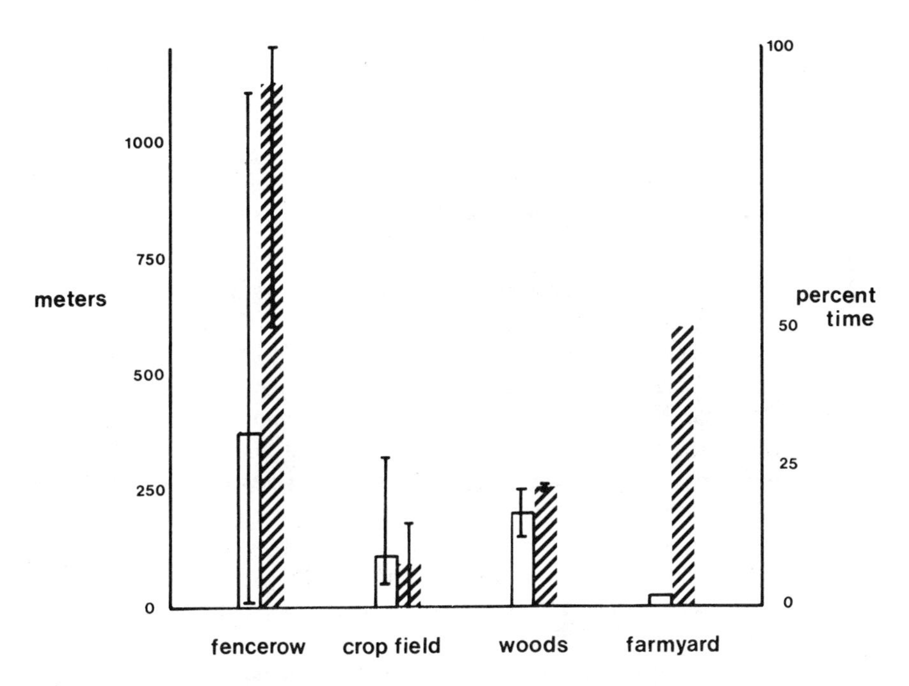

Figure 3. Distances moved and percent of moving time spent in each landscape element, by 18 radio-marked mice released for 48 hours in natural fencerows in unfamiliar farmland. White bars indicate the means of total distances moved by mice, and shaded bars indicate mean proportion of total time used in moving. Vertical lines indicate ranges. See Figure 1.

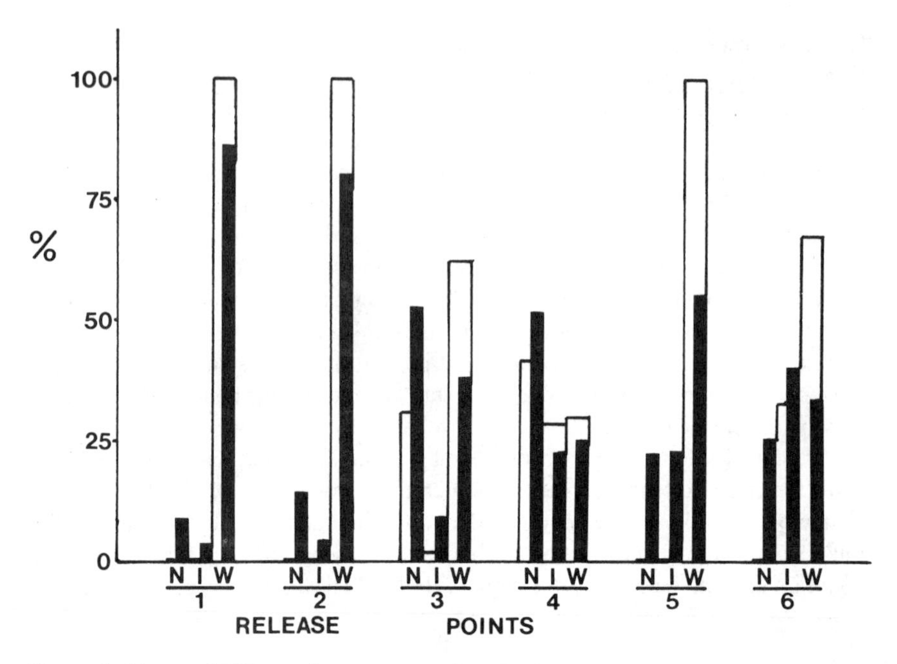

Figure 4. Usage of different fencerow types by 18 mice released in natural fencerows in unfamiliar farmland. N = narrow fencerow with herb and grass cover; I = intermediate-width fencerow with sparse shrub cover; W = wide fencerow with mainly woody cover. Thin shaded bars indicate percent availability of fencerow types within a 200-m radius of each of six release points for mice. Open bars indicate percent of total movement distance by mice in each fencerow type surrounding a release point. Where open bars exceed shaded bars, mice used that fencrow more than expected.

odor that should be obvious to white-footed mice. *Blarina* also have poisonous salivary glands which can paralyze white-footed mice. Therefore a fencerow containing an abundance of *Blarina* and their odor could be a barrier to white-footed mice. We are testing this hypothesis in the laboratory using an analog model that presents a choice of: (1) a *Blarina*-infested (animal or smell) corridor, (2) an overhead runway (containing *Blarina* odor but no shrews) through a *Blarina*-infested corridor, or (3) a *Blarina*-free alternate route simulating a detour by the mouse from the fencerow through an adjacent field (Hawley and Merriam, unpublished). Concurrently, we are using tracking tubes in the field to determine the relative commonness of *Blarina* and white-footed mice, to test whether their distributions in natural fencerows are negatively related. Plastic tubes containing paper, an ink pad, and bait allow visiting small mammals to register their presence by identifiable footprints. Tracking tubes on the ground or tied in trees and shrubs measure species-specific frequency of use of habitat space.

However, many useful questions in landscape ecology are less amenable to replicated controls and treatments, both because of extreme variances associated

Table 1. Landscape area explored and distances moved within landscape elements. Based on 18 radiotelemetrically marked mice released for 48 hours in unfamiliar farmland (Figure 1). Net distance moved includes each meter of fencerow only once, regardless of repeated movement over it.

			Net distance moved (m)			
Mouse	Release point	Landscape explored (ha)[a]	Fencerow	Field	Woods	Farmyard
6	1	12.57	400	17	150	—
2	1	1.54	140	—	—	—
16	1	0.28	60	12	—	—
9	2	2.01	160	27	—	—
3	2	0.05[b]	25	—	—	—
19	2	0.44	75	—	—	—
4	3	0.83	103	—	—	—
8	3	0.31	200	—	—	100
14	3	0.00[b]	4	—	—	—
7	4	28.28[b]	600	150	150	—
12	4	0.00	5	—	—	—
15	4	1.77	150	—	—	—
5	5	22.06	530	—	—	—
10	5	0.05	26	—	—	—
20	5	0.95	110	60	—	—
1	6	4.16	230	—	—	—
17	6	7.07	300	—	—	—
11	6	0.24	55	—	—	—

[a]Area of circle with diameter equal to maximum distance between any two points visited
[b]Killed by predator.

with inherent heterogeneity in space and time, and also because upper limits in time and space scales are too great to manage. This heterogeneity and these hierarchical, and long, time scales are characteristics of landscape ecology. Central questions cannot be set aside because of complications arising from the heterogeneity or time scale or the difficulties of small number systems; some can be dealt with by creativity in research design.

Fragmentation and Functional Role of the Matrix

To measure long-term effects of forest fragmentation on the spatial distribution of woodland species, from invertebrates to trees, we (Middleton and Merriam, 1983) compared agricultural landscape cleared 200 years ago with nearby unfragmented forest. By assembling diagonal transects through a large number of farm woodlots in the computer, we were able to compare two sets of 5-km transects of presence and absence data for over 100 of these woodland species, one set subjected to fragmentation, the other not. We found most species present in both samples, and computer-run tests indicated that the spatial distributions of these species were not significantly different. This means:

1. On a 200-year time scale there is no isolation of these patches for these species, although they are isolated on a 1-year time scale, as evidenced by local extinctions of their populations.
2. Explanations of functional relationships in the landscape may be valid only for the time scale to which they are inevitably constrained by method, paradigm and subject.
3. Most of these woodland species already had the capability to spread and recolonize a heterogeneous mosaic before agricultural clearing, probably because continuous forests are heterogeneous mosaics, too.

Description, analysis, and interpretation of specific spatial patterns without relationship to the biotic processes of the landscape does not provide ecologically convincing answers. Conceptual abstractions and classifications without demonstrable relationships to functioning landscape systems are similarly unsatisfactory. The problem is that "maps" for landscape ecologists must include not only the spatial foundation with its hierarchy of scales, but also the functional overlay of processes with their nest of time scales.

We had accumulated considerable information on wooded fragments, fencerow infrastructure and the *matrix* (a neutral descriptor for the surrounding fields in aggregate) of cropland, when we began questioning the functional role of that matrix in the landscape ecology of *woodland* mice (Merriam, 1984). I wondered whether a field of sweet corn *(zea)* plowed down in October could be useful as winter refuge for these mice, as we knew farm buildings were. A good winter refuge provides food, nest material, and physical cover. A plowed field of sweet corn has corncobs along with corn leaves and stalks, all in tunnels covered by the plowed furrow slice. These thoughts and some observations of mice in such fields at the end of winter prompted reexamination of our maps of forest fragments, corridors, and matrix.

We now find considerable use of this matrix by these mice (Wegner and Merriam, in preparation). Woodland mice enter fields of small grains (oats and barley) when heads are forming on the grain in May, and may stay until after harvest in August (Figure 5). These mice also will enter corn fields when corn is 10 to 15 cm tall and may use these fields until autumn and again after plowing. Individual mice do not just forage out into crop fields from fencerows; some apparently remain in fields and are only caught there for long periods. This species apparently changes its relative use of fencerows and fields in relation to the redistribution of relevant resources caused by seasonal crop growth and land use differences (Figure 5). As agricultural intensity increases, fencerows become less rich in resources relative to crop fields, and fencerow use by mice is low. Early results indicate that this pattern extends to high-intensity crop farms also.

In surviving, these woodland mice use a landscape system including: wooded and brushy patches, fencerows and stream corridors, small grain fields, cornfields, farmsteads, and buildings, but apparently not grassland (i.e., hay fields or pastures). Preliminary data indicate that some individuals move through as much

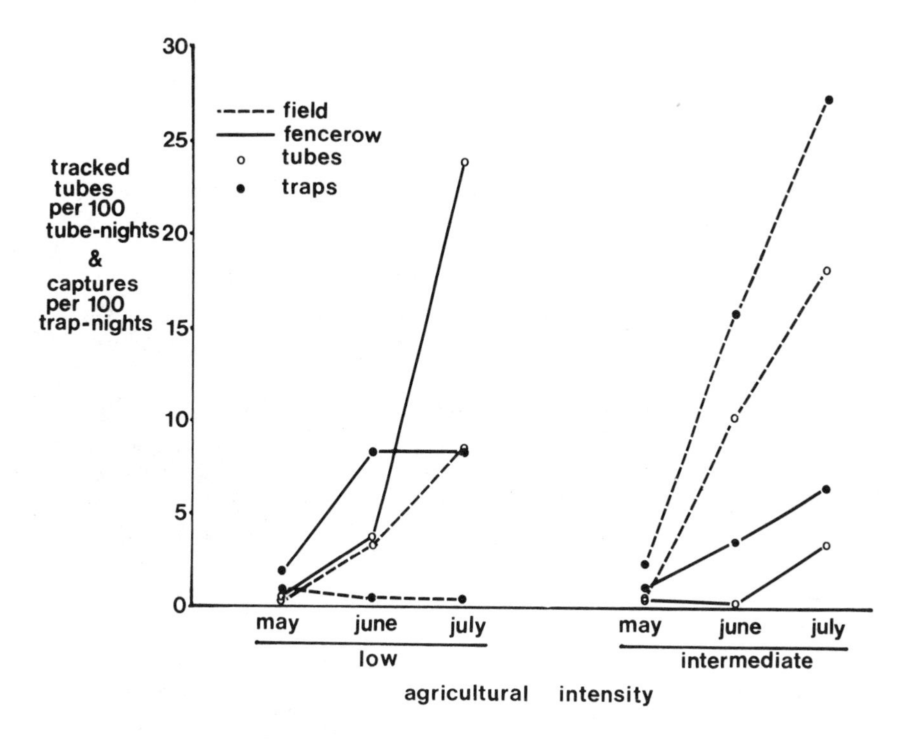

Figure 5. Use of barley and oat fields *(Hordeum, Avena)* compared to use of surrounding fencerows by mice measured over the growing season at two levels of agricultural intensity. Intermediate-intensity farms have less fallow land, higher yields per area, and narrower, less woody fencerows than low-intensity farms. Number of plastic tracking tubes with mouse tracks per 100 tube nights and number of captures per 100 livetrap nights are independent data sets that indicate intensity of use by mice.

as 25 ha of farmland in as little as 2 days. Their spatial scale, in the time scale of one season, may include two or three farms of 40 ha each.

For ecological understanding we need to know what the mice are doing in cornfields, for example, and this requires more behavioral study. So we have built a mouse enclosure over growing corn. White-footed mice released in the enclosure with small luciferin light tags glued to their backs climb to the tops of corn plants at night. Further experiments are underway to show whether the animals will eat arthropods and/or seeds offered experimentally, and whether they will take their food only on the ground or will climb and eat it on the corn plants (Silva and Merriam, unpublished).

From this discussion of current work on woodland mice in crop fields, it is clear that functional processes are critically important to a holistic approach such as landscape ecology. Processes unite the structures and constitute the interactions among spatial patterns. Only on the basis of functional processes can it be proposed that the whole landscape system differs from the sum of its parts. These processes underlie the fundamental holistic tenet of landscape ecology.

The Evolutionary Scale

Full understanding requires that we go beyond the shorter-term ecological processes to the evolutionary dimension. Molecular biology offers opportunities for first steps. We have begun a genetic evaluation to determine whether two closely related species of woodland mice, *Peromyscus leucopus* and *P. maniculatus,* might be in our wooded fragments. Taking salivary samples allowed the mice to be released and allowed us to look at their salivary amylases by electrophoresis (Merriam et al., in press). We have further examined the genetic similarity of *P. leucopus* in about 20 ecologically isolated wooded fragments, spread over several kilometers of farmland. The salivary amylases in these isolated subpopulations were essentially genetically homogeneous. We tentatively conclude that either ecologically isolated patches (with frequent local extinctions) are not isolated when we shift to genetic time and space scales, or we used a genetic analysis too gross to differentiate among subpopulations. We are now resampling these subpopulations to compare their mitochondrial DNA. This requires destructive sampling, isolation of MtDNA, cutting it enzymatically and displaying the fragments by electrophoresis. This will help clarify lines of maternal genetic linkages in the spatial scale of a landscape, and should be on a time scale between the 1-year (local extinctions) and 200-year time scales that we have studied.

Our results to date indicate that a farm landscape is a dynamic system that mice must "solve" seasonally to assemble their resource needs and to avoid the hazards (Merriam, 1988). Spatially, we are working with farm woodlots, fields, fencerows, riverine borders, other corridors, farmsteads, and some barriers. But we know that the mammal species use smaller spatial units at specific times, and we know that individuals use combinations of these landscape elements annually.

This dynamic system functions on a hierarchy of time scales. We have worked from the hourly scale back to 200 years on what might be called the biogeographic time scale, certainly including the evolutionary scale. The genetic time scales controlling the patterns of mitochondrial DNA in the landscape are probably related to the time scale for fragmentation of the forest (agricultural settlements). Farming practice generates a shorter scale of 3 to 10 years by crop rotation and changing land use. The growing season manifested by crop growth and maturation dictates a shorter scale. And within that is nested a sequence of tillage and other farm practices. (Plowing probably *is* an event for mice!) And of course mice have their own time scale, dictated by season, growth, and reproduction, and perforce including survival, if possible, for all 12 months.

Summary and Conclusion

Fragments of deciduous forest set in a matrix of farmland is a common landscape at middle northern latitudes. These mosaics provide models that may be useful in simplifying the study of other heterogeneous and complex environments. In some regions, such forest fragments are the only common form of woodland remaining to the people and to their heritage of woodland species. Understanding how woodland species persist in fragmented forests in agricultural landscapes is worthwhile both ecologically and culturally.

Such understanding is facilitated by adopting a landscape-ecology model incorporating ecological processes and landscape elements. The forest fragments are viewed as part of a heterogeneous mosaic with components that tie together temporal and spatial cascades of ecological processes. These processes functionally interconnect the spatial configuration of landscape elements.

For particular woodland species such as the white-footed mouse *(Peromyscus leucopus)* and the chipmunk *(Tamias striatus),* this means that populations in individual forest fragments can become so small seasonally that local extinctions occur. For the species to continue in the landscape, these local extinctions must be balanced by recolonization. Since the nonforested farmland matrix that intervenes between forest fragments is hostile in varying degrees to woodland species, difficult recolonization might be expected. However, recolonization can be so rapid and so nearly certain that equilibrium biogeographic theory is inapplicable. Ease of recolonization results from use of corridors by woodland species moving between forest fragments, thus avoiding the hazards of intervening cropland. Wooded or brushy fencerows, for example, produce this corridor effect in the movement of small mammals.

Colonists can arise from large source populations in large forest fragments, from small populations in newly recolonized woodlots, or from overwintering populations in anthropogenic refuges such as farm buildings. The supply of colonists is not easily related to density in source populations, and peak of dispersal may not coincide with season of recolonization.

During the breeding season the populations of several forest fragments can be interconnected, so that they operate demographically as a larger *metapopulation*. It is critical that the landscape elements be interconnected so that individuals can move among the small populations composing a metapopulation. However, the model is more general if we define a process parameter called *connectivity,* which may depend on a particular landscape element such as a fencerow, but does not measure structural features. Instead, connectivity measures the potential of the landscape to facilitate recolonization of local extinctions, and therefore measures a behavioral process. Operationally, this requires dispersal for recolonization at the appropriate season without the excessive mortality often associated with dispersal. Beyond recolonization of local extinctions, population increase in isolated patches of a metapopulation is enhanced by connectivity.

This model of the functioning of forest fragments in an agricultural mosaic is based on ecological field experimentation and validated simulation modeling. Results indicate, at least for some rodents, that the heterogeneity and the connectivity of this system are critical, and are best understood in a landscape-ecology model.

The strength of this approach lies in its synthesis of basic ecological processes with natural, agricultural, and technological elements in the normal heterogeneous space emphasized in landscape ecology. Studies using this model produce both fundamental ecological knowledge and planning and management insights in landscape ecology. Results should include knowledge of heterogeneity as a functional element of ecological systems, relationships between connectivity and adaptive values of dispersal mechanisms, demographic-ecological effects of environmental heterogeneity on "evolutionarily stable strategies," and the possibilities of indirect relationships of stability and diversity via heterogeneity. The critical importance of connectivity in survival of a metapopulation is easily applied. Management of isolated habitat units, connecting elements such as fencerows, ravines, and riverine strips, siting of barriers such as highways, and planning for animal populations with known migratory requirements all become more effective when the landscape features affecting connectivity are properly considered.

References

Fahrig, L. and G. Merriam. 1985. Habitat patch connectivity and population survival. *Ecology* 66:1762–68.

Henderson, M. T., G. Merriam and J. Wegner. 1985. Patchy environments and species survival: Chipmunks in an agricultural mosaic. *Biological Conservation* 31: 95–105.

Henein, K. and G. Merriam. In press. The elements of connectivity where corridor quality is variable. *Landscape Ecology*.

Lefkovitch, L. P., and L. Fahrig. 1985. Spatial characteristics of habitat patches and population survival. *Ecological Modelling* 30:297–308.

Levins, R. 1970. Extinction. In M. Gerstenhauber, ed., *Some mathematical questions in biology. Lectures in mathematics in the life sciences.* American Mathematical Society, Providence, Rhode Island, pp. 77–107.

M'Closkey, R. T. 1975. Habitat dimensions of white-footed mice, *Peromyscus leucopus. American Midland Naturalist* 93:158–67.

Merriam, G. 1984. Connectivity: a fundamental ecological characteristic of landscape pattern. In J. Brandt and P. Agger, eds., *Proceedings of the first international seminar on methodology in landscape ecological research and planning.* Roskilde University Center, Denmark, Vol. 1, pp. 5–15.

Merriam, G. 1988. Modelling woodland species adapting to an agricultural landscape. In K.-F. Schreiber, ed., Connectivity in landscape ecology. *Münstersche Geographische Arbeiten* 29:67–8.

Merriam, G., M. Kozakiewicz, E. T. Tsuchiya, and K. Hawley. In press. Barriers as boundaries for metapopulations and demes of *Peromyscus leucopus* in farm landscapes. *Landscape Ecology* 2:227–35.

Merriam, G. and A. Lanoue. In press. Corridor use by small mammals: Field measurements for three experimental types of *Peromyscus leucopus. Landscape Ecology.*

Middleton, J., and G. Merriam. 1981. Woodland mice in a farmland mosaic. *Journal of Applied Ecology* 18:703–10.

Middleton, J. and G. Merriam. 1983. Distribution of woodland species in farmland woods. *Journal of Applied Ecology* 20:625–44.

Morrison, W. O. 1986. *Peromyscus leucopus* in experimental fencerows. M.Sc. thesis, Carleton University, Ottawa, Canada.
Stickel, L. F. 1968. Home range and travels. In J. A. King, ed., *Biology of Peromyscus (Rodentia)*. American Society of Mammalogy Special Publication 2:373–411.
Ulanowicz, R. E. 1986. *Growth and development: Ecosystems phenomenology*. Springer-Verlag, New York.
Wegner, J. F. and G. Merriam, 1979. Movements by birds and small mammals between a wood and adjoining farmland habitats. *Journal of Applied Ecology* 16:349–358.

Part III Natural and Human Processes Interacting to Cause Landscape Change

Landscape dynamics are typically the product of diverse natural processes and socially organized humans that only plan small parcels within a landscape mosaic. Resilience of the landscape depends both on its structure and its history of exposure to particular perturbations. Many human activities are without historical precedent in a landscape, and thus a mosaic of dynamical systems, rather than a steady state, should be expected. Part III explores and elucidates the following patterns and concepts.

River/stream systems are commonly key integrators of landscape processes, and a hierarchical approach is useful because processes differ by scale. At a broad regional scale humans may settle and become dense on flat irrigatable areas, and progress up river valleys where water supply and flood avoidance is predictable. At an intermediate spatial scale, concentric "von Thunen bands" are often visible. At the fine scale of an individual field, (for example, during a normal rainfall year in a Mediterranean climate), crop harvest leaves the field subject to livestock grazing and water erosion, followed by wind erosion during the following dry season. This seasonal pattern results in a reduced resource base for the following year. Then the periodic and predictable drought year arrives causing severe field degradation and little recovery. The effects of time scale are combined, but differ.

The mosaic of ecosystems and the effects of landscape fragmentation result from the combined geologic template of landforms and spatially differentiated human activities. Landforms exert controls on inputs, fluxes and disturbances, and appear to be useful, relatively predictable organizing structures for understanding landscape mosaics and fluxes. In a forested mountainous area a dispersed-patch clearcutting process produced a high contrast, relatively homogeneous grain size over the land, with major implications for all ecological characteristics from fire to biodiversity, wildlife and erosion. In a flatter mainly agricultural area, a half century of change resulted in a major decrease in farmland and a major increase in urban-industrial-transportation land, while forest land remained constant. During this process spatial patterns changed markedly, for example, fragmentation decreased, field size increased, forest tracts became larger and more connected, total edge length decreased, and the fractal dimension decreased.

Synergisms among the four classic, spatially differentiated human harvests (crop farming, livestock grazing, wood cutting, and water use) are striking. An elegant example here weaves together crop production and timing, nonnative annual grasses, goat grazing and movement, water and wind erosion, shrub coppicing and lignotuber removal, the spatial patterning of native versus nonnative plants and wildlife, and temporal and spatial scales. With increasing human population the synergisms lead to an accelerating spiral into poverty. In other landscapes synergisms are seen from rain-on-snow events, and from doubling of crop production while halving cropland area.

Nonmarket values, such as the goods and services provided to people by natural areas, forests and croplands, are woefully underestimated in economic terms. For example, in a large heterogeneous region the total nonmarket value equalled one-third of the market value of timber and agricultural production combined. Energy density (i.e., the fossil fuel energy consumed for human activities per unit of area) for urban areas was 10 times, and cropland 1.5 times, that of forests and other natural areas. Wetlands and stream corridors are especially high in nonmarket values, and public ownership is proposed as important in protecting nonmarket values.

9. The Georgia Landscape: A Changing Resource

Eugene P. Odum and Monica G. Turner

Albert E. Cowdrey (1983) comments that poverty is no friend of natural resources which "typically are devoured piecemeal to sustain existence." Unbridled prosperity is also no friend of natural resources, which tend to be despoiled in the rush to accumulate material wealth. Georgia, along with the rest of the southern U.S.A., has experienced both poverty and prosperity. In the early 1900s, Georgia was poor as measured by most economic and cultural indices and as compared with northern states of the U.S.A. Much of its soil was "devoured" in an effort to scratch out an agrarian existence, and human-resource potential was chronically underachieved as a result of low income and poor education. During the past 50 years the tables have turned. Southern states have made a remarkable recovery from decades of depression, and economic and cultural measures now approach those of the rest of the nation. However, little planning has been done to cope with the changes. Land use, for example, changed dramatically. Once forested, Georgia became primarily agricultural, then made a transition to its present diversity of forests, farms, and cities. The South, including Georgia, now has a relatively large land reserve for which a variety of uses may compete in the future (Healy, 1985).

To provide a background for future planning and management, we studied changes in the landscape and in resource use from 1935 to 1985 in Georgia. We began with the 1930s because the late Howard W. Odum published his masterwork, *Southern Regions of the United States,* in 1936. This book documented the condition of the region's human and natural resources, the reasons for the "chasm between the potentialities and the actualities" (Odum's words), and what needed to be done. Our goal in this study was to compile a sort of "Southern Regions Revisited," with a focus on the state of Georgia as a microcosm of the region. In this chapter, we will describe changes in land and resource use within Georgia and assess some of the ecological implications of these changes.

Georgia is a good microcosm for the southern USA because its topographic gradient from mountains to the sea and its climate (moist, warm temperate) are representative of the region. Human population density, the ratio of urban to rural inhabitants (60/40), and per capita income are also close to the regional average. Accordingly, we can compare Howard Odum's southern regions of the 1930s with Georgia of the 1980s. In the 1930s the chief problems were social and economic; now they are demographic (rapid population growth due to immigration) and environmental. To assess change, we first compiled numerous charts and tables from existing data bases to document changes in the Georgia landscape and its plant, animal, and other natural resources during the past 50 years. For some studies, it was not possible to analyze the whole state during our 2-year project, so representative sample counties were selected for study (Figure 1).

Georgia encompasses three major physiographic regions (Figure 1); the mountains (1,470,310 ha), the piedmont (4,606,139 ha), and the coastal plain (8,971,206 ha). The mountain region ranges in elevation from 183 to 1432 m,

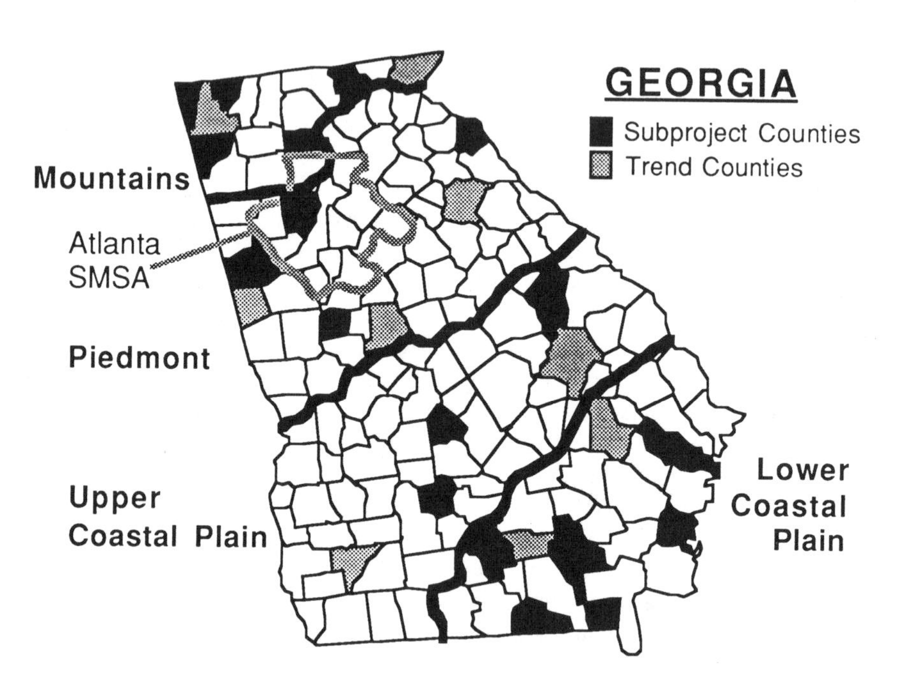

Figure 1. Map of Georgia showing physiographic regions and sample counties used in our study. The nine *trend counties* were used for the analysis of changes in landscape patterns; these plus the other *subproject counties* were used in other portions of the study.

with mean annual temperature ranging from 12.8 to 16.1°C and annual rainfall ranging from 132 to 229 cm. The Georgia piedmont consists of foothills underlain by acid crystalline and metamorphic rock. Elevation ranges from 152 to 457 m. Mean annual rainfall is 112 to 142 cm, and mean annual temperature ranges from 15.0 to 17.8°C. The large coastal plain region has gentle to moderate slopes and sandy soils underlain by marine sands, loam, and/or clays. Elevation ranges from 0 to 300 m; mean annual rainfall ranges from 112 to 135 cm, and mean annual temperatures range from 18.9 to 21.1°C.

Historical Perspective

Forests once covered the state of Georgia, with the exception of coastal salt marshes and grassy areas in the Okefenokee Swamp (Nelson, 1957; Plummer, 1975). The potential natural vegetation of Georgia (Kuchler, 1964) was suggested to be an oak forest *(Quercus)* in the northeastern mountains and an oak–hickory–pine forest *(Quercus–Carya–Pinus)* in the northwestern mountains, the piedmont, and the upper coastal plain. Major forest types in the lower coastal plain are longleaf-slash pine (*P. palustris* and *P. elliotti*) and loblolly-shortleaf pine (*P. taeda* and *P. echinata*), except where southern floodplain forest *(Quercus–Nyssa–Taxodium)* would prevail along the bottomlands. Plummer (1975) used survey maps from the early 1800s to identify the species composition of the early forests. The virgin forest had been modified for centuries by Native Americans, who commonly used fire as a hunting technique, as a means to increase production of browse, berries, and other wild food plants, and to clear fields for agriculture (Stewart, 1956). Extensive clearing and farming accompanied European settlement of the region. Between 1773 and 1823, for example, the Georgia piedmont was converted from stands of virgin timber to agriculture (Bond and Spillers, 1935; Brender 1974), and by 1793 cotton was the dominant crop. Subsequently, nearly all the original topsoil was lost from 47 percent of the uplands, and gullying was apparent on 44 percent of the piedmont (Hartman and Wooten, 1935). Following invasion of the boll weevil *(Anthonomus grandis)* in the 1920s, large amounts of cropland were abandoned. Most abandoned land reverted through natural succession to pine (primarily *Pinus taeda* and *P. echinata*), and this old-field pine comprised more than two-thirds of the total forest area in 1930.

Structure of the Georgia Landscape

Most landscapes, including Georgia, are heavily influenced by human land use. The resulting landscape mosaic is a mixture of natural and human-managed patches that vary in size, shape, and arrangement (e.g., Burgess and Sharpe, 1981; Forman and Godron, 1986; Urban et al., 1987). These spatial patterns can influence a variety of ecological phenomena (Turner, 1989), including animal movements (e.g., Fahrig and Merriam, 1985; Henderson et al., 1985; Freemark and Merriam, 1986), water runoff and erosion (e.g., White et al., 1981; Peterjohn and Correll, 1984; Kesner, 1984), or the spread of disturbance (e.g., Franklin and Forman, 1987; Turner, 1987a; Turner et al., 1989). Thus, our ability to quantify changes in landscape structure through time may be crucial to our understanding of the dynamics of the Georgia landscape. We discuss several different approaches to analyzing these changes.

Changes in Land Use

Land use patterns integrate both the natural and human-developed environments and thus were an important focal point in our study. The extensive changes in land use in Georgia involved not only the *proportion* of land that is farm, forest, urban, and so on, but also the *spatial patterns,* such as patch size and complexity. For the state as a whole, the area in farmland (which includes the fields, pastures, small woods, etc., that are on farms) decreased, with the most rapid change occurring between 1950 and 1970. In the coastal plain, however, cropland increased during the past two decades (Figure 2). Forests of varying age and composition have covered more than 60 percent of the state throughout the half-century period. This included frequent cutovers, but nevertheless there was a tendency for the total forest area to increase. Coverage today is about 65 percent statewide, but varies up to 85 percent in some mountain counties and to as low as 40 percent in some agricultural counties of southwestern Georgia. The greatest changes in land use occurred in the piedmont region (Figure 2), where the majority of Georgians live today.

Since the 1930s, the area in urban-industrial use (including transportation corridors) has increased rapidly. In 1930, Atlanta was the only officially recognized Standard Metropolitan Statistical Area (SMSA, a census bureau term for major cities and their surrounding suburban areas). By 1985, the Atlanta SMSA comprised 15 counties (see Figure 1) and one-third of the state's human population. Now there are seven other SMSAs of two to four counties each, so that 22 percent of Georgia's 159 counties are included in metropolitan districts. The total length of state and federal highways in Georgia has more than doubled since 1935, from 5,400 to 12,000 km (9,000 to 20,000 mi). Southern cities are not as densely populated as many older cities of the Northeast; for example, human density in Atlanta is about 3,000 persons per square mile compared with 10,000 or more in some eastern cities.

Despite the increased urbanization in Georgia, the ground actually occupied by towns and cities remains only about 6 percent of the whole land area of the state (Figure 2), with an additional 2 to 3 percent in transportation corridors. However, because of the very large energy and resource demands of urban-industrial areas, the impact of cities on surrounding cropland, forests, rivers, and aquifers is enormous. In the pie-diagram of Figure 3, we weighted the three major environments (agricultural, natural, and urban) according to **energy density,** that is, how much fossil-fuel energy is consumed for human activities per

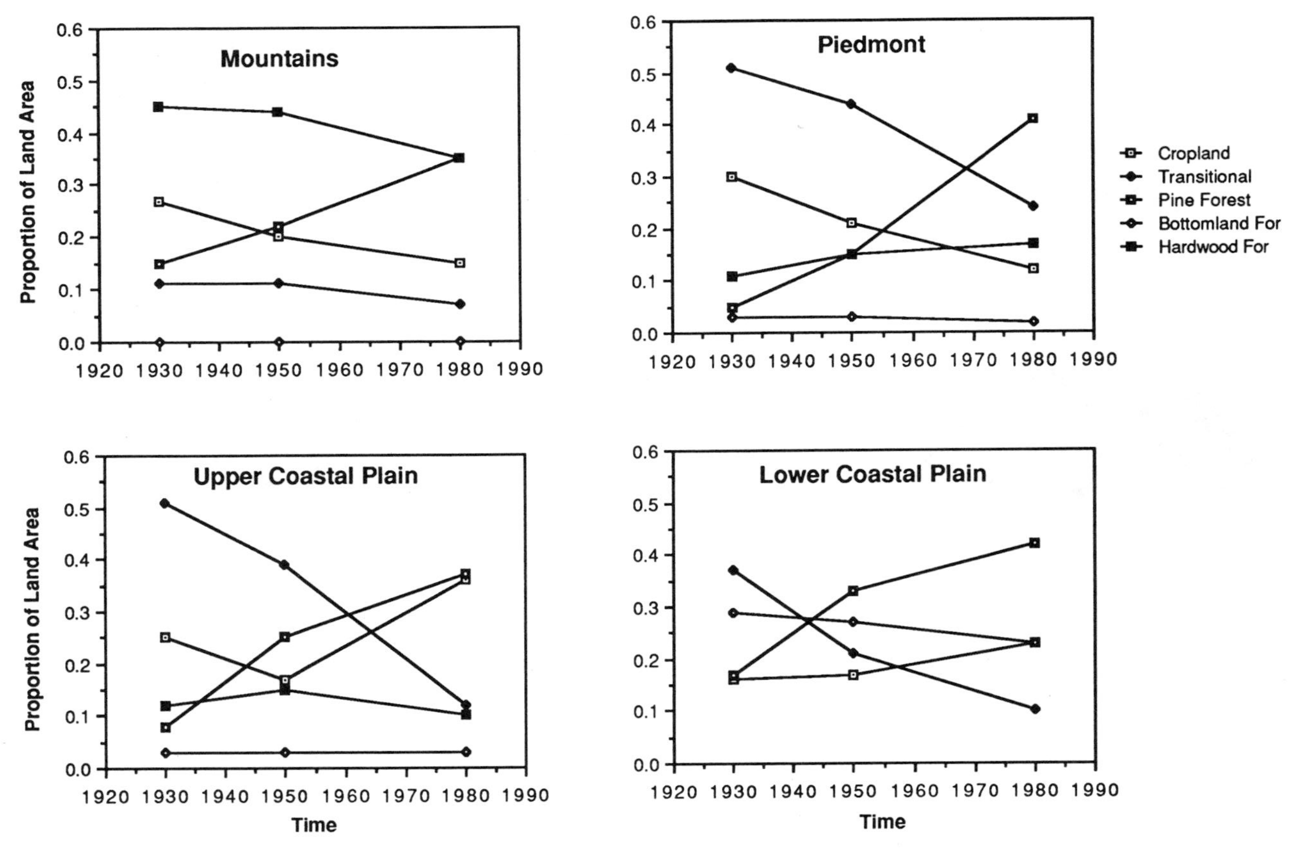

Figure 2. Land use changes in Georgia, 1935 to 1982. (Data reported in Turner and Ruscher, [1988].)

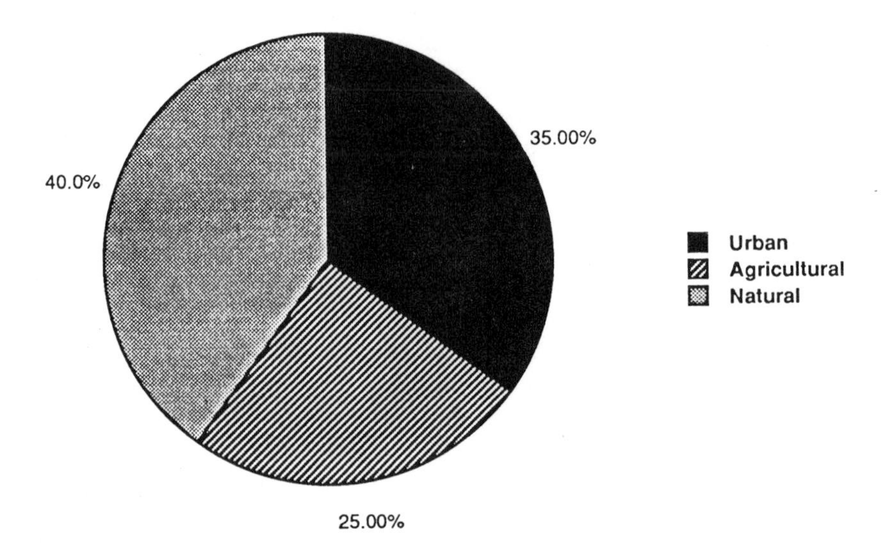

Figure 3. Percent of energy consumed in Georgia in three major land uses. Calculations were based on energy density and the area in each land use. Energy density of agricultural areas is considered to be twice as high as natural environments; that of urban areas ten times as high.

unit of area. Urban areas in Georgia average 10 times the energy density (as estimated from consumption of electricity which is 38 percent of the total), and cropland 1.5 times the energy density of forests and other natural areas. If we multiply these ratios by the area occupied by the three land uses, the percentage of energy consumed in each land use can be estimated (Figure 3). This illustrates the true importance of the urban landscape in terms of energy flow, and, indirectly, its impact on the rural landscape.

Spatial Patterns in the Landscape

The amount of land in various uses has thus changed, but what about the spatial arrangement of the landscape components? The complexity or fragmentation of the pattern may also be important. Historical black-and-white aerial photography was used to analyze the spatial patterns of land use (Turner and Ruscher, 1988). Nine sample counties (included in the trend counties in Figure 1) from different nonurban regions were selected by a stratified random sample: Atkinson, Baker, Emanuel, Heard, Monroe, Oglethorpe, Rabun, Tattnall, and Walker. Photographs encompassing a 12,696-ha rectangular study area in each county were analyzed for three time periods (1930s, 1950s, 1980s), and eight land uses were identified: urban, cropland, abandoned cropland, pasture, pine forest, upland hardwood forest, bottomland hardwood forest, and water. Land use data were digitized manually at a resolution of one ha (see Turner and Ruscher, 1988 for details).

During the 1930s, the landscape was very fragmented, meaning that there were many small patches of different land uses. This fragmentation declined during the past 50 years. For example, the number of forest patches of one hectare in size decreased, but the total amount of forest increased in all regions. Thus, the individual tracts of forest land increased in average size, and forests are now more connected. The size of crop fields changed in different ways in the different physiographic regions. In the piedmont, where cropland has been declining, the average size of crop fields stayed the same. However, in the upper coastal plain, where most of today's agriculture is located, the average size of crop fields increased.

To assess changes in the complexity of the patch shapes, fractal dimensions (Mandelbrot, 1983) of patch perimeters were calculated (Turner and Ruscher, 1988). In general, the fractal dimensions declined during the past 50 years, reflecting a decrease in the complexity of the landscape pattern (Turner and Ruscher, 1988). However, there was some variance among land-cover categories. For example, fractal dimensions in the piedmont generally declined with time (Figure 4), indicating that tracts of cropland, abandoned cropland, and pine forest have become more regular and less convoluted in shape. Hardwood forest was an exception, however, reflecting the increase in patch size and the relatively lower human influence.

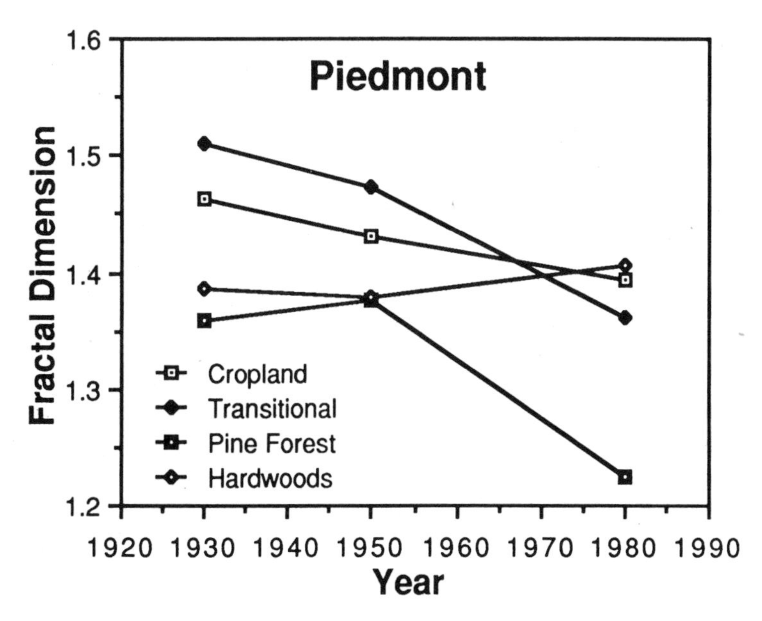

Figure 4. Fractal dimensions *(D)* of patch perimeters of major land uses in the Georgia piedmont. A simple, linear perimeter has $D = 1$, whereas a convoluted perimeter surrounding the same area has D approaching 2. (Data reported in Turner and Rusche, [1988].)

Edges between land types are important for certain species of animals and may affect water runoff and erosion from one land use to another. The total length of edge between cropland and abandoned cropland decreased in all regions, whereas edge between cropland and forest increased in all regions (Turner and Ruscher, 1988). This suggests that the boundaries of crop fields have become more permanent than they were in the 1930s. The total amount of edge between all land uses decreased in all sample counties between 1930 and 1980, again emphasizing that the Georgia landscape has become less fragmented.

Landscape Diversity

Landscape pattern can also be studied by looking at the diversity of units or patches in the landscape. The diversity concept has been applied in many ways to species, but the application of this concept to landscapes is fairly new (Romme, 1982; Romme and Knight, 1982). Commonly used measures of species diversity include species richness, the number of different species in a specified area, and evenness, the equitability of the abundance of each species. By merging the study of species diversity with the geometry of biotic communities, the concept of landscape diversity emerges. For terrestrial studies landscape diversity may be defined as the richness and evenness of the habitats or cover types that make up the landscape mosaic.

Hoover (1986) studied comparative diversity of plant comunities and species across the physiographic regions of Georgia by using the indices developed by Romme (1982). Six sample areas located in counties without extensive agricultural or urban features were studied: two each from the mountains, the piedmont, and the coastal plain. Vegetation maps of the sample counties previously prepared by the state Department of Natural Resources were used to identify the sizes and shapes of natural vegetation communities and agriculture. Vegetation was classified as matrix, corridor, or patch (after Forman and Godron, 1981). Community types were verified in field studies, and species diversity determined from strip censuses conducted in each community type.

Species diversity increased from the coastal plain to the mountains (Hoover, 1986). Landscape diversity, however, exhibited the reverse trend with highest diversity in the coastal plain. The mountains generally exhibited low landscape diversity, having both low evenness and richness. The piedmont landscapes expressed an intermediate level of landscape diversity, with a high degree of evenness but the lowest value for richness. The relatively even distribution of community types might be attributed to the extensive human impacts and prior fragmentation. The coastal plain had the most diverse landscape. The trend in landscape diversity was unexpected, since we had hypothesized that both species and community diversity should increase from south to north because of increasing topographic relief. Hoover's research suggested that the complex drainage patterns of the coastal plain may create more natural diversity than hills and mountains. These results underscore the concept that patterns at one level of scale are not necessarily the same as patterns at another level. Thus, diversity at the species, community, and landscape levels may differ.

For the state as a whole, xeric communities tended to behave as patch components embedded in the matrix of dominant vegetation. In contrast, mesic areas, such as stream courses, behaved as corridors, the most linear feature in the landscape. Corridors join the landscape together and function as vital networks of animal movement and plant dispersal within the landscape. For these and other reasons (water quality, for example) improved statewide protection of stream and river corridors was emphasized among the recommendations that emerged from our research.

Simulating Changes in Landscape Structure

Along with quantifying changes in landscape patterns, the prediction of such patterns would be useful. The application of simulation modeling to landscapes represents a powerful approach. Most ecological modeling, however, has focused on temporal changes and excluded spatial dynamics. Transition models have been used to predict changes in vegetation or land use, but these models have not been spatially explicit. Accordingly, spatial simulation models were developed by Turner (1987b; 1988) to predict changes in land use patterns in the piedmont county of Oglethorpe, adjacent to the Athens metropolitan area. Five land use categories were included: urban, cropland, abandoned cropland, pasture, and forest. Three different types of spatial simulation were compared using cells in a grid:

1. random simulations based solely on transition probabilities
2. spatial simulation in which the nearest four neighbors (adjacent cells only) influence the transition
3. spatial simulations in which the eight nearest neighbors (adjacent and diagonal cells) influence the transition

Models and data were compared using the mean number and size of patches, fractal dimensions of patches, and amount of edge between land types. The random model simulated a highly fragmented landscape having numerous small patches with complex shapes. The two versions of the spatial model simulated the pattern for cropland well, but simulated patches of forest and abandoned cropland were fewer, larger, and less convoluted than those in the real landscape. To improve predictability, several possible modifications of model structure were proposed. This approach has potential generality for simulating human-influenced landscapes and may prove quite useful, particularly if socioeconomic factors can also be included (e.g., Parks, in press).

Ecological Effects of Land Use Changes

With these striking changes in structure of the Georgia landscape quantified, we now turn to the ecological implications of land use change. Here we focus on patterns of primary productivity of the whole landscape, and then more specifically on trends in agricultural production, on forests, and on certain wildlife populations.

Net Primary Production of the Landscape

In natural landscapes, net primary production (NPP) provides the energetic and material basis for all other processes, because every food chain begins with some kind of plant production. NPP also provides an important resource base for human economic activity. NPP includes yield, the merchantable part of plant production, plus all the other organic matter produced by the plants.

A sample of 20 counties, with four to six in each physiographic region (Figure 1), was used to quantify both land use and NPP for the time period 1935 to 1982 (Turner, 1987c). Data on yield of food and fiber harvested from cropland, pastures and forests for each county were converted to estimates of NPP using appropriate conversion factors (harvest ratios, for example). Estimates for NPP of urban areas and wetlands were obtained from the literature.

The NPP for the whole Georgia landscape has more than doubled between 1935 and 1982 (Figure 5), increasing from 2.5 to 6.4 tonnes/ha, a remarkable recovery when one considers the "wasted" land of the 1930s. The pattern of increase varied among land uses and physiographic regions, but followed a general sigmoid growth form with the most rapid increase occurring between 1960 and 1970. The increase in productivity of agricultural lands reflects soil recovery as well as increased fertilization and irrigation, and the increase in forest productivity reflects natural soil-building processes and improved silviculture.

Rates of potential natural NPP predicted by world models are 16 to 18 tonnes/ha for a forested region with the temperature–rainfall pattern of Georgia (Turner, 1987c). The average NPP estimated for Georgia forests was 8.3 tonnes/ha, approximately two-thirds of the predicted rate. Some cropland approximated the predicted rate, but the average was much lower (5.1 tonnes/ha.). Accordingly, Georgia remains below its potential productivity.

Agriculture

Crops

To determine the 50-year trends in Georgia crop production, we converted the yields (bushels and pounds per acre) of all major crops (corn, soybeans, grain, vegetables, cotton, etc.) as reported by the U.S. Agricultural Reporting Service into the common denominator of calories. This allows us to add up all "the apples and oranges as fruit," so to speak. During the five decades since 1935, the area in crops has decreased by half while crop yield (cal/ha) has quadrupled (Figure 6). Thus, total food production in Georgia has doubled. In other words, in the present decade, Georgia grows twice as much food on half as much land as in the decade 1935–1944. This remarkable achievement (in view of the degraded

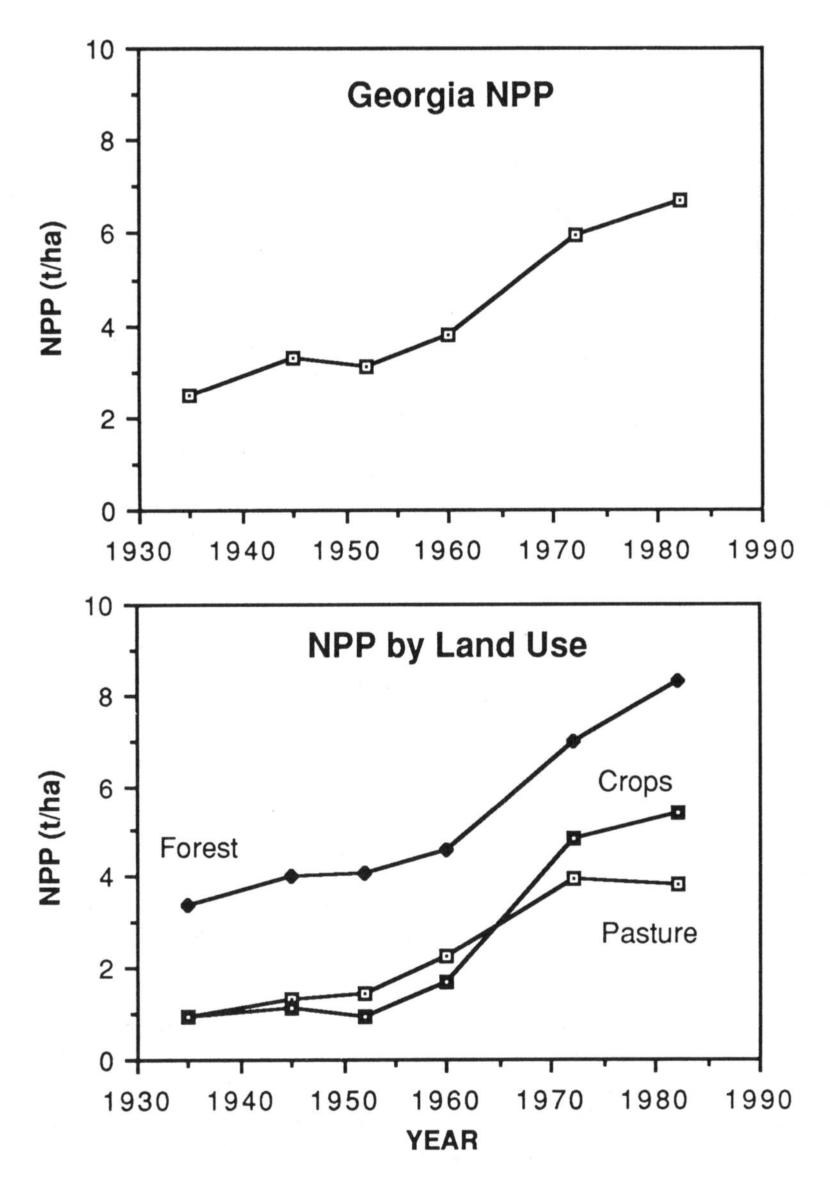

Figure 5. Net primary production of the Georgia landscape, 1935 to 1982. (From Turner, [1987c].)

state of the land in the 1930s) was accomplished partly by development of improved crop varieties (cultivars) and farm machinery, but mostly by increased use of fertilizers, pesticides, herbicides, and (in recent years) irrigation. General mineral fertilizer use increased seven-fold and nitrogen use increased eleven-fold over the half century (Figure 6). Pesticide and herbicide use showed similar trends.

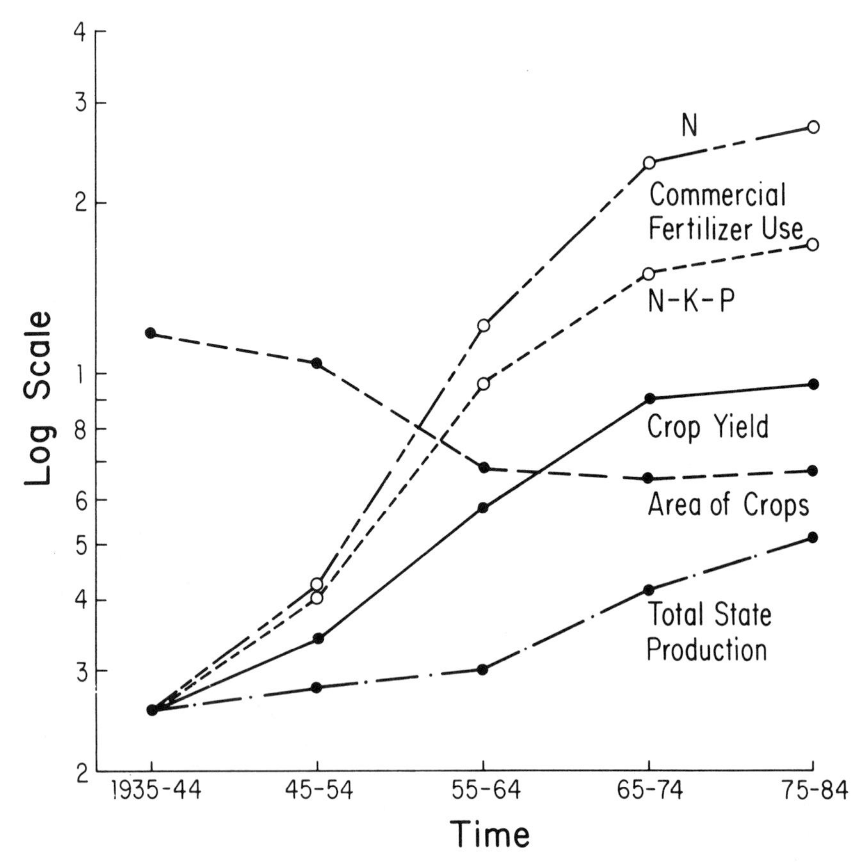

Figure 6. Changes in commercial fertilizer use (10^3 tons/yr), crop yields (cal/ha), area in crops (ha), and total crop production (cal/yr) in Georgia, 1935 to 1985.

The curves in Figure 6 all appear S-shaped, or sigmoid, with the most rapid change occurring between about 1950 and 1970. The fact that yields have essentially leveled off during the past 10 years suggests diminishing returns with agricultural practices that rely on raising yields by increasing chemical subsidies. Much of the fertilizer and pesticide now applied to crops runs off or leaches out of the crop fields, and contaminates surface and ground water. This non-point-source pollution (that is, unwanted and deleterious materials that do not originate from a concentrated or point source, such as a discharge pipe) poses a great threat to our global life-support systems (e.g., effects of acid rain, ozone, greenhouse gases, and so on). Reducing inputs, or input management, may be most important for future non-point-pollution control, but it should also be coupled with "output side" control techniques, such as the best management practices used for erosion control and urban runoff (Odum, 1989). Current research here in Georgia and elsewhere on "reduced-input agriculture," "conservation tillage" and other systems that maintain reasonable yields, but reduce damage to our life-support systems, will contribute to improvements in our agricultural production systems (Odum, 1987; 1989).

In sharp contrast to major cash crops (corn, grains, soybeans), vegetable production in Georgia has declined. The area devoted to growing commercial vegetables declined by about three-fourths since the 1930s, and there has been no improvement in yield. Thus, total production has declined. To an increasing extent, vegetables consumed in Georgia are produced in other states (California and Florida, especially). As transportation costs rise in the future, a revival of agriculture centered on vegetables, especially near large urban areas, seems likely. Also, a greenhouse agriculture, common in European countries such as The Netherlands and Italy, that would provide fresh beans, tomatoes, and other products year round, could be in Georgia's future.

Domestic Animals

Most people are unaware that there are more domestic animals in Georgia than people, and these animals also consume more food. We compared the growth of human and domestic animal populations by converting numbers of cattle, chickens, pigs, turkeys, and so on, to human population equivalents. To do this, we took the calories of food consumed annually by an individual animal (information supplied by animal science nutritionists at the University of Georgia) and scaled this to the one million calories per year required by a human in Georgia. Thus, on an annual basis a beef cow would be equal in energy consumption to three people, while about 73 broiler chickens (which only live a few weeks) would be equivalent to one person. Over the 50-year period, domestic animals in population equivalents tripled while the number of people doubled (Figure 7). The growth of the poultry population has been spectacular, a ten-fold increase since 1935. Broiler chickens as well as laying hens and eggs are major Georgia cash crops. At the present time, chickens alone consume a quantity of corn and soybeans approximately equivalent to the entire state production of these crops. In other words, most of Georgia's crop production could go to feed animals, while most human plant food needs to be imported from out of state!

Aquaculture, chiefly catfish (*Ictalurus* spp.) farming, has increased in the state in recent years. In warm-water regions such as southern Georgia, fish (and shellfish) theoretically convert plant to animal food more efficiently than cattle or poultry, because the fish are cold-blooded and do not divert energy to maintaining a high body temperature. Many people predict that fish farming in artificial ponds will become a major industry in the state; much will depend on the availability of labor and water and the cost of building and maintaining impoundments.

Forests

For the whole state, slightly less than half of the forest land is in pine (less in northern Georgia, more in southern Georgia), a little more than a third is in

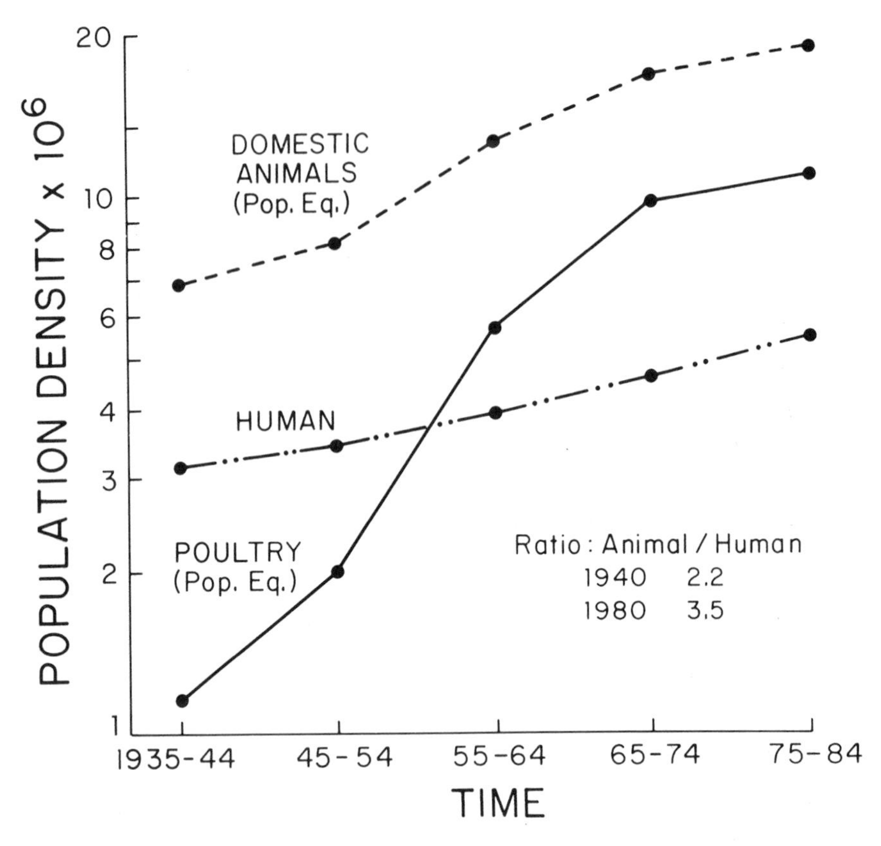

Figure 7. Changes in human population (in millions of individuals) and domestic animals in Georgia, 1935 to 1985. Domestic animals are expressed in human population equivalents calculated on the basis of food consumption in calories. See text.

upland hardwood or in transition to hardwood (oak–pine), and about 15 percent is in bottomland hardwoods (Table 1). Only 15 percent of Georgia forests are in pine plantations, that is, trees planted in rows and harvested and replanted at short intervals, but this type of forest (really a crop) is expected to increase in the future.

Land ownership is an important consideration if we are to project future trends. Less than 7 percent of Georgia forests are in public ownership, about 25 percent are owned by forest industries, and 66 percent are in numerous small to medium-sized tracts owned by individual private citizens (Table 1).

Questions about future forest trends are two-fold: (1) How much of the approximately 400,000 ha (1,000,000 acres) now in private (noncommercial) ownership will remain as such, and how much will be converted to urban or farmland, or be preserved? and (2) How will the competition between pine and hardwood on the Piedmont turn out? Pine currently has a higher market value

Table 1. Forest cover in nine sample counties in Georgia and in the whole state. Data are for 1982 and expressed as acres $\times 10^3$.

County	All land	All forest	Forest Types					Ownership			
			Pine		Oak Pine (transition)	Upland hardwood	Bottomland hardwood	National forests	Other public	Forest industry	Other private
			Plantation	Natural							
Union (MTS)[a]	198	171 (86.7%)	4.4	53.2	8.9	102.2	—	89.1	0.34	3.4	76.0
Chattooga (MTS)	203	149 (73.5%)	2.8	31.8	48.7	65.7	—	15.5	—	17.0	116.5
Cobb (Pied)	222	105 (47.5%)	—	73.0	12.7	12.8	3.2	—	3.0	0.14	98.5
Oglethorpe (Pied)	278	221 (79.3%)	42.5	67.7	11.8	64.7	33.8	3.8	0.34	96.3	120.2
Monroe (Pied)	255	204 (79.9%)	16.3	86.1	28.8	61.2	10.9	—	0.18	59.2	143.9
Emanuel (UCP)	439	285 (64.9%)	61.9	107.7	24.1	29.5	60.9	—	16.0	6.5	48.3
Turner (UCP)	188	82 (44.0%)	10.7	40.1	10.1	6.7	15.0	—	0.08	12.3	70.0
Barrien (LCP	300	181 (60.5%)	15.7	71.2	27.1	14.3	52.9	—	2.6	21.0	157.7
Glynn (LCP)	270	157 (58.2%)	57.1	27.7	11.4	19.9	37.0	—	3.9	112.1	37.1
Totals	2,353	1,555	211.4	558.5	183.5	357.0	203.7	108.4	25.4	327.8	867.2
Percentages		66	14	36.9	12.1	23.6	13.5	8.2	1.9	24.7	65.2
State	37,167	24,242	3,592	7,847	2,960	5,805	3,530	765	819	5,936	16,214
Percentages		65.2	15.1	33.1	12.5	24.5	15	3.2	3.5	26.0	68.3

[a] MTS—Mountain Region; Pied—Piedmont; UCP—Upper Coastal Plain; LCP—Lower Coastal Plain (see Figure 1).

than hardwood, and most of the large Georgia forest industry is based on pine. Johnson and Sharpe (1976) reported that despite silvicultural efforts to maintain pine in the Georgia piedmont, the area occupied by hardwoods increased. Using a 30-year model, they projected this trend to continue, although at a slower rate than if only natural succession was involved. Urbanization and suppression of fire, both of which favor hardwoods over pines, were important factors in the model projection. Healy (1985) noted that urbanites are purchasing an increasing amount of land in the vicinity of cities, and unlike farmers or foresters, they do not obtain their main income from the land. Such owners are more interested in the second home or recreational value of their land, and less interested in managing for pine production; many, in fact, would rather see the hardwoods take over.

Because of these trends, some foresters worry that future demand for pine may exceed supply, because few private land owners are replanting after the very large stands of naturally regenerated pine are cut. Demand projections often are based not only on human population increase, but on an increase in per capita consumption as well. However, paper recycling and more efficient use of wood in home construction could reduce per capita needs. For example, the number of municipalities with paper recycling programs in Georgia has increased from one in 1975 to 17 in 1983. If by the year 2000 most cities and towns in this and other states recycle paper, there should be plenty of pine to meet both local and export demands for paper.

Wildlife Populations

The effect of land use changes on bird populations was examined by Ronald Pulliam (University of Georgia) and associates who analyzed long-term census data collected by the U.S. Fish and Wildlife Servive (Pulliam, 1987, unpublished). Several species that occupy old-field habitat (i.e., recently abandoned crop fields) have declined during the past 20 years. These declines parallel the decline in abandoned or idle cropland (Figure 2). Other bird species have increased in number and extended their ranges southward in Georgia. This group includes the robin *(Turdus migratorius),* house wren *(Troglodytes aedon),* song sparrow *(Melospiza melodia),* house finch *(Carpodacus mexicanus),* and grackle *(Quiscalus quiscula)* that thrive in urban and suburban habitat.

The large growth of white-tailed deer *(Odocoileus virginianus)* populations in Georgia (Figure 8) is a result not only of the shift in land use from agriculture to second-growth forest, but also a successful restocking and management effort by the Georgia Department of Natural Resources. The increase in deer since 1940 has been exponential, and the one million deer now populating the state have detrimental effects on crops, home gardens, and auto accidents. Strong measures (increasing hunting, large scale removal, etc.) will probably be required to halt this growth and establish some desirable equilibrium. Beaver *(Castor canadensis),* which were once almost exterminated in the state, have also made a comeback as a result of state protection and changes in land use.

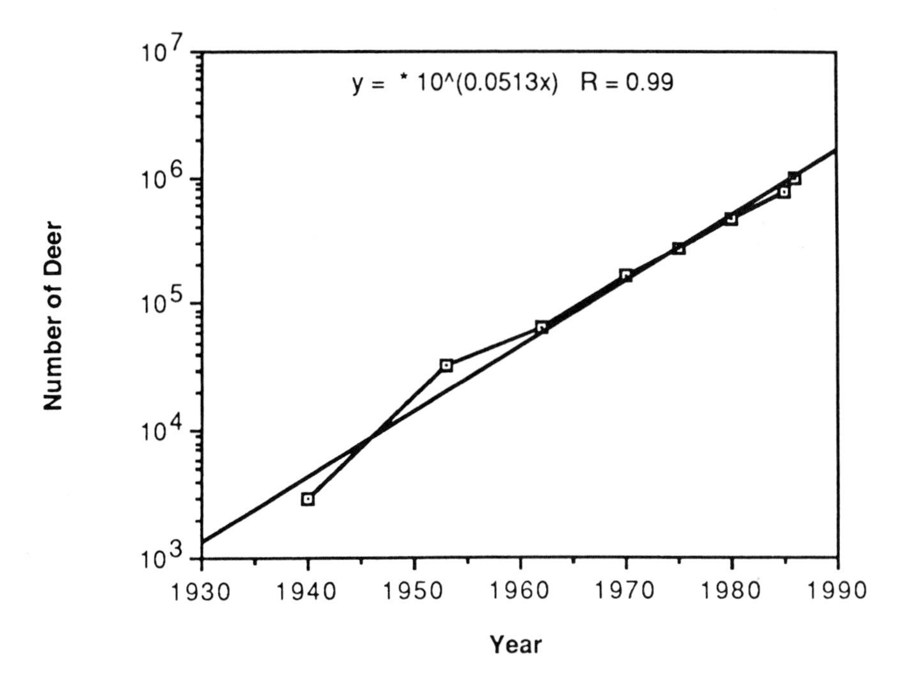

Figure 8. Deer population growth in Georgia, 1940 to 1985. Data provided by the Georgia Department of Natural Resources. The straight line is the exponential curve fitted to the data.

Trends in Resource Use

As land use and human population size both change, the demand for and consumption of resources also changes. In assessing the implications of landscape changes, therefore, it is instructive to assess the trends in resource use and quality. Land planning, or growth management, requires an understanding of such trends and, furthermore, should aim toward maintaining both availability and quality of the resources. Here we focus on two resources, energy and water.

Energy

Per capita energy consumption in Georgia was about 60 percent lower than the national average in 1960, but increased rapidly to approach the national level by 1980 (Table 2, Figure 9). This reflects increasing urbanization and prosperity in the state. Between 1980 and 1985, per capita consumption declined both in the United States as a whole and in Georgia—a result of the rise in energy prices and concurrent energy conservation. A recent report released by the International Energy Agency (cited in Anonymous, 1987) indicated that conservation efforts have increased the efficiency of energy use in western industrialized nations by

Table 2. Energy consumption in Georgia and the United States.

	Total Energy	Population	Per Capita Energy	END USES (BTU × 10^{12})					Sources (% of total)			
	BTU × 10^{12}	× 10^3	BTU × 10^6	Residential	Commercial	Industrial	Transportation	Electric	Coal	Natural Gas	Petroleum	Fossil Fuel
Georgia												
1983	1,666	5,732	291	348 21%	228 14%	524 31%	567 34%	630 37.8%	35.2	18.3	41.9	95.4
1980	1,681	5,463	316	344 20%	207 12%	580 34%	550 33%	600 35.7%	31.1	19.2	44.6	94.9
1975	1,454	4,900	296	297 20%	192 13%	478 32%	487 34%	485 33.4%	21.5	22.9	41.1	92.4
1970	1,252	4,590	273	255 20%	147 12%	450 36%	400 32%	370 29.6%	15.4	27.4	42.6	90.4
1965	876	4,250	206	167 19%	91 10%	361 41%	257 29%	213 24.3%	17.4	25.0	48.6	90.8
1960	651	3,943	165	128 20%	68 10%	247 38%	208 32%	144 22.1%	13.7	29.0	49.4	92.0
United States												
1960	43,806	180,270	243	18.9%	10.8%	46.0%	24.2%	18.8%	22.5	28.3	45.5	96.3
1983	70,508	234,250	301	20.9%	15.2%	36.7%	27.2%	35.4%	22.5	24.6	42.6	89.7

Source: State Energy Data Report.
1960–1983. U.S. Energy Information Administration.

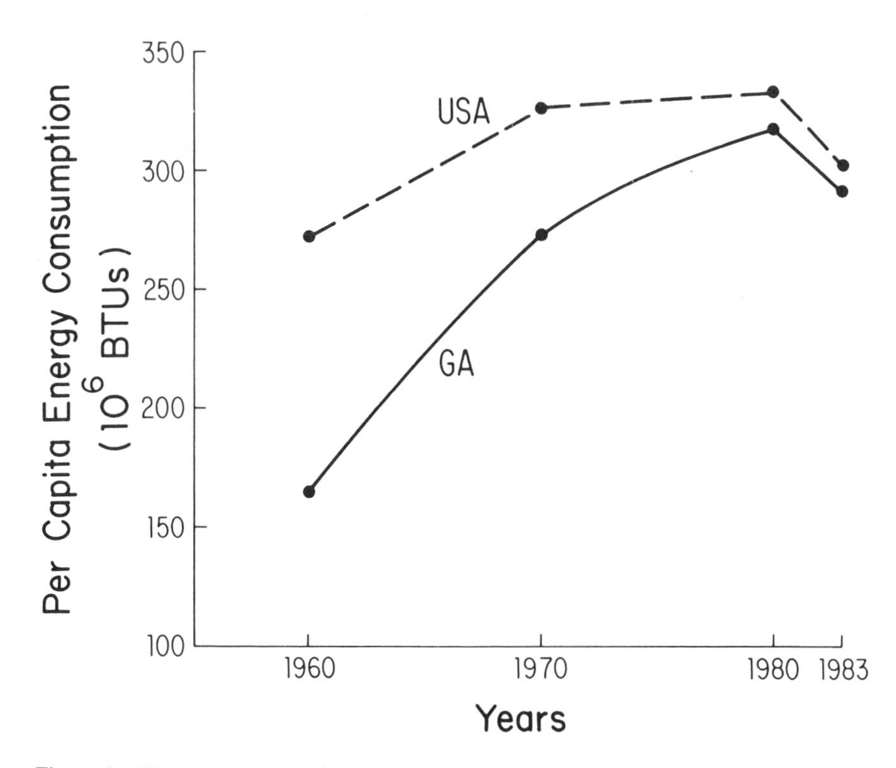

Figure 9. Changes in per capita energy consumption in Georgia and the United States, 1960 to 1983. (From the State Energy Data Report, U.S. Energy Administration.)

about 20 percent since 1970. In the U.S., gross production increased almost 32 percent between 1973 and 1985, but energy consumption increased only 5 percent. The result was a saving of the equivalent of 880 million tons of oil a year—more than the total oil production of western nations (Anonymous, 1987). Accordingly, the downward trend in per capita energy consumption in Georgia parallels the worldwide trend.

Residential, commercial, industrial and transportation energy uses as percentages of the total energy consumption have not changed much over the past 23 years (Table 2). However, electric use increased from 22 percent to 38 percent of the total for Georgia, and 19 percent to 35 percent for the nation. Georgia's dependence on coal as an energy source increased from 14 percent to 35 percent compared to no increase for the nation. Georgia has a 95 percent dependence on fossil fuels (all imported from out-of-state), about average for the nation.

In view of Georgia's dependence on coal, the burning of which in power plants is a major cause of air pollution (e.g., Bormann, 1987), it would seem that power companies should invest in known technology for cleaning coal (removing sulfur and other pollutants) before it is burned. A coal gasification combined power plant rated as the cleanest coal-fired power plant in the U.S. has been operating successfully, and competitively with conventional power plants, in California since 1984 (Spencer et al., 1986). Tax and other incentives could be used to promote this.

The potential for further energy conservation is also very large. The president of a company specializing in designing and retrofitting buildings for energy efficiency estimates that a 40 percent saving in energy consumption can be achieved with today's technology (Milton Bevington, personal communication). If all new buildings in Georgia were 40 percent more energy efficient than conventional construction, much new economic growth could occur with only a modest increase in energy consumption.

While rising costs will undoubtedly result in further reductions in per capita energy use, the state could encourage needed reductions. For example, tax deductions could be provided for the cost of energy-efficient new or retrofitted construction. We strongly recommend that the state become actively involved in energy conservation, not only to save fossil fuels but because reductions in the burning of coal, oil, and gasoline will also reduce the considerable costs of non-point pollution.

Water Quantity

Reliable figures on water consumption were not available before 1950. Water use (both total withdrawals and consumptive use) in Georgia increased rather rapidly from 1950 to about 1970, and has generally leveled off during the past 15 years (Figure 10). Per capita use of water has even declined since 1970 (Figure 10), suggesting that rising costs and droughts have resulted in some water conservation.

The proportions of the public water supply that come from surface water and ground water have not changed markedly over the 35-year period (Figure 10). State-wide, 65 percent is from surface water and about 35 percent from groundwater. In the piedmont, where groundwater is limited, some 90 percent of the public water supply comes from surface water. Water conservation will be extremely important for the future welfare of Atlanta and other rapidly growing urban areas of the piedmont. Water use for irrigation increased about ten-fold between 1970 and 1980, and most of this was from ground water in the prime agricultural areas of the upper coastal plain. The water levels in many shallow wells throughout the state have declined, but water levels in deep wells in the coastal plain have been reduced only slightly. Many artesian wells in the vicinity of paper mills and other heavy water users, however, no longer flow, and salt intrusion is a problem in several coastal counties.

The amounts of water impounded by large reservoirs and smaller lakes and ponds (many built under the supervision of the U.S. Soil Conservation Service) exhibit sigmoid growth patterns, with the most rapid increase between 1950 and 1970. Because many of the best sites on streams and rivers have now been impounded, and because U.S. government funds will be more difficult to obtain for this purpose, the benefit–cost ratio for proposed future impoundments must

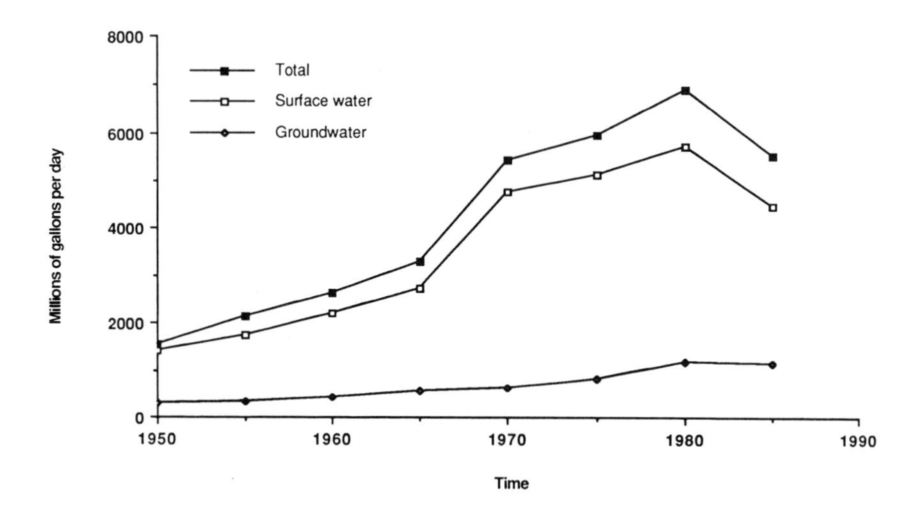

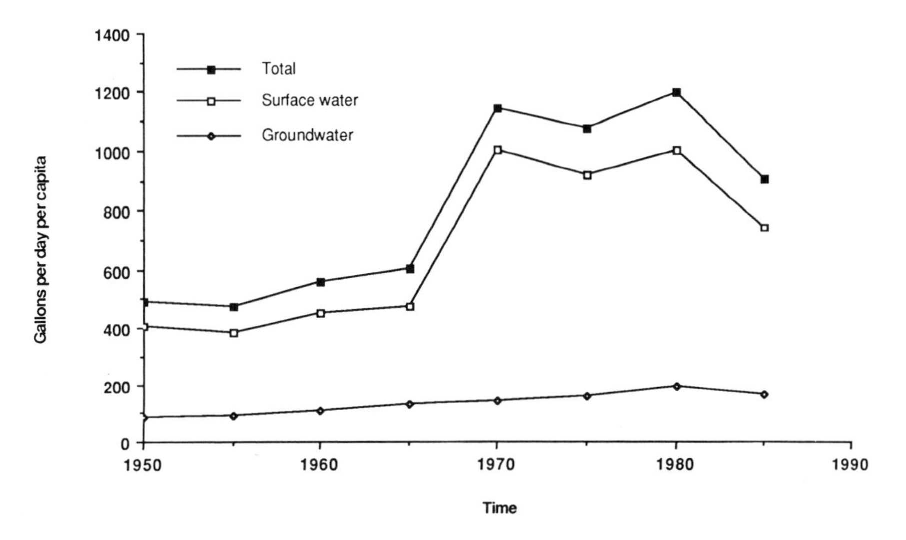

Figure 10. Total freshwater withdrawals (top) and per capita use of freshwater (bottom) in Georgia, 1950 to 1985. (From R. T. James [1987, unpublished report].)

be carefully researched. Many impoundments were originally constructed to promote hydroelectric power and/or recreation, but these are now more important as a water supply for metropolitan areas. Probably the greatest need for new small and medium-sized impoundments will be to store water for cities during droughts. Housing, industrial, and agricultural development on watersheds of such new reservoirs must be restricted if sedimentation is to be minimized and water quality is to be maintained.

Water Quality

Environmental monitoring of Georgia rivers has not been consistent, and information on key aspects was often not available for long periods of time. Nevertheless, some important trends were revealed by data on four rivers (Savannah, Ocmulgee, Oconee, and Yellow Rivers). Coliform bacteria, an index of sewage pollution, have decreased in the Savannah River (Figure 11) and the other rivers since 1970. State and municipal governments have done an excellent job of reducing this point-source pollution, so Georgia rivers in 1985 were more "swimmable and fishable" than they were 15 to 20 years ago. However, non-point pollution as indicated by nitrates and nitrites has increased (Fig. 11), a trend that is nationwide (Smith et al., 1987). Non-point-pollution threats (which include toxic wastes) to water quality can be controlled primarily by reducing the inputs of chemicals into agriculture and industry. As we have already noted, input management as a means of reducing pollution also reduces costs, and is something that mandates statewide (and worldwide) attention.

Nonmarket Values and Natural Area Preservation

Most people are unaware that croplands, forests, and watersheds not only provide material goods like food, fiber, and water, but also provide life-support goods and services. These services (such as air and water purification, recycling and tertiary waste treatment, recreation, and scenic beauty) all are vital to maintaining the economics and quality of life of the urban dweller. It is for this reason that we have included natural-area preservation and nonmarket valuation as important components of this study.

Nonmarket Values in the Georgia Landscape

Landscapes produce many services that historically have been grossly undervalued (Alig, 1983) and presently are not adequately reflected as market values. These nonmarket values may be quite large, but they must be measured indirectly (DeLorme and Wood, 1974) and can be difficult to determine (Westman, 1977; Healy, 1985). Wise and rational management of natural areas requires that these values be incorporated into the decision-making process, and the valuation of the benefits of nonmarketed resources and services has been identified as a major research need (e.g., Odum, 1977; Westman, 1977; Alig, 1983; Costanza, 1984; Loomis and Hof, 1985; Loomis and Walsh, 1986).

Details of our attempts to estimate the nonmarket value of the entire Georgia landscape are reported elsewhere (Turner et al., 1988). Contingent Valuation (an economic approach) and Energy Analysis (an ecological approach) were used, and both approaches yielded similar estimates. The annual value of nonmarketed natural services in Georgia was estimated as $2.6 billion in 1982 dollars, which compares with an annual value of marketed agricultural products of $2.8 billion, and a marketed timber production of $4.5 billion. Assuming a constant stream of

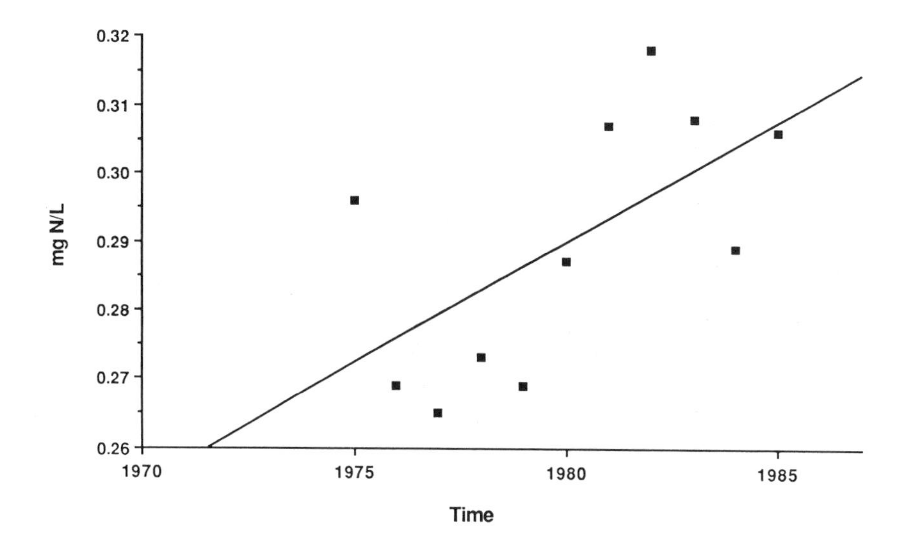

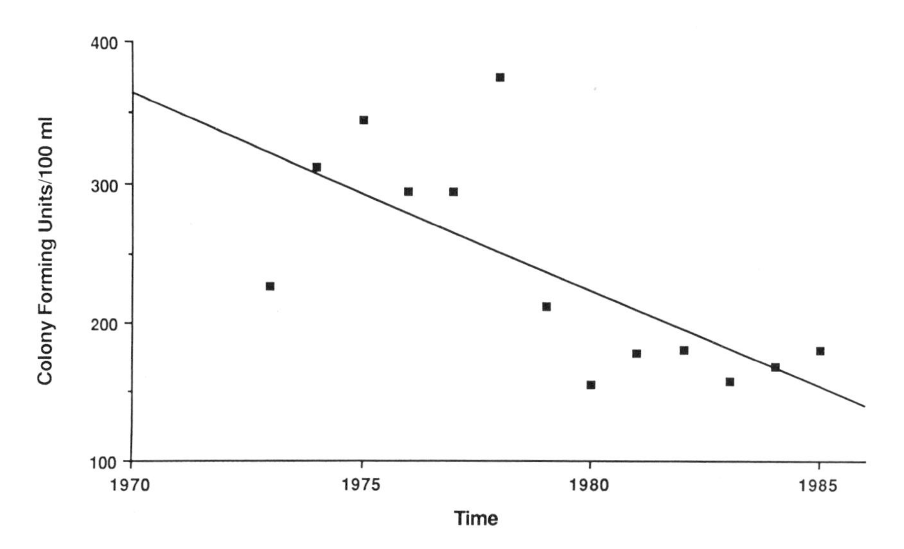

Figure 11. Total nitrate–nitrite nitrogen (top) and fecal coliform bacteria counts (bottom) in the Savannah River, 1970 to 1985. Graphs are five-year moving averages weighted for river flow rate. Data were obtained from the STORET data base of the U.S. Environmental Protection Agency.

services into the indefinite future at a 4 percent discount rate, the total present value of nonmarket goods and services comes to $65.7 billion. This is an estimate of the worth of the state's life-supporting services provided mostly by the natural environment.

Within Georgia, urban lands are of mostly market, whereas wetlands are mostly of nonmarket value (Figure 12). Forests have a high value in both

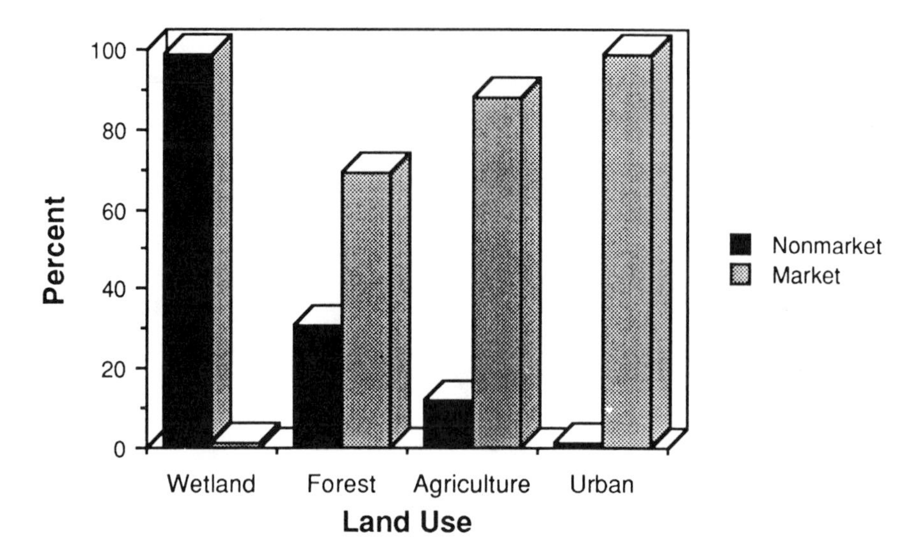

Figure 12. Importance of market and nonmarket values of major land use in Georgia. (Data reported in Turner et al. [1988].)

categories. Thus, changing land use patterns in Georgia and elsewhere will be accompanied by shifts in the relative importances of market and nonmarket values. We believe that the evaluation of nonmarket values is useful, because it demonstrates their magnitude and importance in units (e.g., dollars) that decision makers and the public can readily understand.

Natural Area Preservation

The preservation of large areas in Georgia began in 1836–1837 with the establishment of the Chattahoochee National Forest in the mountains and the Okefenokee Swamp National Wildlife Refuge in the southeastern corner of the state. In both areas, most of the valuable old-growth timber had been removed before the land was acquired by the U.S. government. In the Okefenokee, an unsuccessful attempt was made to drain the swamp to convert it to agricultural use. In the 1950s, abandoned farmland in the eastern piedmont was included in the Oconee National Forest. Between 1965 and 1975, five of the large and unique coastal barrier islands in Georgia were acquired by the U.S. or state government from wealthy families who had purchased the islands after 1865. Acquisition of the Chattahoochee, Okefenokee, and Oconee sites followed resource extraction, which left market values of the lands relatively low. The resilience of these areas has allowed them to recover into valuable resource bases today. Future land acquisition might include currently degraded areas which can be restored.

Despite these acquisitions, only 8 percent of the Georgia landscape is in public ownership, that is, in national, state or municipal forests or parks, wildlife refuges, wilderness areas, etc. For southern states, 8 percent is about average (Parker et al., 1981), but it is less than Florida and far less than all western states in the USA. The Georgia piedmont is especially lacking in preserved areas; less than 2 percent of this physiographic region is preserved. Because large areas of the Georgia landscape are not yet under strong economic pressure for urban-industrial or agricultural development, and because recent opinion surveys indicate strong public support for preserving more natural areas, a once-in-a-century opportunity is here. We suggest a goal of 20 percent preserved land for the year 2000. Accomplishing this would protect critical resources and reduce pollution and congestion, thus improving the quality of life for all Georgians.

Implications for Landscape Management

Based on the results of our research, we offer the following strong recommendations for growth management (i.e., land use planning) in Georgia, giving high priority to maintaining environmental quality:

1. Statewide effort is urgently needed to increase natural-area preservation, especially in the piedmont region, with a goal of placing 20 percent of the state under some form of protection (parks, national and state forests, wildlife refuges, nature preserves, etc.).
2. Statewide effort is needed for the protection of river corridors and freshwater wetlands. Conservation easements and other measures already on the legal books should be encouraged, and new legislation, perhaps similar to the U.S. Coastal Marsh Protection Act of 1970, should be enacted.
3. Educational campaigns are needed to increase public awareness of the: (i) value of the life-supporting environment (nonmarket goods and services of nature); (ii) nature and seriousness of nonpoint pollution; (iii) need for more preserved area; and (iv) need for more effective economic and political infrastructure for planning. Our surveys indicate that large segments of the public are not aware of the significance of changes in the Georgia landscape, and do not perceive the need to take any action to maintain the environmental quality that the state is fortunate to have (Westmacott and Roos, 1987, unpublished).
4. By providing tax and other incentives, state and local governments should encourage power companies to invest in clean coal technology, now economically feasible, to reduce air pollution.
5. In view of the increasing potential for water shortages and declines in water quality, state and municipal governments should actively promote water conservation before there are shortages. To ensure water quantity and quality, all housing, industrial, and agricultural development should be prohibited around all new reservoirs built to store water for urban areas.

6. A central data base for geographical and environmental information should be established, so that decision makers and resource managers will have the continuously updated information needed for managing growth and maintaining environmental quality.

The changes we have documented in Georgia may be useful to those less developed countries that wish to evaluate the processes and effects of changing to a "developed country." In the 1930s, Georgia was largely agrarian, with both low per capita income and low educational levels. Quality industry was lacking, and soil-destroying agricultural practices were prevalent. Today, Georgia is well on its way to becoming developed, with all the problems that come with that status.

Conclusion

In this study, we have followed a *top-down* approach. We examined the patterns of land and resource use for the Georgia landscape as a whole as well as for the major physiographic regions and showed how these patterns changed during the past 50 years. Both spatial and temporal changes were documented. The Georgia landscape has become less fragmented, and forests have increased in area and connectivity. However, although urban areas comprise only 6 percent of the landscape, their impact on energy consumption and use of other resources is several times greater. Landscape diversity varies by physiographic region, but plant species diversity follows a different trend, suggesting that diversity varies with spatial scale. In the Georgia landscape, net primary productivity, as an indicator of functional change associated with land use, is increasing, but still remains below predicted rates for natural land. We conclude that landscape ecological research, which integrates both structure and function, can provide a sound conceptual base for assessing large-scale changes, and in developing strategies for land management.

Acknowledgments

We are grateful for the many contributions and ideas of our colleagues, E. O. Box, J. L. Cooley, F. B. Golley, J. E. Kundell, J. A. Little, V. Meentemeyer, D. Morrison, H. R. Pulliam, and R. Westmacott, who were members of the Physical Resources Task Force. This research was funded by a grant from the Kellogg Foundation to the University of Georgia, as part of a comprehensive study of Georgia conducted by seven task forces. Completion of this manuscript was supported in part by an Alexander Hollaender Distinguished Postdoctoral Fellowship from the U.S. Department of Energy, Office of Health and Environmental Research, administered by Oak Ridge Associated Universities.

References

Alig, R. J. 1983. Impacts of forest land conversion: an overview. *Renewable Resources Journal* 1(4) and 2(1):8–13.

Anonymous. 1987. Conserving energy. *Science* 23:25.

Bond, W. E. and A. R. Spillers. 1935. Use of land for forests in the lower Piedmont region of Georgia. *South Forest Experiment Station Occasional Paper 53*.

Bormann, F. H. 1987. Landscape ecology and air pollution. In M. G. Turner, ed., *Landscape heterogeneity and disturbance*. Springer-Verlag, New York, pp. 37–58.

Brender, E. V. 1974. Impact of past land use on the lower piedmont forest. *Journal of Forestry* 72:34–6.

Burgess, R. L. and D. M. Sharpe, eds. 1981. *Forest island dynamics in man-dominated landscapes*. Springer-Verlag, New York.

Costanza, R. 1984. Natural resource valuation and management: toward an ecological economics. In A. M. Jansson, ed., *Integration of economy and ecology: An outlook for the eighties*. University of Stockholm Press, Sweden, pp. 7–18.

Cowdrey, A. E. 1983. *This land, this south. An environmental history*. University Press of Kentucky, Lexington.

DeLorme, C. D. and N. J. Wood. 1974. Savannah River improvement and environmental preservation. *Land Economics,* August: 284–8.

Fahrig, L. and G. Merriam. 1985. Habitat patch connectivity and population survival. *Ecology* 66:1762–8.

Forman, R. T. T. and M. Godron. 1981. Patches and structural components for a landscape ecology. *BioScience* 31:733–40.

Forman, R. T. T. and M. Godron. 1986. *Landscape Ecology*. Wiley, New York.

Franklin, J. F. and R. T. T. Forman. 1987. Creating landscape patterns by forest cutting: Ecological consequences and principles. *Landscape Ecology* 1:5–18.

Freemark, K. E. and H. G. Merriam. 1986. Importance of area and habitat heterogeneity to bird assemblages in temperate forest fragments. *Biological Conservation* 36:115–41.

Hartman, W. A. and H. H. Wooten. 1935. *Georgia land use problems*. Georgia Experiment Station Bulletin. 191.

Healy, R. G. 1985. *Competition for land in the American South*. Conservation Foundation, Washington, DC.

Henderson, M. T., G. Merriam, and J. Wegner. 1985. Patchy environments and species survival: Chipmunks in an agricultural mosaic. *Biological Conservation* 31: 95–105.

Hoover, S. R. 1986. Comparative Structure of Landscapes Across Physiographic Regions of Georgia. M.S. thesis, University of Georgia, Athens, Georgia.

Johnson, W. C. and D. M. Sharpe. 1976. An analysis of forest dynamics in the northern Georgia piedmont. *Forest Science* 22:307–22.

Kesner, B. T. 1984. The Geography of Nitrogen in an Agricultural Watershed: A Technique for the Spatial Accounting of Nutrient Dynamics. M.S. Thesis, University of Georgia, Athens, Georgia.

Küchler, A. W. 1964. *Potential natural vegetation of the conterminous United States*. Association of American Geographers, Special Publication 36.

Loomis, J. B. and J. G. Hof. 1985. *Comparability of market and nonmarket valuations of forest and rangeland outputs*. USDA Forest Service, Research Note RM-457.

Loomis, J. B. and R. G. Walsh. 1986. Assessing wildlife and environmental values in cost-benefit analysis: State of the art. *Journal of Environmental Management* 22:125–31.

Mandelbrot, B. B. 1983. *The fractal geometry of nature*. Freeman, New York.

Nelson, T. C. 1957. The original forests of the Georgia piedmont. *Ecology* 38: 390–6.

Odum, E. P. 1977. The life-support value of forests. In *Proceedings of the Society of American Foresters National Convention,* Washington, DC, pp. 101–5.

Odum, E. P. 1987. Reduced-input agriculture reduces non-point pollution. *Journal of Soil and Water Conservation* 42:412–14.

Odum, E. P. 1989. Input management of production systems. *Science* 243:177–82.

Odum, H. W. 1936. *Southern regions of the United States*. University of North Carolina Press, Chapel Hill, North Carolina.

Parker, K. C., E. P. Odum, and J. L. Cooley. 1981. *Natural ecosystems of the sunbelt*. Institute of Ecology, University of Georgia, Athens, Georgia.

Parks, P. In press. Incorporating economic considerations in landscape models. In M. G. Turner and R. H. Gardner, eds., *Quantitative methods in landscape ecology*. Springer-Verlag, New York.

Peterjohn, W. T. and D. L. Correll. 1984. Nutrient dynamics in an agricultural watershed: Observations on the role of a riparian forest. *Ecology* 65:1466–75.

Plummer, G. L. 1975. 18th century forests in Georgia. *Bulletin of the Georgia Academy of Science* 33:1–19.

Romme, W. H. 1982. Fire and landscape diversity in subalpine forests of Yellowstone National Park. *Ecological Monographs* 52:199–221.

Romme, W. H. and D. H. Knight. 1982. Landscape diversity: The concept applied to Yellowstone Park. *BioScience* 32:664–70.

Smith, R. A., R. B. Alexander, and M. G. Wolman. 1987. Water quality trends in the nation's rivers. *Science* 235:1607–15.

Spencer, D. F., S. B. Alpert, and H. H. Gilman. 1986. Cool water: Demonstration of a clean and efficient new coal technology. *Science* 232:609–12.

Stewart, O. C. 1956. Fires as the first great source employed by man. In W. L. Thomas, ed., *Man's role in changing the face of the earth*. University of Chicago Press, Illinois. pp. 115–84.

Turner, M. G., ed. 1987a. *Landscape heterogeneity and disturbance*. Springer-Verlag, New York.

Turner, M. G. 1987b. Spatial simulation of landscape changes in Georgia: A comparison of three transition models. *Landscape Ecology* 1:29–36.

Turner, M. G. 1987c. Land use changes and net primary production in the Georgia, USA, landscape: 1935–1982. *Environmental Management* 11:237–47.

Turner, M. G. 1988. A spatial simulation model of land use changes in a piedmont county in Georgia. *Applied Mathematics and Computation* 27:39–51.

Turner, M. G. In press. Landscape ecology: The effect of pattern on process. *Annual Review of Ecology and Systematics* 20.

Turner, M. G. and C. L. Ruscher. 1988. Changes in landscape patterns in Georgia, USA. *Landscape Ecology* 1:241–51.

Turner, M. G., R. H. Gardner, V. H. Dale, and R. V. O'Neill. In press. Predicting the spread of disturbance across heterogeneous landscapes. *Oikos*.

Turner, M. G., E. P. Odum, R. Costanza, and T. M. Springer. 1988. Market and nonmarket values of the Georgia landscape. *Environmental Management* 12:209–17.

Urban, D. L., R. V. O'Neill, and H. H. Shugart. 1987. Landscape ecology. *BioScience* 37:119–27.

Westman, W. E. 1977. How much are nature's services worth? *Science* 197:960–3.

White, F. C., J. R. Hairston, W. N. Musser, H. F. Perkins, and J. F. Reed. 1981. Relationship between increased crop acreage and nonpoint-source pollution: A Georgia case study. *Journal of Soil and Water Conservation* 36:172–7.

10. Landscape Change in Mediterranean-Type Habitats of Chile: Patterns and Processes

Eduardo R. Fuentes

In Chile the subtropical winter-rain and summer-dry climate ranges from about 27° to 40° latitude south. The more restrictive conditions for a Mediterranean type of climate (Aschmann, 1973; di Castri, 1973) are found on the lowlands only between approximately Combarbala (31° 11′) and Linares (35° 51′). The altitudinal limits of these climatic conditions are more difficult to delineate, because of the steep altitudinal gradients, and because meteorological stations in mountainous areas are scarce in Chile. At the latitude of Santiago (33° 27′), the altitudinal limit is about 1500 m, but this varies with latitude and general topography. In this chapter I will focus on areas between Combarbala and Linares below 1500 m, thus excluding a few peaks of the coastal ranges plus large areas of the high Andes.

Within this region the emphasis will be on non-flat areas, where interactions between people and the biological-physical dimension are less capital-intensive and extreme than those typically found on the flat and well-irrigated lowlands. That is, I will examine landscapes conditioned by two basic factors: (a) an overall Mediterranean type of subtropical climate, and (b) a generally mountainous topography. Mountains, water, and people are central to understanding all landscapes of this region, and linkage of these three forces is a central theme of this chapter.

Since about 80 percent of the 12 million inhabitants of Chile live between Combarbala and Linares, and a major driving force for recent landscape change is related to human activity, I will focus on three main questions:

1. How have people coped with the mountainous character of the landscapes? We will be particularly interested in microgeographical differences in human density and land use that reflect topographical restrictions and possibilities.
2. How have people coped with the within-site variability of water supply? Here we will focus on the seasonal, as well as the between-year, variability in rainfall, and the modes by which people have adapted to this variability.
3. What are the ultimate landscape-dynamical consequences of these variability patterns in water supply and human activity?

Landscapes are dynamic entities, resulting from continuous interactive processes between socially organized humans and the current state of the biological-physical frame. In this dynamic, not only are planned and conscious actions important, unconscious and unplanned side effects are also often critical. The actual changes produced and their significance depend on overall landscape structure and its inherent resilience, and therefore on the history of the system. Moreover, there is no a priori reason why the system should absorb, or even recuperate the old dynamics after a perturbation, particularly one that is without historical precedent in the system (van Dobben and McConnel, 1975).

These considerations are particularly relevant for the dynamics of the landscapes in this region of Chile. On the one hand, human presence here is relatively new (ca. 11,000 yr, Montane, 1968), and since the Spanish arrival, the area has become the most densely populated region of the country (Garcia-Vidal, 1982). On the other hand, many human needs, expectancies, perceptions and capacities to act are culturally determined, and in central Chile these modes and attitudes have a Spanish-European origin. The result of the interaction of two such worlds, the cultural and the biological-physical, each with partially separate histories and driving forces, will not necessarily be a steady state. In fact, as I will show, a whole mosaic of dynamical situations can be expected, depending on the local conditions of topography and climate and the extent to which humans have interacted with natural systems.

In this chapter, description of geomorphology and climatic variability will provide the background for a presentation of key functional couplings between the physical and biological landscape dimensions. Once these bases are established I will analyze the three focal questions presented regarding the "strategies" people have used, and the consequences of these strategies in molding Chilean landscapes and in providing insight into landscape pattern and process.

The Physical Components

Landforms

Two mountain ranges running parallel to the Pacific coast are primary structuring factors of the central Chile landscapes (Figure 1). Within the study area these ranges reach altitudes of 6000 m, and in general their peaks are above 3000 m.

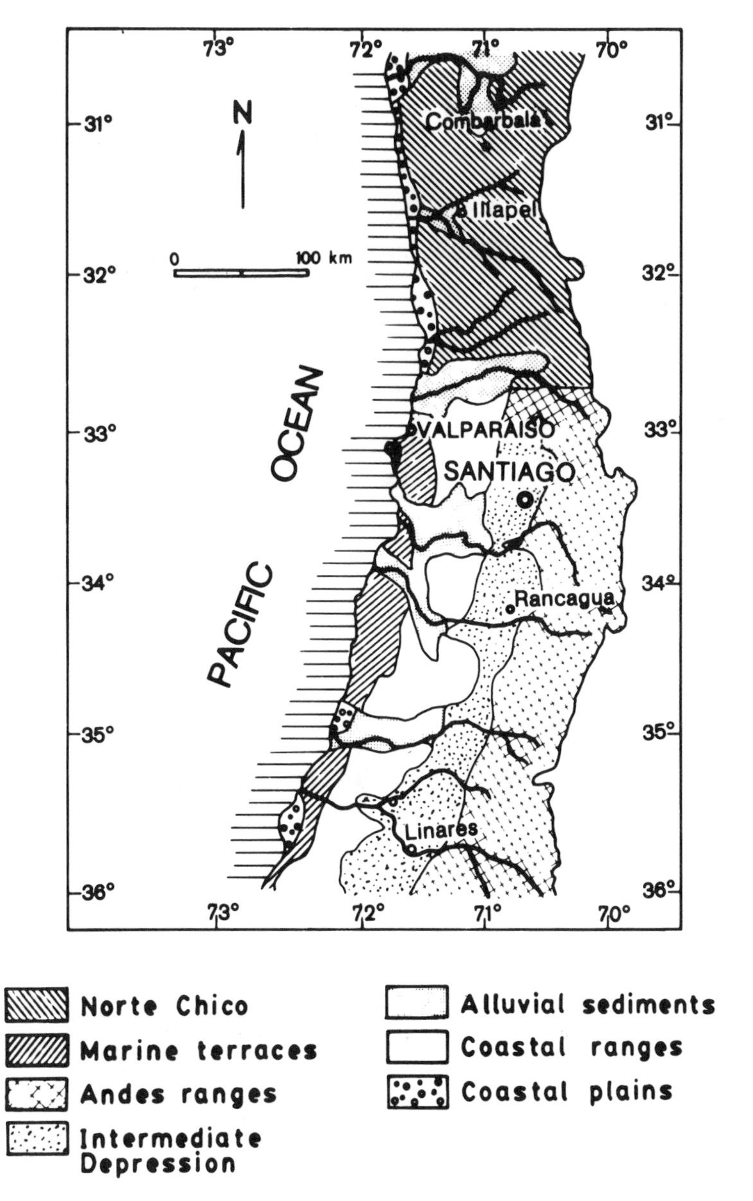

Figure 1. Major geomorphologic units in central Chile. Notice that the Intermediate Depression extends from essentially Santiago southward.

The higher, geologically complex Andes are: (a) generally constituted of crystalline rock, (b) mountains with characteristically steep slopes, (c) covered by glaciers on their high peaks, and (d) topped by a chain of recent active volcanoes (Weischet, 1970).

The coastal ranges are generally much lower (1000 m or less) and more rounded and accessible to people (Figure 2). The coastal ranges have a grano-

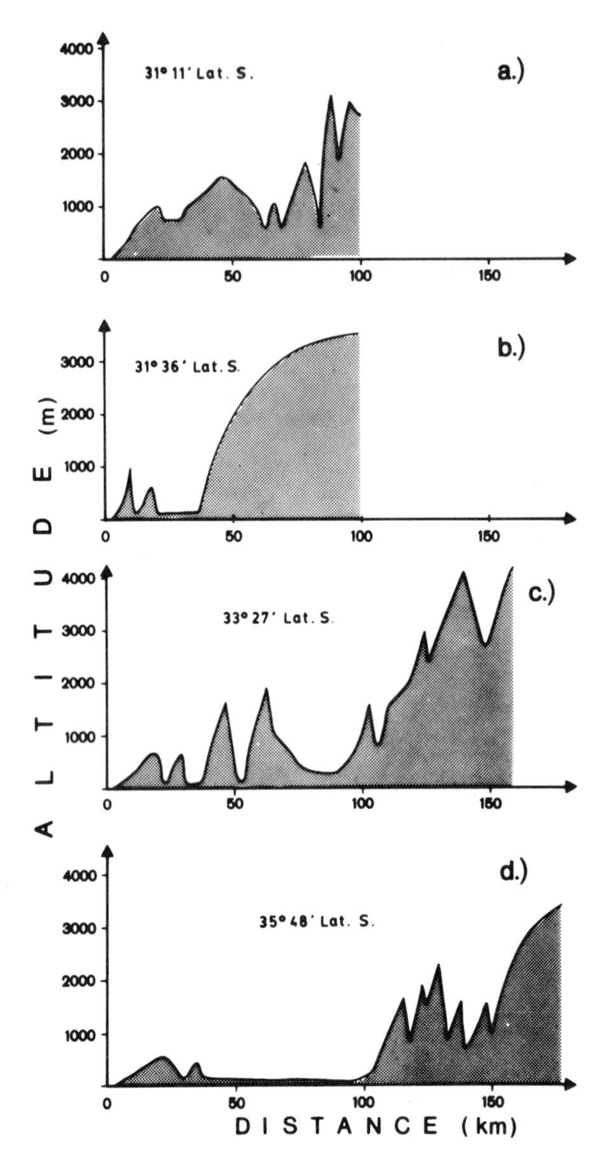

Figure 2. Smoothed east-west topographic profiles across central Chile. (a) Slightly south of Combarbala; (b) by Illapel (notice the large river valley near the coast); (c) near Santiago, where the Intermediate Depression starts; (d) close to Linares. Vertical scale is exaggerated approximately 20 times. Distances are from the Pacific Ocean to the left. Notice that the total land mass west of the water divide, as well as the amount of flat area, becomes greater toward the south. Compare with Figure 1 and see text for further discussion.

dioritic composition with signs of deep (ca. 80 m) weathering (Wright, 1959–1960).

Other important landforms in the area are the various coastal plains and terraces, the fluvial valleys and terraces, and the Intermediate Depression (Figure 1). The Intermediate Depression is a Tertiary graben that reaches an elevation of about 600 m near Santiago and extends southward, where it becomes wider and lower. Between 30° and 33° latitude there is no Intermediate Depression, and only a few large rivers coming from the Andes dissect the overall rugged and generally east-west inclined topography (Figures 1 and 2).

As a consequence of the relatively steep slopes the best developed soils are formed on deposition areas, that is, along the fluvial forms and on the Intermediate Depression. On the Andean slopes, soils tend to remain young, skeletal, and chemically diverse, due to the variety of parent materials and complex geology of these ranges (Wright, 1959–1960). By contrast, soils on the weathered coastal ranges tend to be the oldest and best developed in the region (Wright, 1959–1960; Rovira, 1984).

These geomorphic characteristics provide several important constraints on human land use, that is, the area north of Santiago and the two mountain ranges constrain human use more than the remaining flatter forms (Figure 1). However, in addition to restrictions, relief provides opportunities for use of landscapes. One opportunity relates to water supply for the lowlands. Watercourses originating in the low-altitude ranges tend to have sporadic discharges, whereas rivers from the high Andes have more dependable year-round flows (Weischet, 1970; Niemeyer, 1984). Discharges of these latter rivers are composed of rainfall, seasonal snow, and glacier melting, whereas watercourses originating within the study area (with a Mediterranean type of climate) depend only on current rainfall. The presence of the high Andes thus provides water for irrigation, industrial, and urban uses in the nearby lowlands.

A second opportunity provided by the high-energy relief of Central Chile is transhumance, here the movement of goats and sheep from the lowlands in the winter towards the highlands in the summer. Over short distances (ca. 100 km), the progressive shift of growing seasons with altitude provides the opportunity to reduce grazing loads in the lowland at precisely the moment at which shrub growth rates stagnate (Kummerow et al, 1981).

Based on these geomorphically determined constraints and opportunities, at least three interacting and altitudinally differentiated landscape types are recognized (Figures 1 and 2): (a) the low, irrigated or potentially irrigated flatlands, (b) the intermediate hills, where evergreen or drought-deciduous shrublands and woodlands are found today, and (c) the highlands capped with ice and snow.

Climatic Variability

Although the region has a general Mediterranean type of climate, there are important differences between subareas.

Rainfall and Water Supply. Annual rainfall is about 950 mm in the southern portion and decreases exponentially towards the north, where it is about 100 mm (van Husen, 1967). This same pattern is observable in the discharges of rivers originating in the higher Andes (Niemeyer, 1984). In addition, the interannual variability of both rainfall and river flow decreases at higher latitudes (Figure 3).

Autocorrelation analyses of the last 41 years in 10 meteorological stations within the study area show no significant increase or decrease in total annual rainfall over time ($p > .20$). These patterns indicate that the average water supply and its reliability, both for rain-fed farming and for irrigated agriculture, is much lower to the north. Weischet (1970) claims that, in spite of the smaller flatland surfaces north of Santiago and even in irrigated areas, water can be in short supply.

Since rainfall is produced by advective movement of air masses coming from the Pacific Ocean, local topographic configurations account for important geographical differences in average rainfall (Hajek et al., 1985). The coastal ranges have higher rainfall values than the Andes ranges at the same latitude and altitude, and the latter have generally higher values than localities on the Intermediate Depression. The altitude of the coastal ranges also influences the amounts of rainfall on the slopes of the Andes at the same latitude (Hajek et al.,

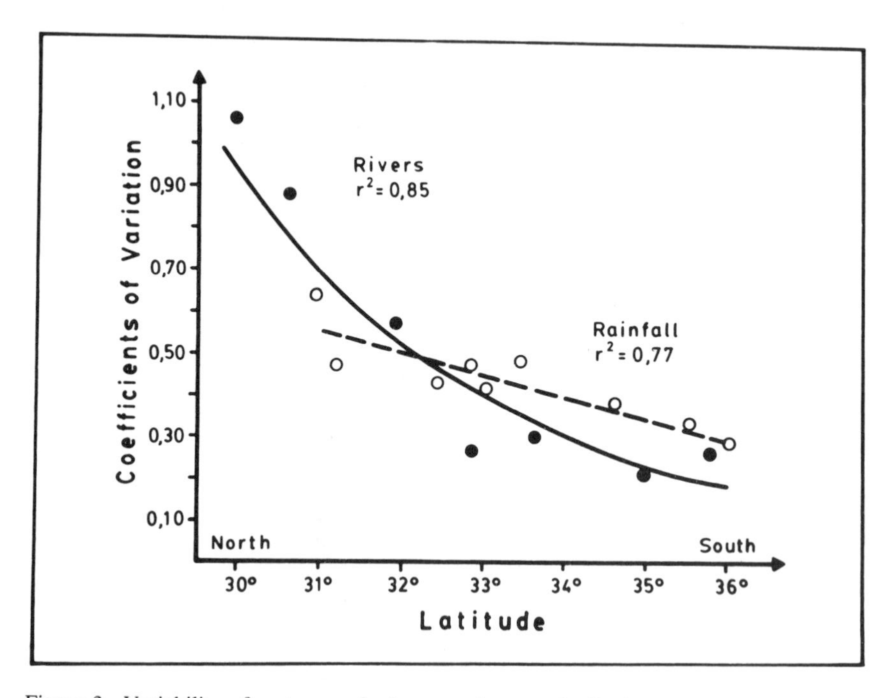

Figure 3. Variability of water supply from north to south. Variability was measured by the relative between-year differences in rainfall and river discharges, i.e., the coefficients of variation at each latitude.

1985). Thus, besides the north-south precipitation gradient, there are east-west trends due to local topography.

High, short-term rainfall events, important as a potential trigger for soil erosion (Rovira, 1984), seem to correlate only with east-west tendencies in average rainfall. An analysis of the maximum rainfalls in 24 hours using the combined data of van Husen (1967) and Peralta (1976) shows that: (a) meteorological stations on the coastal plains tend to exhibit these higher rainfall events more frequently than inland stations (chi-square, $p > .01$); and (b) by contrast, there is no clear latitudinal tendency for these events, either among coastal stations or among stations on or near the Intermediate Depression. Only for the high-rainfall events during April is there a positive rank correlation with latitude ($r^2 = .83, p > .05$). Moreover van Husen's (1967) data for the 17 meteorological stations in the study area show that the number of rainfall events with more than 50 mm increases irregularly towards the south, to the point that its correlation with latitude is not significant ($p > .10$). All this means that in spite of average rainfall values, high-rainfall events are infrequent, but not unusual in the northern and drier area.

Temperature Patterns. Average temperatures vary relatively little between years and tend to follow topography more than latitude (Rovira, 1984). Mean annual temperatures within our area vary between approximately 14 and 18°C. The between-year standard deviation in mean temperature for La Serena (29° 53′) is .84 ($N = 70$), whereas for Santiago (33° 28′) to the south it is only .42 ($N = 70$). In relative terms, between-year variability in mean temperature, as measured by coefficient of variation, is only 7 percent of the equivalent coefficient for rainfall at Santiago, and 9 percent at La Serena. Therefore, rainfall more than temperature is doubtless the main factor accounting for between-year differences in primary productivity. Geographically this means that since rainfall variability decreases with latitude (Figure 3), the more northern localities have greater year-to-year variance in plant growth. In other words, to keep a production-to-consumption ratio on the safe side, demand by humans and grazing herds should be more finely tuned in the North, where production is on the average smaller and more variable.

Both mean annual temperature and its between-year variability differ with altitude and location (Hajek et al., 1985). These differences result in greater household heating requirements and surrounding wood extraction on the two mountain ranges than on the Intermediate Depression.

Functional Coupling of the Biological and Physical Components

Within the study area, the relationship between water availability and water demand by organisms plays a major role in differentiating vegetation and landscape types (Mooney, 1977, and references therein), as well as animal abundance and distribution (Cody et al., 1977; Le Boulenge and Fuentes, 1978; Fuentes and Campusano, 1985).

In addition to the latitudinal and altitudinal differences, water supply varies with slope exposure position in the soil catenas and with soil depth. We will use distribution of the shrubs and trees to exemplify these differences. Our main reason in choosing this example is that composition and cover of the shrub and small trees are the original matrices (Forman and Godron, 1986) of landscapes in our area, and as such, play major roles as resources and as indicators of past history.

In the northern parts of the region where summers are dry, and episodically the winters as well (van Husen, 1967), xeric shrublands predominate. They are mostly composed of shallow-rooted cacti (*Trichocereus* spp., *Eulychnia* spp.) and drought-deciduous shrubs (*Flourensia thurifera, Cordia* spp., *Proustia* spp., *Oxalis gigantea, Gochnatia* spp., etc.). Towards the south, where winter rainfall becomes more reliable and sometimes summer rains are present, evergreen shrublands composed of deep-rooted plants tend to dominate the undisturbed landscape (Figure 4). These evergreen sclerophyllous shrubs and small trees include *Lithraea caustica, Quillaja saponaria, Kageneckia oblonga* and *Colliguaya odorifera* (Mooney, 1977; Quintanilla, 1983). Still further south, but outside our study area, winter-deciduous trees (i.e., *Nothofagus oblicua*) become more abundant.

In the northern part of this north-south transition zone the short, relatively unpredictable growing season favors drought-deciduousness and shallow root systems as a survival strategy (Schmithüsen, 1956; Walter, 1968; Mooney and

Figure 4. Evergreen shrublands east of Santiago. Note that the open spaces between shrubs are colonized by annual grasses.

Dunn, 1970). Towards the south, as droughts become shorter and rainfall more predictable, evergreen sclerophyllous plants outcompete drought-decidous ones. On the average evergreen sclerophyllous plants do not have a positive energy balance in the northern areas and are therefore absent. Still further south beyond the study area, cool winters accompanied by equable spring and summer periods impose a selective regime in which winter-deciduousness is favored.

This overall north-south sequence of vegetation belts is somewhat distorted along east-west axes due to average rainfall, slope and soil depth differences between the coastal and Andean ranges. Thus, evergreen sclerophyllous shrubs penetrate further north along the moister and generally more rounded coastal ranges, than along the Andes. On the other hand, xerophytic formations tend to penetrate south along the drier, steeper Andean ranges and *inselbergs* (Schmithüsen, 1956; Quintanilla, 1983). This broad-scale pattern is modified at finer scales by: (a) altitude (Hoffmann, 1982), (b) slope exposure (Armesto and Martinez, 1978), (c) local soil depths (i.e., alluvial fans versus the bordering rocky slopes, or the steep versus gently sloped sides of monoclinal uplifts), and (d) the presence of running watercourses (Mooney, 1977; Rundel, 1981). Thus for example, equator-facing slopes and shallow soil conditions typically favor the more xerophytic and even drought-deciduous vegetation throughout the study area. But polar-facing slopes, ravines, and deep alluvial fans tend to support lusher, sclerophyllous vegetation. In this way overall broad scale latitudinal differences are modified by progressively finer scale phenomena to produce a mosaic of landscape types. The complexity of the mosaic increases when human activity is considered. Human intervention, among other things, frequently favors the more xerophytic vegetation types (Armesto and Gutierrez, 1978).

Five functional couplings of shrubland ecosystems on the slopes are now presented, and where possible, related to potential human use.

Production and Rainfall

A strong coupling is evident between annual rainfall and the phenology and production by herbs and shrubs (Mooney, 1977; Kummerow et al., 1981). At Santa Laura (33° 10′) on the coastal range, herbaceous biomass production is low, and varied from 55 g/m^2y in 1973 (when total rainfall was only 412 mm) to 202 g/m^2y in 1977 (when total rainfall was 851 mm) (Kummerow et al., 1981). Evergreen shrubs seem to have less between-year variation than drought-deciduous shrubs or herbs, but unfortunately no appropriate data exist for a comparison. Also, for the Santa Laura site it has been shown (Kummerow et al., 1981) that annual production of leaves (in 1973) was 283 g/m^2y, of young stems 86 g/m^2y, and of stems 44 g/m^2y. The aboveground standing crop at that time was 1200 g/m^2. Thus production is relatively low and quite variable from year to year.

The between-site variation in production within the region is very high. As yet no data exist to determine the potential carrying capacities of natural shrublands for domestic grazers (i.e., goats and sheep).

Vegetation and Erosion

There is a positive relation between plant standing crop and sediment yield. Annual grasses and shrubs have distinctive effects on erosion in different years. The effects have been measured for grasses (Figure 5), drought-decidous shrubs such as *Flourensia thurifera,* and cactus species such as *Trichocereus chilensis* (Figure 6). Although methods used for the grass and shrub studies differed (Fuentes, unpublished), in both cases sediment transport was reduced by plant cover.

In addition, experiments in which all *F. thurifera* shrubs were pollarded (coppiced) in parcels of 50 × 15 m, at a site with 35° slope (Fuentes and Espinoza, unpublished) showed that the total sediment yield increased by a factor of about six. That is, the experimental parcels collected about six times more sediments than neighboring control areas where the *F. thurifera* cover remained intact. In the year these experiments were performed (1984), 95 percent of all the sediments collected were eroded out during a single rainstorm of 140 mm, before

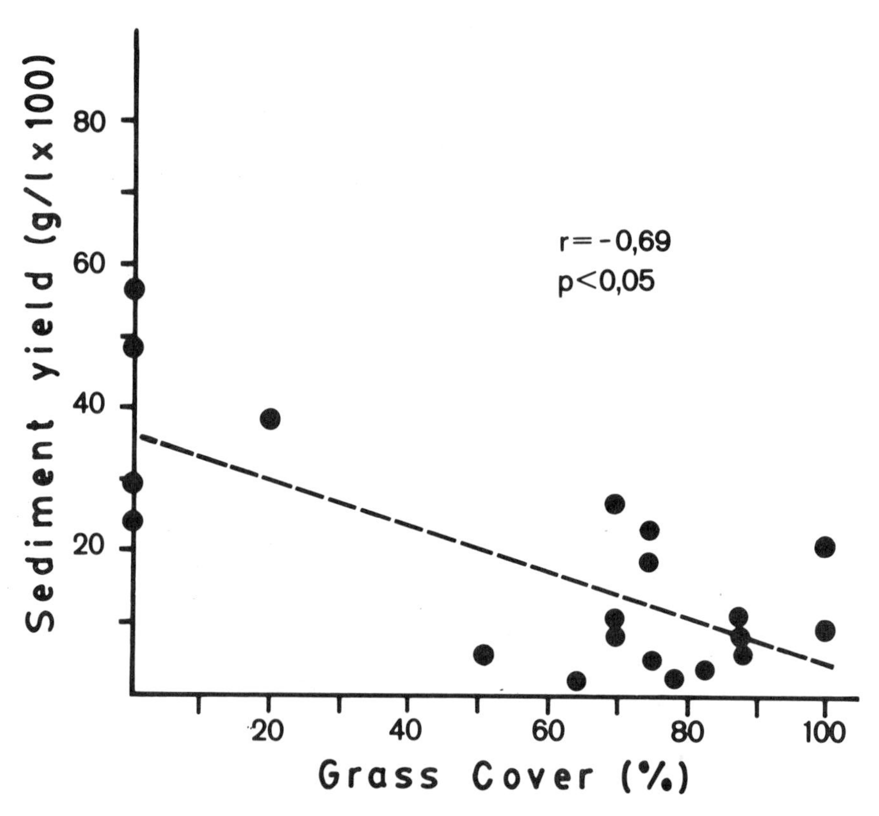

Figure 5. Average sediment related to cover of annual grasses. Ordinate is concentration of sediments in the water discharge from parcels with different percent grass cover. Note that if a high grass cover is available, sediment yield from erosion can be negligible.

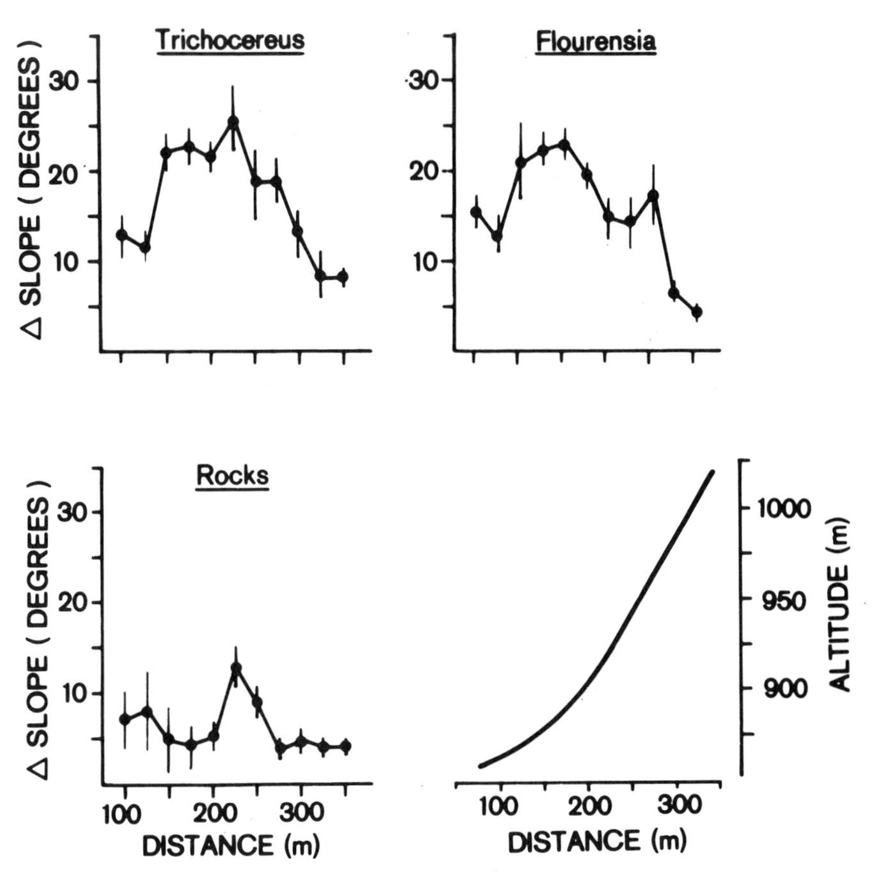

Figure 6. Soil retention capacity of cacti, shrubs, and rocks. The vertical axis is the difference in degrees of slope just above and below an object: a cactus *(Trichocereus)*, a shrub *(Flourensia)*, and rocks of about 40 cm in diameter. Greater differences in slope indicate greater soil retention capacity by the object. The horizontal axis indicates locations of samples taken 50 m apart up a steep slope (see diagram in lower right) from ca. 860 to 1020 m elevation. Note that all these objects have a soil-retention effect, although the effect varies according to position of the object in the mountains.

the annual grass cover had developed and exerted a soil retention effect. In contrast, the following year was dry, with negligible differences in sediment yields between vegetation removal parcels and controls.

Thus, in years with heavy rainfall early in the season, even drought-deciduous shrubs can provide better soil protection than grasses. But in years with a protracted rainy season the annual grasses have enough time to develop and play an important role in reducing soil losses. However, as part of the high between-year variability, heavy-rainfall episodes are likely to occur before July (van Husen, 1967; Peralta, 1976), and the total amount of precipitation in this period can be highly variable. All this variability produces an erosion pattern that is irregular in time and space.

Vegetation Change

Complex couplings exist within as well as between vegetation life forms. In habitats where the invasive shrub *F. thurifera* dominates, there is evidence of allelopathic effects of this species on both shrub seedlings and herbaceous plants (Fuentes et al., 1987). Thus, at least on certain slopes, the alternative of grasses or shrubs can be an almost all-or-none issue, with major implications for erosion control.

Evergreen shrubs do not seem to have this chemical-inhibition effect (Montenegro et al., 1978; Fuentes et al., 1987), but they provide shrub seedlings with protection from rabbits (Fuentes et al., 1983) and desiccation (Fuentes et al., 1986), and also segregate native herbaceous plants under shrubs and exotic herbaceous plants between shrubs (Keeley and Johnson, 1976; Jaksic and Fuentes, 1980).

On the other hand, the opposite effect also seems to exist. Grasses are capable of reducing the life expectancy of seedlings of the common shrubs *Lithraea caustica* and *Quillaja saponaria* (Fuentes, unpublished). The common introduced European grasses can probably prevent, or at least retard, the establishment of some native shrubs through competition for available water in the first few centimeters of topsoil. In other words, a conversion from shrubland to grassland or from evergreen shrubland to *Flourensia* stand could produce quasi-irreversible landscape change.

The chances of actually managing these landscapes and converting them from one type to another are highly dependent on the mechanisms of vegetation change, particularly interactions between species and between types of life forms. Vegetation changes within this area are, in general, slower on drier areas and slopes than on wetter ones (Fuentes et al., 1986). This is particularly valid for the transitions that require the arrival of species with low dispersal distances. Early colonization is generally by European grasses. Woody plants colonize later by rapid occupation of a few wind-dispersed plants such as *Baccharis* spp. and *Gochnathia* spp. that can tolerate summer droughts and some herbivory. Later a slow diffusion process from neighboring mature areas, if present, takes place.

Alternatively, large woody nurse plants such as the cattle-dispersed *Acacia caven* can invade the European grass cover. This then facilitates colonization by sclerophyllous species (Fuentes et al., 1986).

On the other hand, primary succession on road-cuts, where soils are nutrient-poor and European grasses scarce, occurs by the simultaneous colonization of a high diversity of evergreen shrubland species (Guinez and Fuentes, 1989). In short, there are several possible directions of vegetation change after large disturbances, and in general recovery will be slow or may not occur at all.

The Role of Herbivores

The evolutionarily important browsers in the study area are defoliating insects (Fuentes and Etchegaray, 1983). Although on the average they remove only a small percentage of the total photosynthetic surface of each shrub, this amount may be enough to change the competitive equilibrium in the shrub community (Fuentes and Etchegaray, 1983, and references therein). In the new equilibria that emerge following defoliation, the shrubs with maximum capacity to compensate for the removed leaf surface predominate (e.g., *Lithraea caustica*). Evidence suggests that the plants with this capacity most developed are least palatable to goats (Torres et al., 1980). In short, even small numbers of domestic goats can change the equilibrial composition of shrublands by favoring species like *L. caustica*. At high densities, goats eliminate all woody species except cacti. Consequently, in areas of low goat density a change in shrub species composition and a reduction can be seen, whereas in localities with high goat density drastic reductions in cover and diversity are found.

In addition, the recent introduction and naturalization of European rabbits *(Oryctolagus cuniculus)* and hares *(Lepus capensis)* has produced significant effects on the vegetation by increasing the mortality of seedlings of evergreen shrubland species (Fuentes et al., 1983) and of *Prosopis chilensis* (Fuentes et al., in press). Thus the landscapes we see today are at least in part a by-product of a complex interaction with an introduced biota.

Vulcanism and Vegetation Resilience

The Chilean evergreen sclerophyllous vegetation recovers rapidly from pollarding and human-caused fires (Araya and Avila, 1981). This recovery has two components: individual shrub resprouting and subsequent species succession. Although there are other plausible explanations, this resilience of the landscapes is likely due to a long history of vulcanism and vulcanism-caused fires in the region (Fuentes and Espinoza, 1986), that has selected for resprouting, but not for enhancement of seed germination after fire (Muñoz and Fuentes, 1989). It therefore seems that the coupling between the geologically young vulcanism of the Andes and the evolutionary history of Chilean shrublands ultimately permits a greater and more diverse human use of these ecosystems than if this inherent resilience had not evolved.

Landscape Patterns and Processes

The central linkage among mountains, water, and people will now be addressed more directly. The two "strategy" questions originally posed provide the perspective. How have people coped with the mountainous character of the landscapes, and how have they coped with the within-site variability of water supply? In the context of the physical and biological components elucidated above, I will also address the landscape-dynamical consequences of variability in water supply and human activity, particularly with respect to soil-conserving and soil-depleting activities.

Regional Controls on Landscapes

Broad-Scale Spatial Patterning and Human Settlement

On a broad scale (1:1,000,000 to 1:5,000,000) the distribution of the human population tends to coincide with the product of two sets: the distribution of flatter areas and the distribution of water supply. That is, where slopes are steep or the water supply is low and irregular, the established human population is small, and vice versa. In the northern sector (Figures 1 and 2) between 31 and 32.5° latitude, because of the rugged topography and the low, fluctuating water supply, population density (ca. 8.6 inhabitants/km^2) is lower than further south (20 to 220 inhabitants/km^2) where water is in greater supply (Carrera de Cartografia, 1983). Moreover, within this northern sector human densities are high and agricultural activities intense only near large rivers originating in the high Andes (Fuentes and Campusano, 1985). On the rain-fed interfluves or on the Andes ranges themselves population is low and pastoral activities are extensive. Between 33 and 36° latitude the flat and now irrigated Intermediate Depression supports the largest populations, especially near larger rivers (Carrera de Cartografia, 1983). The relatively wet and moderately steep coastal ranges rank second in population density, and the high Andes ranges with their comparatively steep slopes rank third. In fact, density on these latter slopes is similar to that on those north of Santiago.

In general, human densities and economic well-being tend to be correlated at this regional scale. Flatter irrigated areas have not only more population, but also more productive and capital-intensive activities. The mountainous and dry areas tend to support populations in a near-subsistence economy. This general pattern, largely a consequence of the geoecological restrictions described earlier, seems to be old. The 30 to 70 times less dense aboriginal populations (I.G.M., 1983) doubtless were familiar with these restrictions, and also tended to concentrate on flatter areas near water sources (Aschmann and Bahre, 1977). There is no evidence that these residents had permanent settlements above 1500 m. The later European colonization was also largely of the wetter lowlands (Garcia-Vidal, 1982), and not even a long war with the aborigines forced the Europeans to build settlements on higher locations.

In the eighteenth century, when an explicit urbanization policy was designed to remove people from rural areas (Garcia-Vidal, 1982), more than 40 settlements were established on flatter areas near larger water courses (I.G.M., 1983). Although it formed a sharper contrast with surrounding unsettled land (irrigation of the Intermediate Depression began only in the nineteenth century [Ellis 1960]), the overall spatial pattern was the same as we see today. In sum, occupancy of the land had its focus on or close to the flatter lowlands and later progressively expanded towards the surrounding mountains.

Intermediate-scale Patterning and Human Settlement

On an intermediate geographical scale (1:5,000 to 1:10,000) the principles behind human density and use of the landscapes are similar. Traditionally, rural property, either singly or communally owned, has been concentrated in the wet lowlands (Borde and Gongora, 1956; Baraona et al., 1961). Land ownership is less clear and buildings have been less prevalent in the surrounding hills.

Along watercourses, bands of land use similar to patterns described by von Thunen (Haggett, 1979) can be seen. On flatter areas close to the water are houses and irrigated fields. Beyond these are the fields where rain-fed agriculture is practiced and sometimes goats fed. Still further away from the water source are grazing grounds, which slowly give way to less disturbed areas where wood is harvested and animals such as European rabbits *(Oryctolagus cuniculus),* hares *(Lepus capensis)* and viscacha *(Lagidium viscacia)* are hunted. Today, even around large cities like Santiago, human population density declines towards the steeper slopes, and irregular concentric rings of activity types can be distinguished (Fuentes et al., 1984).

In all these cases, the same principles seem to hold. First, a central place is selected close to a flatland associated with a dependable water supply but where flooding risks are minimized (Weischet and Schallhorn, 1974). Second, the presence of more or less defined land use rings or bands, ultimately associated with the costs of movement, generally progress up the slope. These two classical principles of geography (Haggett, 1979), in conjunction with the distribution of mountains and rainfall, explain the current distribution of agricultural and grazing lands, hunting areas, introduced plants and animals (Fuentes, in press), the best conserved shrublands, and, as we will see, the distribution of the most degraded lands.

Human Land Use Processes

With these locational patterns in mind, the historical trends in human use of the landscapes will now be considered. Diaguitas and Picunches, the two aboriginal groups living here before European arrival, are believed to have had fine-scale agriculture of the slash-and-burn type on the lower interfluves, and irrigated agriculture on the floodplains. In addition, they had llamas *(Lama lama)* and alpacas *(Lama pacos),* two domestic species of camelids (Aschmann and Bahre, 1977). Although landscape modifications are thought to have been relatively concentrated close to the water sources, the actual changes produced by agricultural and pastoral activities, the use of fire, and the spreading of plants is unknown (Bahre, 1979). Generally, these changes are considered to have been localized in comparison with later changes.

During colonial times there was a large increment in the types of use of landscapes and in the surfaces involved. Initially, during the sixteenth and seventeenth centuries, the flatter areas were used for livestock and small agricultural parcels. During the eighteenth century the vice-royalty of Peru, on which Chile depended, determined that central Chile would grow wheat *(Triticum),* due to local pest problems to the north (Garcia-Vidal, 1982). This production was obtained from the flatter areas, north of Santiago first, but later from the more southern plains also (Garcia-Vidal, 1982). Meanwhile livestock was pushed towards the hills (Baraona et al., 1961; Cunill, 1971). Throughout this period the mountainous lands were used as a source of wood and various plant products, and as hunting grounds (Cunill, 1971). Later, during the second half of the nineteenth century and the beginning of the twentieth century, the high price of wheat in the international market led to an increment in the land surface plowed (Sepulveda, 1959). This time slopes were cultivated, and livestock had to be pushed onto even steeper slopes.

An additional important resource emerged that affected the woodlands and shrublands originally covering the slopes, namely mining and ore smelting. Under Chilean law, mining activities have always had priority over agriculture. Thus since colonial times, the effects of wood extraction for ore smelting and household cooking and heating needs have played a major role in eliminating woody cover on the hills (Cunhill, 1971; Bahre, 1979). By 1580 woodcutting around Santiago had already been severe and wood was considered scarce. By the eighteenth century there were more than 40 big smelters between 27 and 32° latitude, and since not enough wood was locally available to feed them, it had to be brought from Chiloe (approximately 42° latitude). Not even the so-called woodminers (named for their wood searching capability) could find enough wood on the drier slopes north of Santiago (Cunhill, 1971).

Fine-Scale Site Controls on Landscapes

The description of land use up to now has been necessarily broad and general, as it is mainly based on historical records. Some of the major cultural forces acting on the landscape are clear, but the finer dynamics that underlie and to some extent drive the macroscopic pattern now need to be addressed. Unfortunately, this cannot be done with the historical evidence available; therefore, the common assumption in historical studies that the forces and principles we see today are similar to those acting in the past under similar circumstances will be utilized. Given the current land use patterns, this assumption seems reasonable. In this manner one can examine how people have adapted their behavior, or reacted to the climatic variability within the study area. Here agricultural, pastoral, and woodcutting activities will be sequentially linked to site characteristics. These linkages provide insight into the synergisms present, and underlie understanding of macroscopic landscape dynamics.

Rain-Fed Agriculture

Steep slopes of up to 50 percent on hills formerly covered with shrublands and woodlands have been plowed (Figure 7) to cultivate wheat, barley, or cumin

Figure 7. Land use zonation with irrigated and rain-fed cultivation on contrasting sites. Houses are near irrigated fields being plowed. Steep hillsides in the background were plowed and seeded with cumin *(Cominum)* three months before.

(Triticum aestivum, Hordeum vulgare, Cominum cyminum) (Etienne et al., 1983), for subsistence requirements or to supply external wheat markets. Typically farmers plow and plant wheat, barley, or cumin after the first storms of the season, when the soil is wet and not too hard. This practice leaves the sometimes very steep slopes without a protective plant cover during an important part of the rainy season. When all the yearly rains are concentrated in the first few months, this means that practically all rain falls on these steep, little-protected slopes. Water-caused erosion is very high. After harvest, with a yield of about 600 to 1000 kg/ha or less (Contreras et al., 1986), goats are brought into the fields. The soil, is again left uncovered, this time to erosion-causing winds in the dry season.

After 8 to 10 years of repeating this cycle, the typical 1- to 7-ha parcels of land are generally abandoned for 20 or more years. During this time some recovery of the woody plant cover can occur, depending on grazing intensity. The usual cover, however, is of grasses from the Mediterranean region and weedy wind-dispersed shrubs (e.g., *Baccharis* spp.) or the livestock-dispersed shrub, *Acacia caven*. A whole new multiannual cycle can be repeated if the site recovers and if subsistence-farming requirements make it necessary.

The overall patterns and yields are variable from place to place, but the important alternating phases involving heavy water- and wind-caused soil erosion are constant. On a microgeographical scale such an activity pattern can explain the progressive widening of the agricultural belt, as lands closer to a central place are degraded, and new shrubland areas on steeper slopes are incorporated into the local agricultural economy. It also explains how some secondary grazing lands can develop close to the central place and thus temporarily invert the original scheme of parallel bands.

Grazing and Transhumance

Grazing can be coupled with agricultural rotations and thus generate about 50 percent of the total family income in the northern part of the region (UNCOD, 1977; Etienne et al., 1983, Contreras et al., 1986). Annual average income for families of 5 or 6 people in some of this area is about US $1100 (Etienne et al., 1983). A goat herder usually has fewer than 40 animals (Contreras et al., 1986) which, during winter and spring, feed freely on the slopes surrounding the house.

The within-year variability in rainfall is traditionally overcome by moving a fraction of the goats and sheep to high pastures on the Andes (Aranda, 1971). This transhumance process acts as a conservation measure and a way to increase the carrying capacity of the lowlands, because the removal of the animals occurs precisely when the sustaining capacity of lowland pasture begins to decrease. However, the number of animals actually taken to the uplands is highly variable between years and between subareas (Aranda, 1971), and does not seem to follow only from productivity criteria. Using data from Aranda (1971) and later unpublished information, no significant correlations ($p > .20$) were found between transhumance and rainfall in the lowlands (a measure of local productivity according to Fuentes and Campusano, 1985), either of the current year or of preceding years. Whatever the complex economic and social reasons are that determine the number of animals involved in transhumance, between 13 and 50 percent of the sheep and goats are moved on any given year (Aranda, 1971).

During years when the winter carrying capacity of pastures remains low due to regional drought, there is no relief for the resource base. An equivalent mechanism to the above-mentioned seasonal east-west transhumance would involve, for example, movement of the herds either south (e.g., to 38° latitude) or south-southeast to areas with higher, more predictable rainfall. This has occurred infrequently in the recent past. Normally, during drought years goats and sheep simply remain in the same field (Fuentes and Hajek, 1978). In other words, animals trample and eat everything in sight, during a period when plant productivity is almost nil. During the ensuing massive starvation periods (Baraona et al., 1961), both the herbaceous and woody plant material is removed, making erosion high and recovery difficult during the following rainy season (Fuentes and Hajek, 1978). During these drought periods the prices of animals are low, and people are forced further into a spiral leading to more poverty.

From the perspective of landscape dynamics, this type of management of the ratio of pasture carrying capacity to grazer density can produce severe range degradation, and with an accelerated rhythm. This is because in each drought goats are confronted with a progressively deteriorated resource base.

Wood Extraction in Shrublands

Woodcutting and charcoal making are also important activities within the study area. Bahre (1979) estimated that in the Coquimbo province (in the Norte Chico) about 70 percent of the people used wood and charcoal for cooking and household heating in 1967 (Contreras et al., 1986). Each family consumes the equivalent of about two large shrubs, or 60 to 80 kg of wood per week (Contreras et al., 1986), with some estimates as high as 100 kg/week (Baraona et al., 1961). In Chile as a whole about 15 percent of the total energy supply presently comes from biomass (Garcia-Vidal, 1982). Wood extraction from shrublands is heavy, not only around rural settlements, but also near urban centers.

When shrubs are repeatedly pollarded they usually resprout if cutting pressure is released. But under conditions of high population density or during drought years, when people use every possible resource to obtain cash, even the subterranean lignotubers and roots are removed. With this practice, regeneration of shrubland during the following high rainfall years is impeded. This practice has far-reaching consequences, given the high erosion rates that can occur in the following rainy season and the previously mentioned obstacles to range recovery in the absence of nursing plants or nearby mature stands.

Year-to-year variability in rainfall and primary productivity has an aggravating effect on this clearing process. Following an idea by Hajek et al. (1972), minimum monthly temperatures were correlated with total annual rainfall in Santiago for the period from 1910 to 1980. A positive association ($r^2 = .33, p > .05$) between these variables was found, with an average decrease of 1°C in the minimum monthly temperature for every 100-mm decrease in annual rainfall. In other words, in drier years vegetation has a lower productivity not only because of a lower water supply, but also because the air is colder. Simultaneously, in these dry years, demand for wood is also likely to be higher because of the somewhat lower air temperatures. For the same reason goats also must eat more food, just to maintain their weight.

Land Use Synergisms

The three sources examined, agriculture, grazing/trampling and woodcutting, act synergistically, not separately. This is because the land is the same, and the subsistence requirements of human populations are fairly constant, with few effective choices. The natural oscillation of the carrying capacity of this land, in the face of a constant human demand, results in accelerating land degradation. All three major human activities have in common that they progressively reduce vegetation cover, and thus when combined, have the capacity to trigger heavy soil erosion, extremely slow recovery, and a significantly reduced resource base for humankind (Figure 8).

The reduction of woody cover and topsoil tends to favor annual plants that have evolved the capacity to evade overgrazing and overexploitation of land in general. However, as mentioned previously, annuals provide relatively little soil protection in areas where sometimes all rain falls in a few storms early in the

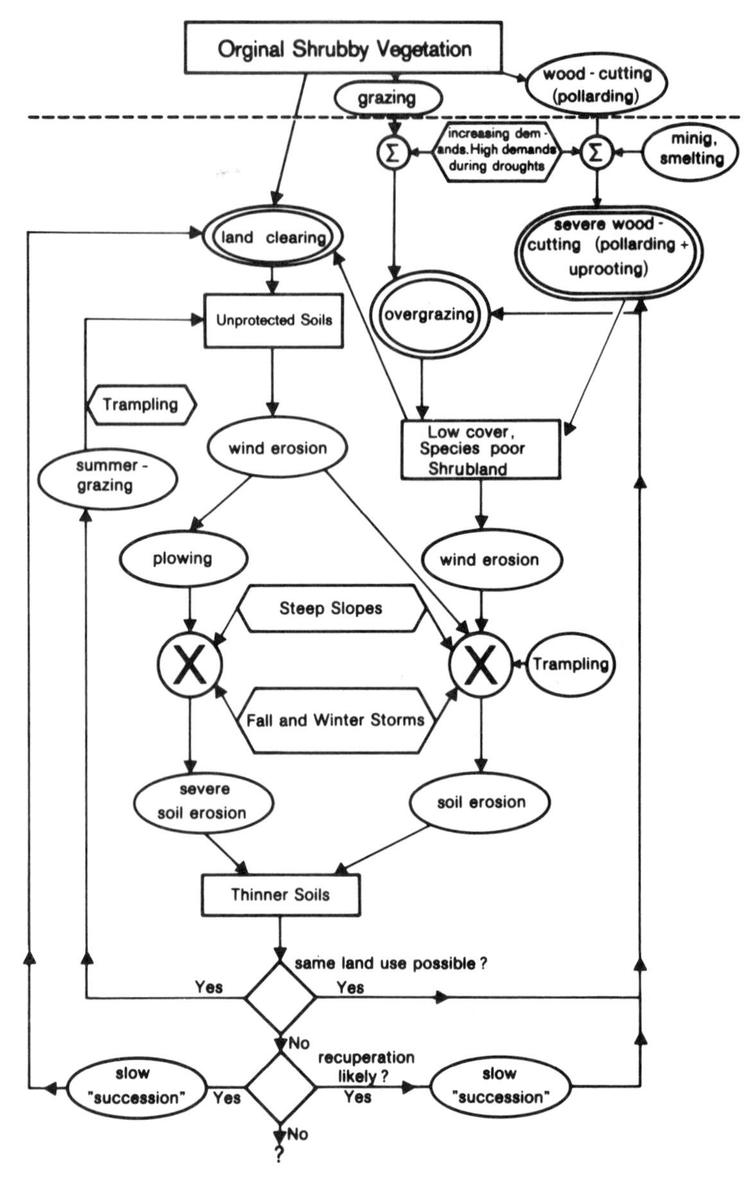

Figure 8. Integrated model of landscape degradation. Three main sources of impact are shown near the top (each encircled with two lines): land clearing for plowing (left), overgrazing (middle), and woodcutting (right). All three sources, especially the first, decrease plant cover and thus enhance soil loss and land degradation. Agriculture and grazing can intermingle somewhat during fallow periods or when areas are rotated between grazing and dry-farming. The last alternative at the bottom of the figure indicates the uncertain future of badly degraded lands, in which their sustaining capacity for woody plants is extremely reduced. See text for further explanation.

season. In addition, conversion of shrubland to annual grassland has the negative effect of leaving people too dependent on current-year precipitation. Productivity of annuals tends to correlate with rainfall of the current season, whereas larger shrubs with greater biomass and root systems (Kummerow et al., 1981) possess the capacity to tap deep water sources and even out two or more years of variable rainfall. People in these areas have few opportunities to capitalize on and store production from good rainfall years. Therefore a shrub-to-grass conversion ultimately leaves them vulnerable to dry years and highly dependent on a progressively deteriorating resource base.

To what extent these human behavior patterns were the same in previous centuries is unknown. All evidence available suggests that the behavioral patterns and adaptations were at least similar to those described here, and may have underlain the historical trend of expansion of land use from wet lowlands to drier highlands. One difference might be that wood is no longer used for ore smelting. However, it is likely that today the high wood-consumption rates for household purposes play a similar role.

Spatial Trends in the Land Use Process

The effect of the localized, synergistic land use processes varies geographically in landscapes across the region. On the relatively accessible and rounded coastal ranges south of Valparaiso there is a long history of wheat cultivation and grazing that, together with the relatively high rainfall, has produced landscapes with severe mantle and gulley erosion (Weischet, 1970; Peralta, 1978; Fuentes and Hajek, 1979). In contrast, the relatively flat Intermediate Depression satisfies only the high-population-density criterion, and has the lowest soil degradation in the area. Within the same general subarea, near Santiago, the Andes and nearby slopes have been used to a limited extent (compare Borde and Gongora, 1956 with Baraona et al., 1961), and, coupled with the moderate rainfall, this has produced relatively little erosion (Peralta, 1978; Fuentes and Hajek, 1979).

In the area north of Santiago, land use has been intensive (Figure 9), terrain is generally mountainous and average rainfall low. High rainfall years and events are infrequent, but not unusual. Perhaps they are infrequent enough to be forgotten by people, but nevertheless of importance in a long historical process of soil erosion. The result of these conditions coinciding has unfortunately been the genesis of one of the showcase areas of desertification in the world (UNCOD, 1977; Gasto and Contreras, 1979). Here it is possible to see large areas with little remaining topsoil, almost no vegetation, and very poor human populations with an uncertain future (Fuentes and Hajek, 1979). Using aerial photographs of selected sites, it has been estimated (Etienne et al., 1983; Valdes, 1983) that in this subarea, cover by woody species has been reduced at a rate of 0.5 to 1.5 percent per year for at least 30 to 40 years. This reduction in shrub cover has been found to be significantly more extreme in communally owned areas, than in nearby areas with individually owned properties (Valdes, 1983).

Figure 9. Synergisms among land uses, leading to erosion and land degradation. Here wood removal combined with grazing (and perhaps former cultivation) on steep slopes produces strong contrasts with the productive fields on valley bottoms. Superimposing these interacting human processes onto the seasonal and year-to-year drought regime results in a significantly diminished resource base for these inhabitants.

At a finer geographical scale (1:5,000 or more), the interactions of relief, high rainfall and land use intensity have produced rings of degraded land near human settlements. It can be frequently observed that around towns or even small households, the nearby areas are the most degraded, and as one moves away from them, the shrubby matrix appears. Only relatively inaccessible slopes and peaks maintain vegetation similar to what must have been the original communities. Dry-farming, grazing, woodcutting, and hunting activities may be present as concentric rings or bands surrounded by the vegetation matrix. For example, south of Santiago, where the land has not been over-exploited, and beyond a band of no woody vegetation, invasive European grasses and shrubs, such as *Baccharis* spp., can dominate the landscape. Further away, shrub cover increases and more mesophytic plants predominate. Here, shrub cover is initially composed of species like *Lithraea caustica,* and only the more distant areas show higher cover and diversity. Where human impact has been relatively long and severe, bands close to settlements can show the strongest degradation, with little or no woody cover except for cacti. Further away, the invasive shrubs and sometimes the more mesophytic elements are present.

North of Santiago, where the shrub cover is lighter and the climate harsher, degradation around settlements has led to bare soils. More distant surroundings have only a sparse cover even in the best cases.

Conclusion

Dynamic tendencies and current spatial patterns of the mountainous landscapes with a Mediterranean-type climate in Chile, are largely explained by two major groups of factors:

1. The innate fragility of the local ecosystems, partly a consequence of the rugged topography and partly due to high year-to-year climatic variability. The generally low and fluctuating productivity that leads to overexploitation of the resource base is part of this factor.
2. The progressive geographical and numerical expansion of the human population from the lowlands towards the mountains.

The overall human land use approach has been inadequate given the topographic and climatic constraints present. The question remains whether any other outcome was possible in view of these constraints. That the Chilean outcome is only one of many possibilities can be seen by comparing (as far as such a comparison is possible) the Chilean case with what Elsasser (1985) describes for the mountainous, Mediterranean area of the Cevennes in southern France. There, a complex multiresource strategy has been developed and maintained through generations with little land degradation. Alternatively, in the mountains of Andalusia, southern Spain, where many of the inhabitants of Chile originated, several conservative land use practices include contour plowing, the use of terraces, wooded strips and patches of natural (i.e., *Quercus ilex*) or planted (i.e., almond and olive) species. The heavier dependence on trees in Andalusia than in Chile takes better advantage of the seasonality in water supply and of the higher resilience of these woody plants compared with the annuals favored in Chile (Fuentes, 1988). Here again, landscape degradation has been relatively less than in Chile. The Chilean patterns thus seem to result from a unique combination of human and natural factors that produced the landscapes we see today.

Acknowledgments

I am grateful to J. Filp, R. T. T. Forman, H. Romero, H. Steinlin, and W. Weischet for their suggestions regarding earlier drafts of this contribution.

I would also like to express my gratitude to the Universidad Catolica de Chile, the Institut für Physische Geographie (Universität Freiburg), and the Alexander von Humboldt Foundation for providing me with opportunities to work on the ideas presented. Most of my own research cited here was done with the support of the Universidad Catolica (DIUC), PNUMA, UNESCO-MAB and FONDECYT; their help is gratefully acknowledged.

References

Aranda, X. 1971. Algunas consideraciones sobre la transhumancia en el Norte Chico. *Informaciones Geográficas. Universidad de Chile*. Santiago. Numero Especial, pp. 141–70.

Araya, S. and G. Avila. 1981. Rebrote de arbustos afectados por fuego en el matorral chileno. *Anales Museo Historia Natural (Valparaiso)* 14:107–13.

Armesto, J. J. and J. R. Gutiérrez. 1978. El efecto del fuego en la estructura de la vegetación de Chile central. *Anales Museo Historia Natural (Valparaiso)* 11:43–8.

Armesto, J. J. and J. A. Martinez. 1978. Relations between vegetation structure and slope aspect in the mediterranean region of Chile. *Journal of Ecology* 66:881–9.

Aschmann, H. 1973. Distribution and peculiarity of Mediterranean ecosystems. In F. di Castri and H. A. Mooney, eds., *Mediterranean type ecosystems. Origin and structure*. Ecological Studies 7. Springer-Verlag, New York. pp. 11–19.

Aschmann, H. and C. Bahre. 1977. Man's impact on the wild landscape. In H. A. Mooney, ed., *Convergent evolution in Chile and California. Mediterranean climate ecosystems*. Dowden, Hutchinson and Ross, Stroudsburg, PA, pp. 73–84.

Bahre, C. 1979. *Destruction of the natural vegetation of north-central Chile*. University of California Publications in Geography *23* University of California Press, Berkeley.

Baraona, R., X. Aranda, and R. Santana. 1961. *Valle de Putaendo. Estudio de estructura agraria*. Editorial Universitaria, Universidad de Chile, Santiago.

Borde, J. and M. Gongora. 1956. *Evolución de la propiedad rural en el Valle del Puanque*. Editorial Universitaria, Universidad de Chile, Santiago.

Carrera de Cartografia. 1983. Distribución de la población. In *Atlas de la República de Chile*. Instituto Geográfico Militar, Santiago, pp. 88–9.

Cody, M. L., E. R. Fuentes, E. Glanz, J. H. Hunt, and A. R. Moldenke. 1977. Convergent evolution in the consumer organisms of Mediterranean Chile and California. In H. A. Mooney, ed., *Convergent evolution in Chile and California. Mediterranean climate ecosystems*. Dowden, Hutchinson and Ross, Stroudsburg, Pennsylvania, pp. 144–92.

Contreras, D., J. Gasto, and F. Cosio., eds. 1986. *Ecosistemas pastorales de la zona Mediterránea de Chile. Estudio de las comunidades agricolas de carquindaño y yerba loca del secano costero de la región Coquimbo*. Publicaciones UNESCO, Montevideo.

Cunill, P. G. 1971. Factores en la destrucción del paisaje chileno: Recolección, caza y tala coloniales. *Informaciones Geográficas*. Universidad de Chile, Santiago. Número Especial, pp. 235–64.

di Castri, F. 1973. Climatographical comparison between Chile and western coast of North America. In F. di Castri and H. A. Mooney, eds., Mediterranean type ecosystems. Origin and structure. *Ecological studies*. Vol. 7. Springer-Verlag, New York, pp. 21–36.

Ellis, M. 1960. *La división de la tierra en Chile central*. Editorial Nascimiento. Santiago.

Elsasser, K. 1985. Analyse intégrée d'un espace montagnard Nord-Méditérranéen à l'exemple de la vallée de Taleyrac en Cevennes. *Schriftreihe des Instituts für Landespflege der Universität Freiburg*. Heft 4.

Espinoza, G. and E. R. Fuentes. 1984. Medidas de erosión en los andes centrales: Efectos de pastos y arbustos. *Terra Australis* 3:75–86.

Etienne, M., E. Caviedes, and C. Prado. 1983. *Bases ecologiques du développement de la zone aride Méditérranée du Chili*. C.N.R.S./C.E.P.E., Montpellier, France.

Filp, J., E. R. Fuentes, S. Donoso, and S. Martinic. 1983. Environmental perception of mountain ecosystems in central Chile: An exploratory study. *Human Ecology* 11:345–51.

Forman, R. T. T. and M. Godron. 1986. *Landscape ecology*. Wiley, New York.

Fuentes, E. R. 1988. Landscape development in mountainous habitats with similar climates and cultural backgrounds: Central Chile and Andalusia. *Mountain Research and Development* 8:75–7.

Fuentes, E. R. In press. How do introduced plants and animals fit into the landscape? In R. Groves and F. di Castri, ed., *Biogeography of Mediterranean invasions*. Cambridge University Press, United Kingdom.

Fuentes, E. R. and E. R. Hajek. 1978. Interacciones hombre-clima en la desertificación del norte chico Chileno. *Ciencia e Investigación Agraria* 5:137–42.
Fuentes, E. R. and E. R. Hajek. 1979. Patterns of landscape modification in relation to agricultural practice in central Chile. *Environmental Conservation* 6:265–71.
Fuentes, E. R. and J. Etchégaray. 1983. Defoliation patterns in matorral ecosystems. In F. J. Kruger, D. T. Mitchell, and J. V. M. Jarvis, eds., *Mediterranean-type ecosystems*. Ecological Studies, Vol. 43. Springer-Verlag, New York, pp. 525–42.
Fuentes, E. R. and C. Campusano. 1985. Pest outbreaks and rainfall in the semiarid region of Chile. *Journal of Arid Environments* 8:67–72.
Fuentes, E. R. and G. Espinoza. 1986. Resilience of central Chile shrublands: A volcanism-related hypothesis. *Interciencia* 11:164–5.
Fuentes, E. R., F. M. Jaksic, and J. Simonetti. 1983. European rabbits versus native rodents in central Chile: effects on shrub seedlings. *Oecologia* 58:411–14.
Fuentes, E. R., G. Espinoza, and I. Fuienzalida. 1984. Cambios vegetacionales recientes y percepción ambiental. El caso de Santiago de Chile. *Revista de Geografía Norte Grande* 11:45–53.
Fuentes, E. R., G. Espinoza, and G. Gajardo. 1987. Allelopathic effects of the Chilean shrub *Flourensia thurifera*. *Revista Chilena de Historia Natural* 60:57–62.
Fuentes, E. R., A. J. Hoffmann, A. Poiani, and M. C. Alliende. 1986. Vegetation change in large clearings: Patterns in the Chilean matorral. *Oecologia* 68:358–66.
Fuentes, E. R., R. Aviles, and A. Segura. In press. Landscape change under indirect effects of human use: The savanna of central Chile. *Landscape Ecology* 2:73–80.
Garcia-Vidal, H. 1982. *Chile. Esencia y evolución.* Instituto de Estudios Regionales, Universidad de Chile, Santiago.
Gastó, J. and D. Contreras. 1979. *Un caso de desertificación en el norte de Chile. El ecosistema y su fitocenosis.* Boletin Técnico 42. Facultad de Agronomia, Universidad de Chile, Santiago.
Guiñez, R. and E. R. Fuentes. 1989. Sucesión primaria en el matorral de Chile central: Los cortes camineros como modelo cuasi experimental. *Revista Chilena de Historia Natural.*
Haggett, P. 1979. *Geography. A modern synthesis.* Harper and Row, New York.
Hajek, E. R., M. Pacheco, and A. Passalacqua. 1972. Analisis bioclimático de la sequía en la zona de tendencia mediterránea de Chile. *Publicaciones Instituto Geografía Universidad Católica de Chile* 45.
Hajek, E. R., E. R. Fuentes, and G. A. Espinoza. 1985. Bases para una geografía climática de Chile central. *Revista de Geografiía Paralelo* 37:327–36.
Hoffmann, A. J. 1982. Altitudinal ranges of Phanerophytes and Chamaephytes in Central Chile. *Vegetatio* 48:151–63.
I.G.M. (Instituto Geográfico Militar). 1983. *Atlas de la República de Chile.* Santiago.
Jaksic, F. M. and E. R. Fuentes. 1980. Why are native herbs in the Chilean matorral more abundant beneath bushes: Microclimate or grazing? *Journal of Ecology* 68:665–9.
Keeley, S. and A. W. Johnson. 1976. A comparison of the patterns of herb and shrub growth in comparable sites in Chile and California. *American Midland Naturalist* 97:120–32.
Kummerow, J., G. Montenegro, and D. Krause. 1981. Biomass, phenology and grow. In P. Miller, ed., *Resource use by chaparral and matorral. A comparison of vegetation function in two Mediterranean-type ecosystems.* Springer-Verlag, New York. pp. 69–98.
Le Boulenge, E. and E. Fuentes. 1978. Quelques données sur la dynamique de population chez *Octodon degus* (Rongeur Histricomorphe) du Chili central. *La Terre et la Vie* 32:325–41.
Montenegro, G., O. Rivera, and F. Bas. 1978. Herbaceous vegetation in the Chilean matorral. Dynamics of growth and evaluation of allelopathic effects of some dominant shrubs. *Oecologia* 36:237–44.

Montane, H. J. 1968. Paleo-Indian remains from Laguna de Tagua-Tagua, central Chile. *Science* 161:1137–8.

Mooney, H. A., ed. 1977. *Convergent evolution in Chile and California. Mediterranean climate ecosystems*. Dowden, Hutchinson and Ross, Stroudsburg, Pennsylvania.

Mooney, H. A. and E. Dunn. 1970. Photosynthetic systems of mediterranean climate shrubs and trees of California and Chile. *American Naturalist* 104:447–53.

Muñoz, M. and E. R. Fuentes. In press. Does fire induce shrub germination in the Chilean matorral? *Oikos*.

Niemeyer, H. 1984. *Hidrología*. Instituto Geográfico Militar, Santiago, Chile.

Peralta, M. 1976. *Uso, clasificación y conservación de suelos*. Ministerio de Agricultura, Santiago, Chile.

Peralta, M. 1978. Procesos y áreas de desertificatión en Chile continental. *Ciencias Forestales* (Septiembre):41–4.

Quintanilla, V. 1983. *Biogeografía*. Instituto Geográfico Militar, Santiago, Chile.

Rovira, A. 1984. *Geografía de los suelos*. Instituto Geográfico Militar, Santiago, Chile.

Rundel, P. W. 1981. The matorral zone of central Chile. In F. di Castri, D. W. Goodall, and R. C. Specht, eds., *Mediterranean-type shrublands. Ecosystems of the world*. Elsevier, Amsterdam. Vol 11. pp. 175–201.

Schmithüsen, J. 1956. Die räumliche Ordnung der chilenischen Vegetation. Forschungen in Chile. *Bonner Geographische Abhandlungen*. 17:1–86.

Sepúlveda, S. 1959. *El trigo chileno en el mercado mundial*. Editorial Universitaria, Universidad de Chile, Santiago.

Torres, J. C., J. Gutiérrez, and E. R. Fuentes. 1980. Vegetative responses to defoliation of two Chilean matorral shrubs. *Oecologia* 46:161–3.

UNCOD (United Nations Conference on Desertification). 1977. *Case-study on desertification, region of Combarbalá, Chile*. September, 1974, Nairobi, Kenya.

Valdés, J. 1983. Dinámica de la desertificación en tres áreas del secano interior de la IV Región. Tesis, Escuela Ciencias Forestales, Universidad de Chile, Santiago.

van Dobben, U. H., and R. H. L. McConnel, eds. 1975. *Unifying Concepts in Ecology*. Junk, The Hague.

van Husen, C. 1967. Klimagliederung in Chile auf der Basis von Häufigkeitsverteilungen der Niederschlagssummen. *Freiburger Geographische Hefte* 4.

Walter, H. 1968. *Die Vegetation der Erde in ökophysiologischer Betrachtung*, Band II. Die gemässiöten und arktischen Zonen. Gustav-Fischer Verlag, Stuttgart, Federal Republic of Germany.

Weischet, W. 1970. *Chile. Seine länderkundliche Individualität und Struktur*. Wissenschaftliche Länderkunden. Band 2/3. Wissenschaftliche Buchgesellschaft. Darmstadt, Federal Republic of Germany.

Weischet, W. and E. Schallhorn. 1974. Altsiedlerkerne und frühkolonialer Ausbau in der Bewässerungskulturlandschaft Zentralchiles. *Erdkunde* 28:295–302.

Wright, C. 1959–1960. Observaciones sobre los suelos de la zona central de Chile. *Agronomía Técnica*. Chile.

11. Landscape Patterns, Disturbance, and Management in the Pacific Northwest, USA

Frederick J. Swanson, Jerry F. Franklin, and James R. Sedell

The ecology of landscape-scale processes is richly expressed in forested mountain landscapes of western Oregon and Washington in northwestern North America (Figure 1). The mosaic of landscape patterns is especially dynamic in these geomorphically active areas of high relief, heavy precipitation, and frequent disturbance by fire, wind, and other processes. Indeed, the time scales of geomorphic and ecosystem change overlap in these areas of active volcanism, unstable hillslopes, and long-lived trees. Mount St. Helens in Washington State, for example, has had eruptive episodes over the last 2500 years, interspersed with dormant periods lasting 200 to 700 years (Mullineaux and Crandell, 1981). The eruptions have altered the surrounding conifer forests, dominated generally by Douglas fir *(Pseudotsuga menziesii),* which can live well beyond 1000 years.

Only a few decades ago, much of the Pacific Northwest was blanketed with forests that originated after wildfires during the twelfth through the early nineteenth centuries. This was before nonaboriginal peoples entered the area (about A.D. 1800) or had significant effect on fire ignition (beginning about 1840) and suppression (beginning about 1910). Aboriginal people may have influenced fire history during much of the Holocene Epoch, but their influence is mostly unknown and may not have been widespread in the massive, generally wet forests west of the crest of the Cascade Range (Morris, 1934; Burke, 1979; Teensma, 1987) (Figure 1).

Today, privately owned forest lands have been almost entirely cut over, and extensive tracts of Federal forest land are now being logged and converted to

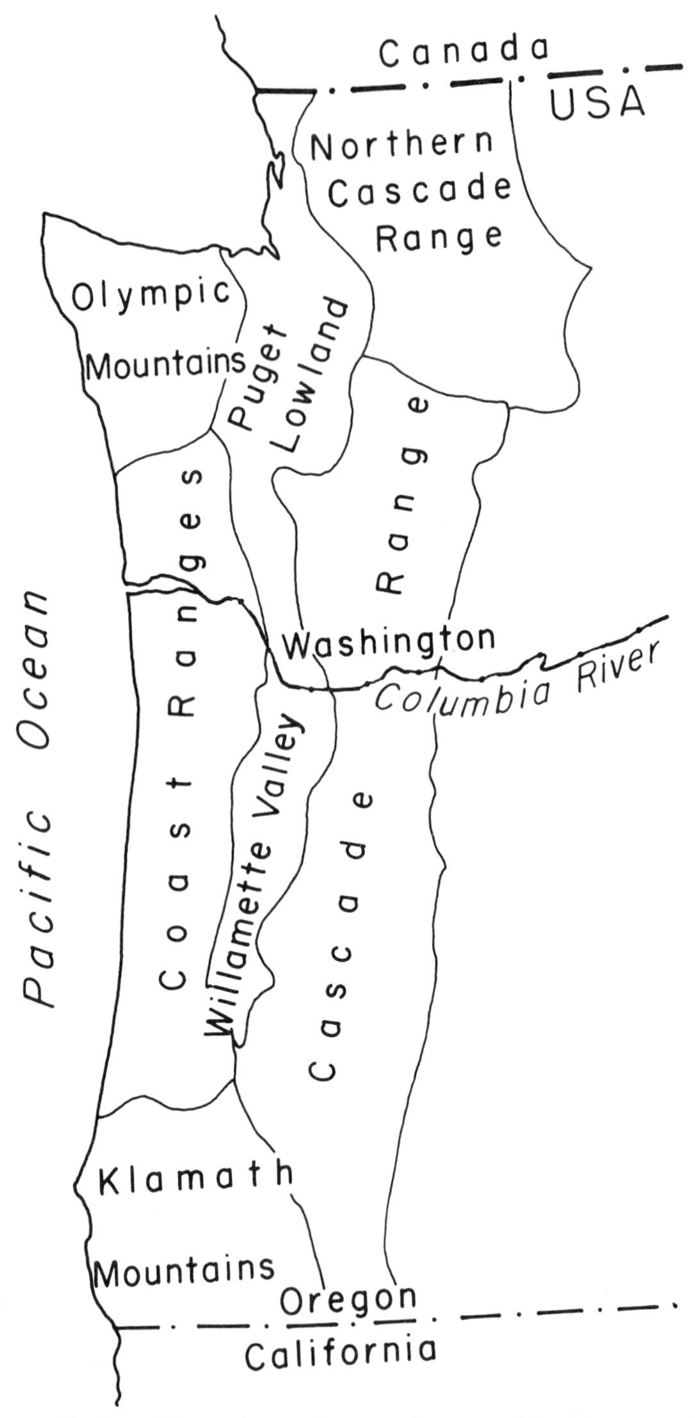

Figure 1. Physiographic provinces of western Oregon and Washington, Pacific Northwest, United States.

managed forests. Despite recent forest cutting, the natural patchwork of forest stands before the influence of nonaboriginal people and the history of disturbances over the past 500 to 800 years can be reconstructed from aerial photographs and dendrochronological records. The patterns can be compared with the landscape pattern imposed by management activities. Many current, politically charged issues in forest management concern the degree to which altered landscape structure and forest fragmentation influence wildlife, hydrology, water quality, and susceptibility of forests to catastrophic disturbance by agents such as wind, fire, and pests.

In this chapter, we present several examples of key landscape-ecology processes operating in forest and stream systems of the Pacific Northwest before and after the conversion from wildland to managed forest land. We do not offer a general model for these landscapes, but instead examine some of the elements that might go into such a model. We discuss in particular the relation among forest and flowing-water ecosystems, disturbances, and geomorphology. Geomorphology concerns both landforms and the geomorphic processes that sculpt them.

Landscapes of the Pacific Northwest

The ecological, physiographic, and geologic complexities of the Pacific Northwest (Franklin and Dyrness, 1973) have created highly varied landscapes. The landscape pattern is strongly influenced by environmental factors that vary spatially either as discrete patches or as broad gradients. Discrete variation in landscape pattern is controlled primarily by land use, wild disturbances (i.e., by natural processes as distinguished from management-imposed disturbance), and geology. Rock types on opposite sides of knife-sharp geologic contact may contrast may strongly in physical and chemical properties, thereby causing abrupt changes in vegetation composition and in dominant geomorphic processes that can trigger disturbances. Superimposed on this geologic template are gradients of important environmental variables: (1) precipitation, which varies with elevation, latitude, and mountain-created rain-shadow effects; (2) snowpack accumulation and timing of snowmelt which, in turn, control the timing and magnitude of flood flows; (3) wildfire, which is controlled primarily by effects of climate on vegetation type, fuel mosaic, ignition, and spread; (4) high winds, including frequency, magnitude, and seasonal timing; (5) glacial imprints on topography and soil; and (6) volcanism, which produces zones of air-fall tephra and mudflow deposits extending outward from stratovolcanoes in the Cascade Range.

Topography changes conspicuously across the region. At one extreme are areas with long, steep slopes in highly glaciated, hard-rock terranes in the northern Cascade Range and Olympic Mountains of Washington (Figure 1). In these areas, landform relief (height of ridge crest above adjacent valley floor) may exceed by 20 times the typical heights (50 to 70 m) of the once widespread old-growth forest. The unglaciated Coast Ranges of Oregon and Washington, underlain by moderately indurated Tertiary sedimentary rock, have low relief (landform relief of two to six old-growth tree heights in many areas), although slopes may be steep and cut by a high density of stream channels. Broad inland valleys—e.g., the Willamette Valley and the Puget Lowland (Figure 1)—and coastal marine terraces are among the landforms of lowest relief in the region. Major landforms are mainly the result of erosion, but the Quaternary stratovolcanoes in the Cascade Range exceeding 2300 m in elevation are imposing features of constructional origin (Figure 2).

Disturbance regimes, influenced in part by landforms (Swanson et al., 1988), also differ across broad gradients in the region. Climate creates a great diversity of wildfire regimes. Essentially fire-free ecosystems occur along a narrow coastal strip from Washington to southeastern Alaska. At the other extreme, a complex regime of frequent, low- to high-severity fires of highly variable patch

Figure 2. High-relief landscape on the southeast flank of Mount Rainier, Washington. Near the northern end of the Cascade Range (Figure 1).

sizes is characteristic of southern Oregon and northwestern California. The Cascade and Coast Ranges of Oregon and Washington experience an intermediate fire regime that includes large (100,000-ha) stand-replacing fires (Teensma, 1987). The importance of wind as an agent of disturbance is greatest near the coast and through the Columbia River Gorge (Franklin, in preparation). Landslides create fewer but larger disturbance patches in the Cascade Range than in the highly dissected Coast Range of southern Oregon (Swanson and Lienkaemper, 1985). Snow avalanches are common at higher elevations of the Cascade Range but do not occur in the Coast Ranges.

Flowing-water and valley-floor ecosystems of the region also show great variation; they range from headwater channels a few meters wide flowing among 70-m-tall conifer trees to broad, multichanneled rivers bordered by shrubs and hardwoods that are frequently disturbed by floods. Bedrock outcrops and landslides from hillslopes create constrictions—local sites of narrow, steep channels—and alter the magnitude of shading, litter production, and other effects of streamside vegetation on stream ecosystems. Channel reaches along valley floors that are free of these constraints may be meandering or braided; lateral channel migration is common, and floodplain forests can significantly affect the dominant processes and food base of the aquatic ecosystem. Geologic conditions, including the resistance of rock to erosion and the rate and recency of uplift, affect the type and geographic arrangement of constrained and unconstrained stream reaches.

This diversity in landform and disturbance regimes has created a complex natural mosaic of forest patches ranging from less than 0.1 to 100,000 ha. Fire, wind, landslides, snow avalanches, patches of root-rot mortality, changes in river channels, and other natural processes are responsible for the mosaic of disturbances. Further complexity in the form of environmental-resource patches (Forman and Godron, 1986) is created by extremely shallow soil, bedrock outcrops, wetlands, talus fields, and other nonforested sites that persist through forest disturbances.

Forest cutting, road construction, and maintenance of forest vegetation for stream protection and other purposes are creating new patchworks that differ markedly in their structure and rate of creation. On U.S. Government land, for example, a *staggered-setting* or dispersed-patch approach to timber harvest uses clear-cut areas of about 15 ha dispersed through the forest (Figure 3). In this system, cutting occurs progressively over the entire rotation (time between successive complete harvests of a site). The rotation in Douglas fir is commonly 50 to 100 years. The staggered-setting system is an effort to disperse the effects of management activities across time and space (Franklin and Forman, 1987). On private industry lands, on the other hand, cut areas commonly form much larger patches, and an entire drainage basin of hundreds of hectares may be cut within a few years; consequently, management activities are altering landscape patterns over large areas and in dramatically different ways with little regard for the natural landscape mosaic or the processes that created it.

Figure 3. Early stage in the staggered-setting or dispersed-patch system of forest cutting used on U.S. Government land.

Concepts of Landscape Ecology

Existing landscape concepts emphasize terrestrial landscapes as a system of patches of different types, shapes, and functions and also the interactions among patches (Forman and Godron, 1986). This perspective has great use in analyzing landscapes of forested mountain regions. Further useful landscape-ecology concepts include: (1) the effects of landforms on ecosystem patterns and processes, such as disturbance and the dynamics of terrestrial mosaics (Swanson et al. 1988); (2) link between terrestrial and aquatic systems; (3) models of flowing-water ecosystems from small to large channels; and (4) structure and function of landscapes that are constantly changing as a result of forest management. We now explore these concepts in sequence.

Effects of Landforms on Ecosystem Patterns and Processes

Many aspects of landscape pattern in mountainous areas reflect the influence of landforms on disturbance and resources for ecosystem development. Landform effects on ecosystems can be viewed as occurring in four ways (Swanson et al., 1988):

1. Slope gradient, elevation, and aspect affect the quantity of solar energy, water, nutrients, pollutants, and other materials received by a site.
2. Landform position and slope gradient affect the flow of materials (water, dissolved material, organic and inorganic particulate matter), organisms, propagules, and energy across landscapes by their influence on gravitational gradients, by guiding flowpaths of wind, and by forming barriers to movement.
3. Landforms and slope gradient influence the frequency, spatial pattern, and intensity of disturbances, such as fire and wind, which are strongly and positively influenced by the presence of vegetation.
4. Landforms influence the spatial pattern and frequency or rate of disturbance by geomorphic processes.

The first three of these classes of landform effects assume topography to be static; only the fourth class accommodates the dynamic geomorphic character of many mountain landscapes.

Patterns in most landscapes represent superimposed effects of several classes or all four classes of landform influences. At Mount Rainier in Washington State, for example, repeated volcanic activity and glaciation have formed a steep, high-relief landscape with slope lengths locally exceeding 1000 m (figures 2 and 4) (Hemstrom, 1982; Hemstrom and Franklin, 1982). Vegetation habitat types therefore occur in zones defined by altitude, because of strong effects of temperature and moisture (class-1 effect) (Figure 4). Downslope movement of moisture and nutrients superimposes a pattern of higher productivity on lower slope locations (class-2 effects). Frequency, size, and geographic patterns of fire appear to be significantly affected by landforms (Hemstrom, 1982). Major ridges and valley bottoms impede fire movement. Snow avalanches repeatedly sweep along bedrock-controlled paths, thereby creating treeless areas that may serve as fire breaks. In addition, all major valleys that drain the mountain have had mudflows that began on the upper flanks of the mountain (class-4 effect).

The resulting landscape mosaic at Mount Rainier is complex but predictable to a degree. Gross landforms are relatively invariant on a time scale of several thousand years; where landforms strongly control the pattern of disturbance, boundaries between patches along ridge lines and valley floors are likely to persist through disturbance. Other segments of boundaries between patches may be ephemeral and readily overridden by subsequent disturbances. This process may occur in midslope positions where fire boundaries have been determined by shifts in wind direction or the onset of rainy weather. Landscape dynamics can be analyzed in terms of the persistence of boundaries between patches, based on the degree of landform control on boundary locations as a measure of ecotone stability.

The influences of landforms on disturbance patterns (class-3 and class-4 effects) pose interesting questions about the importance of the relative scales of

(A)

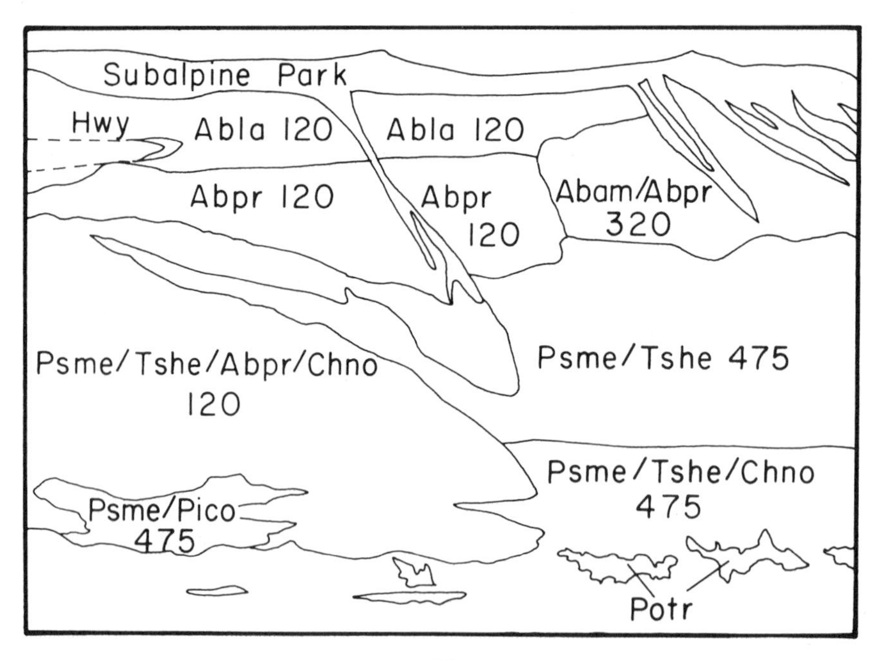
Subalpine Park
Hwy
Abla 120
Abla 120
Abpr 120
Abpr
120
Abam/Abpr
320
Psme/Tshe/Abpr/Chno
120
Psme/Tshe 475
Psme/Tshe/Chno
475
Psme/Pico
475
Potr

(B)

vegetation and landforms. We hypothesize that in steep mountain landscapes where landform relief does not exceed several tree heights, disturbance agents such as fire and wind can readily move through the forest with little regard for topography. Landforms may have greater effect on the spread of disturbance and mosaic structure (i.e., patch size and edge location with respect to landform feature) where relief substantially exceeds tree height.

Links Between Terrestrial and Aquatic Systems

Major ecological links occur where terrestrial and aquatic systems meet in the riparian zone, the three-dimensional zone of interaction between terrestrial and aquatic systems (Swanson et al., 1982). Analyses of landscapes containing aquatic systems should consider both terrestrial and aquatic components and their zone of interaction. The riparian zone is a distinctive element of the landscape commonly subject to disturbances characteristic of both fluvial and terrestrial systems and having mosaic structure and some species not found in either terrestrial or aquatic areas.

From a biological point of view, forest–stream interactions include transfers of organic matter and nutrients and the regulation of energy flow. In steep lands with narrow valley floors and wet climates, the dominant direction of influence is from forest to aquatic system. The composition of aquatic communities and rates of stream ecosystem processes are strongly influenced by the structure and composition of streamside vegetation. The vegetation produces litter and shading, which regulates water temperature and light available for instream primary production. The opposite interaction (from stream to forest) may dominate in areas of broad floodplains and especially in arid and semiarid areas where streamwater flows outward into the groundwater system, recharging it. In virtually any system bordered by forests, large woody debris from adjacent stands can influence the structure of aquatic habitat and the ability of the aquatic system to retain organic detritus, making it available for consumption by aquatic organisms (Harmon et al., 1986).

The interactions between riparian vegetation and aquatic ecosystems differ in response to disturbance history, vegetation succession, and geomorphic setting. A key factor is the height of vegetation relative to the widths of valley-floor geomorphic surfaces. We are now studying these relations in valley floors of

---

Figure 4. Photograph (A) of mosaic of forest stands on Sunrise Ridge in the White River drainage of Mount Rainier. This east-southeast facing slope extends from 900 to 1800 m elevation. Drawing (B) shows forest-stand types and ages reflecting landform influences on environment and disturbance. Snow avalanche tracks cut through the upper elevation forests. The White River (bottom of photograph) has experienced major mudflows generated on Mount Rainier and extensive lateral channel migration. Tree species dominating stands are: Abla = *Abies lasiocarpa*, Abam = *Abies amabilis*, Abpr = *Abies procera*, Psme = *Pseudotsuga menziesii*, Tshe = *Tsuga heterophylla*, Chno = *Chamaecyparis nootkatensis*, Pico = *Pinus contorta*, Potr = *Populus trichocarpa*. Numbers indicate the stand age in years.

third- to fifth-order (stream-ordering system of Strahler, 1957) mountain streams to determine the rates and patterns of change in stream-habitat structure and streamside vegetation in response to disturbance by fire, channel change during floods, streamside landslides, and debris flows from tributaries. This analysis recognizes a critical hierarchy of structural scales in the fluvial system: (1) single particle; (2) subunits, such as a patch of like-sized particles; (3) channel units (e.g., pool or riffle); (4) reach type (defined below); (5) sections composed of multiple reaches (e.g., high-gradient mountain streams, meandering rivers in major valleys); and (6) the full drainage network (Figure 5). Furthermore, in forested mountain systems some individual scales are not sharply distinguished from adjacent scales. Single particles, for example, can be as large as whole, fallen old-growth tree stems, which are the size of and control formation of channel units two scales larger.

To interpret and predict the long-term (centuries to millennia) ecosystem behavior of valley-floor environments, scientists are examining the geomorphic structure and behavior of the fluvial system across the first five scales. These

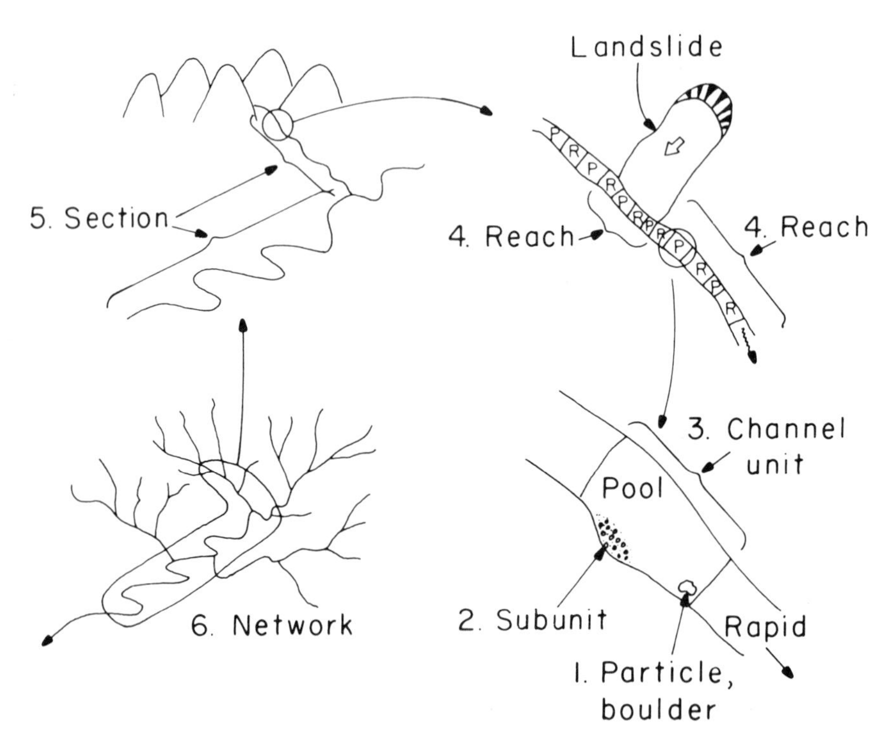

Figure 5. Six scales of the hierarchical structure of river systems from single particle to drainage network. Scales are numbered from the single particle (1) to the full drainage network (6). At the channel-unit (3) and reach (4) scales, P = pool and R = riffle. Scale 5, the section, shows mountain and lowland valley examples.

channel conditions must be interpreted in the context of their valley-floor (floodplain, terrace, alluvial fan) and valley-wall environments (landslide, earthflow, bedrock, soil-mantled slope). A key to predicting long-term valley-floor dynamics is to understand the geomorphic behavior of reach types. Reach types are defined in terms of the type and degree of constraint on the stream system by factors exogenous to the mainstem river system. Three reach types common in mountain streams are: (1) reaches with passive constraint by bedrock exposed in the bed and bank of the channel, (2) reaches with active constraint of slow-moving landslides (termed earthflows) from a hillslope (Figure 5), and (3) areas free of these constraints where the channel has greater opportunity for lateral shifting. We think that fluvial disturbances in bedrock-confined areas will be infrequent and limited in areal extent. Reaches constricted by earthflows have frequent, small, streamside slides (Swanson et al., 1985) that create a fine-scale mosaic of stands of streamside vegetation in various successional stages. Areas free of these constraints have larger, elongated disturbance patches where recent channel changes have provided fresh substrates for establishment of streamside vegetation. Many ecosystem characteristics differ among these reach types, including channel-habitat structure, the spatial distribution and geomorphic functions of large woody debris, and valley-floor wildlife habitat.

In the high-gradient, fourth- and fifth-order mountain streams of the western Cascade Range in Oregon, broad valley floors occur predominantly upstream of earthflow constrictions because of a local damming effect (Figure 6). These areas may be several hundred meters wide, several kilometers long, and contain a

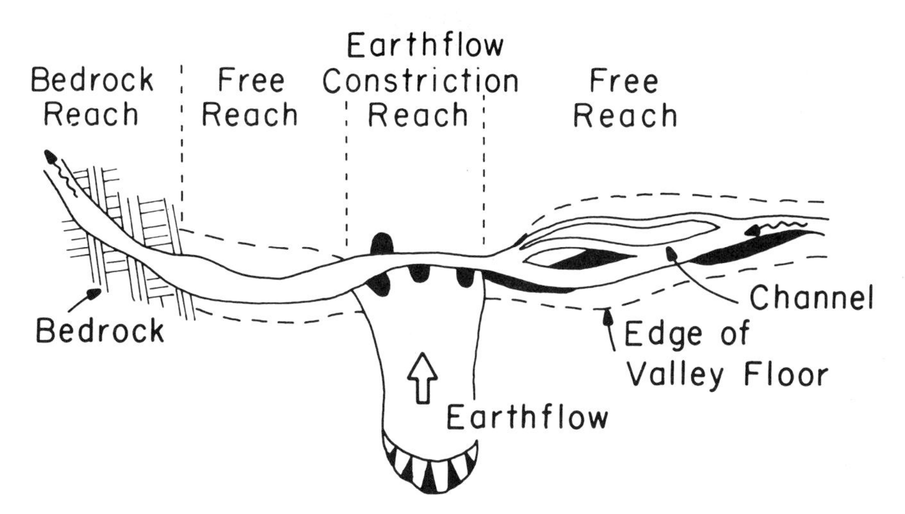

Figure 6. Stream reach types near an earthflow constriction. Dark patches are areas of recent disturbance caused by channel change (in the free reach upstream) and by streamside slides (in the earthflow reach). Interactions between terrestrial and aquatic systems are greatest in the upstream free reach.

complex braided network of perennial, intermittent, and ephemeral channels flowing through shrub and forest stands of different ages (Vest, 1988). The extent of floodplain inundation at high discharges, and hence interaction between terrestrial and aquatic systems, is greatest in these unconstrained reaches.

The size and extent of patches created by disturbances in riparian zones differ by type of disturbance and the type of reach. In French Pete Creek, a wilderness stream drainage in the western Cascade Range in Oregon, a major 1964 flood with associated debris flows in disturbed patches of streamside forests (Grant, 1986) now characterized by stands of alder *(Alnus rubra)* established after that event. The width of these patches is greatest where debris flows from tributaries entered stream reaches free of bedrock and earthflow constraint (Figure 7). Along the free reaches unaffected by debris flows in 1964 and in an earthflow-constricted reach, more than 70 percent of the channel length is bordered by vegetation patches of 1964 origin that are less than 10 m wide (Figure 7). The

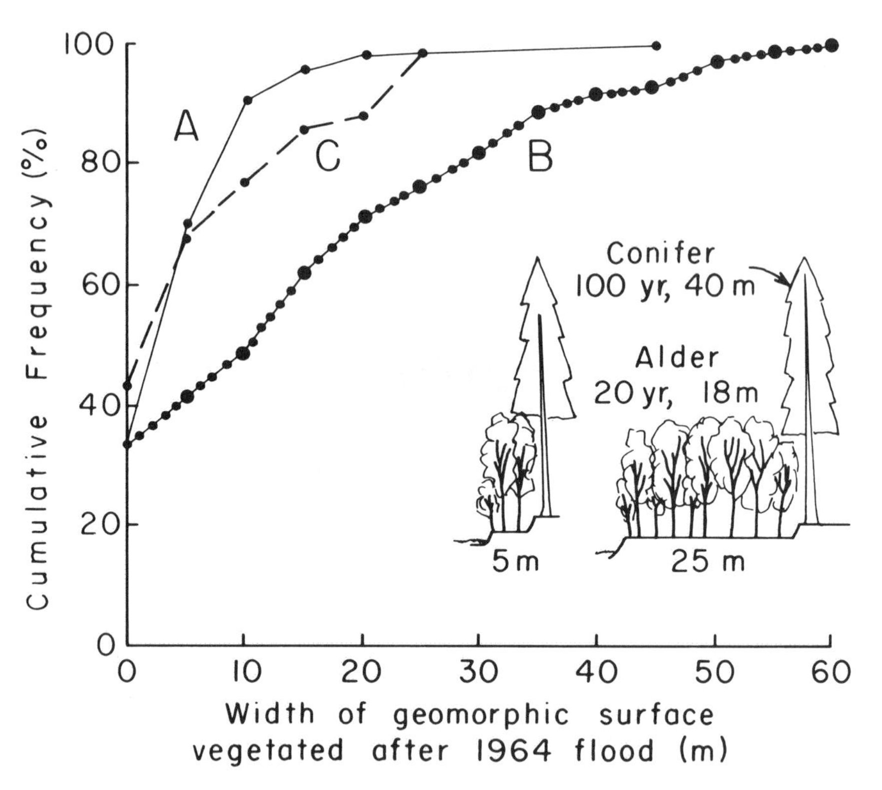

Figure 7. Widths of streamside geomorphic surfaces with vegetation dating from a major flood in 1964, distinguished by type of stream reach. A = reach free of earthflow and bedrock constraints; B = same type of reach, but affected by debris flow in 1964; and C = reach constricted by earthflow. Distance is measured perpendicular to the channel axis. Sketches show relations between surface width and stand structure 20 years after establishment of an alder stand.

effects of patch width on forest–stream interactions is determined in part by the ratio of patch width to tree height. These disturbed areas along French Pete Creek are generally narrow in relation to the heights of the alders established after the 1964 flood and to the 100- to 500-year-old conifers on the adjacent, older surfaces (Figure 7). This low ratio of disturbance-patch width to tree height results in continuity of shading and some litter input to streams through fluvial disturbances of riparian vegetation. Periodic disturbance maintains a narrow streamside band of deciduous vegetation bordered by taller coniferous forest. This mosaic of stands produces a mixture of coniferous and deciduous litter for the stream.

The rate of geomorphic and, hence, ecological change of valley-floor ecosystems differs greatly between high-gradient, gravel-bed streams and low-gradient rivers (Table 1). Along the steep, straight mountain channels of French Pete Creek, Oregon, and North Boulder Creek, Colorado, fluvial and debris-flow events in the past century have disturbed vegetation in an area equivalent to 1.0 times the channel width or less (Furbish, 1985; Grant, 1986). In contrast, lateral channel migration of a reach of the low-gradient, meandering Little Missouri River in North Dakota, USA, reset floodplain vegetation over an area of 5.9 channel widths over the past century. Lateral channel change in mountain streams is limited by high gradients, coarse sediment, bedrock outcrops, and hillslope movement into channels. These contrasting disturbance regimes are manifest in the age distributions of streamside forests. Streamside stands along the Little Missouri have a broad spectrum of ages up to 300 years, and all are primary stands established on deposits resulting from channel migration. Forest stands along the steep channel of French Pete Creek are of only a few age classes as a result of establishment after wildfire, floods, and debris flows.

In summary, the structure, function, and disturbance regime of valley-floor

Table 1. Extent of valley floor disturbance, expressed as even-aged patches of trees less than 100 years old, in two high-gradient, straight mountain streams (French Pete Creek, Oregon [Grant, 1986] and North Boulder Creek, Colorado [Furbish, 1985]), and a low-gradient, meandering river (Little Missouri River [Everitt, 1968]).

		Extent of valley floor disturbance in previous 100 yr	
Site	Mean stream channel slope (m/m)	Hectares per km of valley length	Channel widths (mean disturbance width divided by mean channel width)
French Pete Creek	0.042	1.5	1.0
North Boulder Creek	0.030	<0.05	<0.1
Little Missouri River	0.00085	54	5.9

ecosystems are best analyzed within a geomorphic context. The type and degree of geomorphic constraints, including reach-level phenomena, strongly influence landscape dynamics and forest–stream interactions.

Models of Flowing-Water Ecosystems

River networks and their riparian zones are integral parts of many landscapes. The river-continuum concept (Vannote et al., 1980; Minshall et al., 1985) is useful for integrating flowing-water systems at the drainage-network scale into analyses of landscapes. The initial river-continuum concept predicted downstream changes in physical and biological properties of stream systems based on increased discharge and depth and decreased influence of streamside vegetation. Challenges to the simple predictions of the concept have come where geomorphic conditions, such as changes in channel and valley-floor slope, have caused changes in the degree of interaction between streams and floodplain vegetation. Changes in bedrock type, for example, may cause a river valley to change abruptly from a bedrock-confined gorge to a broad floodplain with multiple channels and extensive exchange of organic matter between terrestrial and aquatic systems. Major changes in the biological properties of the aquatic system at the section or reach scales may disrupt network-scale patterns predicted by the river-continuum concept (Minshall et al., 1985; Sedell et al., in press). Landscape ecology would benefit from further development and incorporation of concepts such as the river continuum.

Management-Created Landscapes and Landscape Management

Management practices in the Pacific Northwest are creating new landscapes with little regard for the ecological design of management at a landscape scale. Issues of landscape management are emerging rapidly through litigation, legislation, and growing concerns of managers and the public. The cumulative effects of management activities is an important issue (which must be assessed for significant U.S. Government actions as stipulated in the National Environmental Protection Act). Cumulative effects can be considered as resulting from: (1) effects of multiple actions through time at a site, and (2) downstream, off-site effects of multiple activities at one or more sites within a drainage basin or airshed. This second case—cumulative watershed effects—is most fruitfully examined in the context of landscape ecology.

A decade ago, when water quality was the major concern in watershed management in the United States, the pivotal issue in forestry was *how* to manage individual patches of land to meet water-quality objectives. New issues are now centered on the *where* and *when* aspects of management activities.

Many major issues in forest land management in the Pacific Northwest today concern landscape-scale problems, such as the effects of clear-cutting on floodplains, designing old-growth forest reserves to protect a rare owl and other old-growth-dependent species, the interaction between management-caused landslides and fish, and maintaining a management strategy in the face of disturbances by wind, fire, and other processes. Brief discussions of these issues pinpoint some applications of the perspectives of landscape ecology in forested mountain ecosystems.

Hydrology of Rain-on-Snow Systems

Warm, wet snow has a major role in the hydrology of the Cascade Range. Much of the Cascades is in a transient snow zone extending from 350 to 1200 m in elevation in Oregon. In this zone, snow accumulates and melts several times each year, mainly during rain-on-snow events when water stored in the warm, transient snowpacks melts during heavy rainfall (Harr, 1981). Precipitation at lower elevations falls mainly as rain, and a seasonal snowpack accumulates at higher elevations. The major floods in basins draining the Cascade Range have resulted from such rain-on-snow events, which are particularly important in the transient snow zone (Harr, 1981; Christner and Harr, 1982).

Harr (1981) argues, based on theoretical considerations, that changing forest structure in the snow zone by clear-cutting increases snow accumulation and also the rate of melt during rain-on-snow events. Warm snow falling in forested areas may catch in the canopy and melt during the snowfall, whereas in nearby clear-cut areas snow accumulates in a pack. During rainy periods, energy exchange between the air mass and snow surface is greatest in nonforested areas because of greater wind speed and turbulence. These relations were observed in field studies contrasting the snow hydrology of adjacent forest and clear-cut patches in the transient snow zone (Berris and Harr, 1987). Runoff records from small experimental watersheds suggest that the size of flows caused by rain-on-snow events is increased by clear-cutting (Harr, 1986). Furthermore, Christner and Harr (1982) observed increases in size of peak streamflow in fourth- and fifth-order basins; they interpret these increases to be the result of the hydrologic effects of clear-cutting in the transient snow zone.

The observations suggest that the timing and location of clear-cut areas in drainage basins within the transient snow zone can affect peak streamflow at downstream points. A high rate of cutting in the transient snow zone could have a major, near-term impact on size of peak flows. These impacts might be minimized by distributing cutting units across a range of elevations.

Landslides and Fish

The practice of forestry in unstable, landslide-prone areas is a major challenge in landscape management in the Pacific Northwest. Landslides are common in forested areas, and their occurrence may be increased by clear-cutting and road construction. Landslides may cause direct and indirect impacts on fish and their habitat many kilometers away from the site of landslide initiation. Therefore, forestry–landslide–fish interactions must be examined in a landscape context.

This can be approached by zoning landscapes in terms of the natural frequency of slides and the magnitude of increased incidence of sliding in response to clear-cutting and road construction (Swanson and Dyrness, 1975). Slope steepness, soil properties, and slope form, as they affect the concentration of water at slide-prone sites, are principal variables in evaluating hillslope stability. Clear-cutting is believed to increase sliding primarily by causing a period of reduced root strength after logging (Ziemer, 1981). The main effect of roads on slope stability is the alteration of surface and subsurface hydrology. In most areas of the Pacific Northwest, road rights-of-way have a substantially higher frequency of landslides than do clear-cut or forested areas (Swanson et al., 1987).

An important consideration in managing landslides is the geographic and temporal distribution of activities that reduce slope stability. Good management in the near term ideally would minimize landslides associated with clear-cutting and roads. This could be accomplished by first cutting land with low susceptibility to sliding; however, this strategy leaves the most unstable land until late in the cutting cycle, when practically all cutting and road construction would be on unstable land. From this perspective, good land management distributes cutting and road construction in the most slide-prone lands through time, which means cutting some unstable sites every decade, even though a recognized high probability of sliding exists. Another alternative is to cut no forest at all in the most unstable areas; this is the practice in parts of the Oregon Coast Range where landslide incidence and fisheries values are particularly high.

Many of the effects of landslides on fish habitat are obvious and adverse; however, some effects of landslides may be beneficial. The evaluation of landslide effects must be done at a landscape scale and with the recognition that fish in the Pacific Northwest have evolved in an environment with frequent and diverse disturbances. Scientists and managers face a tough challenge in distinguishing effects of human actions from those of natural events.

The negative effects from management-induced landslides on habitat of anadromous fish are a major land use issue (Kessel, 1985). Slides from roads and freshly clear-cut slopes enter streams, commonly forming debris flows that move rapidly down channels. Debris flows, typically containing several thousand cubic meters of soil, alluvium, and organic matter, affect fish habitat by altering channel structure, damaging riparian vegetation, and blocking fish passage.

Assessing the effects of landslides on fish habitat involves: (1) the distribution of slide-prone areas in a drainage basin, (2) the probable travel path of debris flows through the stream network, (3) the geomorphic effect of debris-flow deposits at the terminal deposition site, and (4) the response of fish to altered habitat along the path of the slide and at the site of final deposition. Slide-prone areas can be identified by topographical criteria: steep, concave slopes have high susceptibility to sliding. Slides from these sites can enter channels and flow downstream as debris flows, entraining additional soil, vegetation, and alluvium along the flow path. Debris flows tend to stop where the channel gradient decreases to about 4 to 6 degrees and at abrupt changes in channel direction (Benda, 1985). These two conditions are commonly met at the junction of small (first- to third-order) and large (third- to fifth-order) streams. The effects of debris-flow deposits on fish habitat in the receiving stream depend on the relation of volume of debris-flow material to the size of the channel. Small streams may be blocked by deposits; large streams may wash the debris-flow material downstream. Intermediate-size streams may be partially blocked by the debris-flow deposit and a pool may form, benefiting fish habitat in basins where fish production is limited by the quantity and quality of pool habitat. In the 52-km^2 Knowles Creek basin in the Oregon Coast Range, for example, two pools representing just 0.5 percent of the available habitat provided overwintering habitat for over 40 percent of the coho salmon *(Oncorhynchus kisutch)* smolts in one winter (Rodgers, 1986).

This type of analysis of slide potential, debris-flow-runout characteristics, and fish habitat for a drainage basin has led to landscape-scale zoning of basins (Swanson et al., 1987). Such a landscape/drainage basin perspective is essential in planning cost-effective efforts to mitigate management-induced slides. These management activities represent one form of synthesis of the spatial patterns of processes that propagate cause-effect relationships across landscapes.

Forest-Cutting Patterns

The creation of new forest patchworks in the natural landscapes of the western United States is a major management issue. Cutting patterns are creating new, managed mosaics that contrast markedly with the natural forest patterns in types and sizes of patches. In managed landscapes, a strong tendency exists for uniformity in patch sizes and homogeneity of structure within forest patches. The patch patterns created during the first cycle of forest cutting will have long-term consequences because they will be perpetuated through subsequent rotations.

Landscape issues that are consequences of specific forest-cutting patterns are increasingly being recognized. The issues include how cutting patterns affect susceptibility of forests to damaging agents (e.g., wildfire, wind, and pests and pathogens), overall ecological diversity, production of game and other wildlife species, fish production, water yields, flood levels and frequencies, and sediment yields. All are related to the rate and arrangement of forest cutting.

A theoretical analysis was done of the effects of the dispersed patch system of clear-cutting that is widely used in northwestern North America (Franklin and Forman, 1987). In the dispersed-patch system, 10- to 20-ha patches are interspersed with areas of uncut forest and older harvest units (Figure 2). The objective is to delay cutting adjacent forest patches for as long as possible. Forest regeneration, residue disposal, and development of road systems were considerations in the selection of this system (Smith, 1985).

The dispersed-patch-cut model results in some distinctive geometric patterns when the model is systematically applied to a forested grid (Franklin and Forman, 1987). Rapid changes in the average size of remnant forest patches occur between the 30- and 50-percent cutover points, as the forest matrix becomes increasingly fragmented by continued cutting (Figure 8). After the

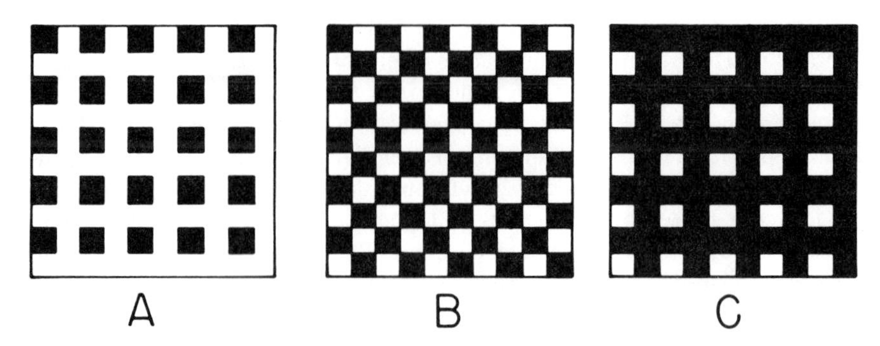

Figure 8. Progression of clear-cutting in a grid pattern using the dispersed-patch model, in which areas are selected for cutting so as to be regularly distributed through the landscape. Shading indicates the (A) 25 percent, (B) 50 percent, and (C) 75 percent cutover points.

50-percent cutover point, all remaining forested patches are the size of the grid segments (10 ha in the hypothetical model). This may effectively eliminate forest interior environments in these tall forests, where edge effects penetrate substantial distances, perhaps two to three tree heights, into a stand. The amount of edge or forest-cutover ecotone is maximized in this system as compared with most other standard cutting patterns.

The dispersed-patch-cut system affects many ecological characteristics of the landscape, such as species diversity, game populations, and abundance of species requiring interior forest conditions (Franklin and Forman, 1987). For example, consider some effects of developing a landscape dominated by the old-growth forests (generally conifer-dominated stands more than 200 years old) that were widespread in the region a few decades ago. When cutting begins, an initial loss of species is predicted because of fragmentation of an old-growth forest and the resulting loss of large blocks of interior forest environment and the species dependent on such blocks, such as the northern spotted owl *(Strix occidentalis)*. A second episode of loss of animal species is predicted when the last patch of old-growth forest is cut. Old-growth-related species, such as some salamanders with very small home ranges, that could survive in a 10-ha forest patch but not in a young, managed forest will be lost. Dead-wood structures (i.e., standing dead trees and fallen logs) are keys to the survival of many species; the magnitude of total species loss will depend on whether such structures are maintained in the managed landscape.

The forest patterns created with a dispersed-patch-cut system contribute substantially to increased risk from damaging agents (Franklin and Forman, 1987). Windthrow potential increases dramatically in the partially cutover landscape because of high densities of forest edge and long wind fetches in cleared areas. The potential for ignition and spread of wildfire in residual forest patches also increases as dispersed patches are progressively cut. Responses of pests and pathogens are highly variable depending on the distribution and abundance of suitable forest patches (e.g., specific ages of forest) and the biology of the particular pest or pathogen (e.g., method and dispersal).

Indeed, patch-cut landscapes do demonstrate high susceptibility to disturbance by agents, such as wind and fire. In the 37,000-ha Bull Run River management unit in western Oregon, for example, 482 and 899 ha of old-growth forest blew down in 1973 and 1983 windstorms. In these blowdown events, 48 and 81 percent, respectively, were directly associated with clear-cuts and road clearings. Similarly, many of the larger wildfires in this and neighboring areas have been associated with escaped slash burns.

Managers can select their cutting methods to achieve specific landscape-spatial patterns and consequent ecological effects (Franklin and Forman, 1987). For example, progressive strip clear-cutting can be used to reduce the amount of edge and therefore the potential for catastrophic windthrow. Forest patch sizes can be altered to fit the needs of interior species. What is essential is that the long-term, landscape-scale consequences of specific cutting programs be considered when plans are developed.

Wildlife and Habitat Fragmentation

Production of wildlife has always tended to be a landscape issue because many wildlife species are wide ranging and make use of several habitats. Large ungulates and many top predators range over large areas of diverse vegetation and topography, and many game species make heavy use of edge or ecotonal habitats.

An emerging global concern is maintaining nongame wildlife species. In northwestern North America, the focus has been primarily on the wildlife species associated with specialized habitat, such as old-growth forest. Such concern is legally mandated in the United States by legislation such as the National Forest Management Act (on National Forest lands) and the Endangered Species Act (on all land).

The northern spotted owl is an outstanding example of the many species identified as old-growth related and requiring a landscape approach to habitat management. This species appears to be strongly dependent for its survival on the special conditions in old-growth forest patches of 400 to 1200 ha per pair of owls, the area varying among physiographic provinces (Carrier, 1985; Forsman and Meslow, 1985). Some scientists believe the owl is potentially endangered by the reduction and fragmentation of this habitat (Gutierrez and Carey, 1985; National Audubon Society, 1986). Measures taken to preserve this bird have important ecological and economic consequences.

Maintaining viable populations of the northern spotted owl involves biological questions from the levels of the gene and population to the landscape (Marcot and Holthausen, 1987; Ruggerio et al., 1988). Critical landscape questions include the size of old-growth forest patch needed to maintain a pair of breeding birds, and the geographic arrangement of the patches across the bird's range. Island biogeographic concepts have been useful but do not address the central issues of spatial configuration in a changing mosaic. The viability of patches, especially their susceptibility to damage and to penetration of edge influences (e.g., microclimate effects), are also critical because the owl and other old-growth-related species are basically interior-forest species. Indeed, the interspersing of clear-cut patches within an old-growth forest matrix may alter habitat in favor of a competing owl species. Landscape issues in maintaining the northern spotted owl are complex and must go beyond most current thinking on habitat fragmentation.

Several management proposals have been developed; all involve some level of old-growth forest-patch preservation. One plan under consideration by Forest Service land managers would establish a network of 400- to 1200-ha old-growth patches (the size varying between physiographic provinces) where timber harvest is prohibited. These habitat patches would be arranged in a network with 10- to 20-km spacing.

Harris (1984) developed an innovative proposition for protecting old-growth forest patches, based partially on island biogeographic concepts. It involves identifying and protecting reserved old-growth islands that are surrounded by much larger buffer areas, and using riparian zones as corridors. The buffer areas would be managed on long cutting rotations (250 years) and cut in such a way that only a small segment of the edge of an old-growth island would be exposed to a clear-cut opening.

In summary, managing many animal species requires a landscape perspective. However, the science and application of landscape ecology are in early, exploratory stages.

Conclusions

To address critical ecological issues in forested mountain lands, it is essential to move from the traditional scales of research on forest plots and stands to mosaics, and from pools, riffles, and stream reaches to the drainage network. This is the perspective of landscape ecology. A further challenge is to integrate landscape-scale views of upland, riparian, and aquatic systems. In physically dynamic, forested, mountain land, the analysis of landscape ecology must be placed in the context of geomorphology. Geomorphic and nongeomorphic processes frequently change the landscape patchwork in upland and stream corridor environments. Landforms constrain the location and areal extent of disturbance and produce predictable spatial patterns in terrestrial, riparian, and aquatic systems. These considerations lead to many questions about the long-term development of landscapes in the face of change imposed by land management, acid deposition, climate change, and other factors.

Steep, forested landscapes present many opportunites to further the science and application of landscape ecology. Major issues in forest management are increasingly centered on landscape-scale problems—the cumulative effects of landscape modification on watershed conditions and functions (e.g., hydrology and sediment production) and on wildlife and stream habitat. These legislatively mandated concerns of forest managers are pushing land management to become landscape management.

Acknowledgments

This work was supported in part by grants BSR-8414325 and BSR-8508356 from the National Science Foundation for support of the Long-Term Ecological Research (LTER) program and riparian research at the H. J. Andrews Experimental Forest.

References

Benda, L. E. 1985. Delineation of channels susceptible to debris flows and debris floods. In *Proceedings of the 1985 international symposium on erosion, debris flow and disaster prevention*. The Erosion-Control Engineering Society, Tsukuba, Japan, pp. 195–201.

Berris, S. N. and R. D. Harr. 1987. Comparative snow accumulations and melt during rainfall in forested and clear-cut plots in the western Cascades of Oregon. *Water Resources Research* 23:135–42.

Burke, C. J. 1979. Historic fires in the central western Cascades, Oregon. M.S. thesis, Oregon State University, Corvallis, Oregon.

Carrier, W. D. 1985. In R. J. Gutierrez and A. B. Carey, eds., *Ecology and management of the spotted owl in the Pacific Northwest*. USDA Forest Service, General Technical Report PNW-185, Portland, Oregon, pp. 2–4.

Christner, J. and R. D. Harr 1982. Peak streamflows from the transient snow zone, western Cascades, Oregon. In *Proceedings of the 50th western snow conference,* Colorado State University, Fort Collins, Colorado, pp. 27–38.

Everitt, B. L. 1968. Use of the cottonwood in an investigation of the recent history of a floodplain. *American Journal of Science* 266:417–39.

Forman, R. T. T., and M. Godron. 1986. *Landscape ecology*. Wiley, New York.

Forsman, E. D., and E. C. Meslow. 1985. In R. J. Gutierrez and A. B. Carey, eds., *Ecology and management of the spotted owl in the Pacific Northwest*. USDA Forest Service, General Technical Report PNW-185, Portland, Oregon, pp 58–9.

Franklin, J. F. and C. T. Dyrness. 1973. *Natural vegetation of Oregon and Washington*. USDA Forest Service, General Technical Report PNW-8, Portland, Oregon.

Franklin, J. F. and R. T. T. Forman. 1987. Creating landscape patterns by cutting: Ecological consequences and principles. *Landscape Ecology*. 1:5–18.

Furbish, D. J. 1985. The stochastic structure of a high mountain stream. Ph.D. Dissertation, Department of Geological Sciences, University of Colorado, Boulder.

Grant, G. E. 1986. Downstream effects of timber harvest activities on the channel and valley floor morphology of western Cascade streams. Ph.D. dissertation, Johns Hopkins University, Baltimore, Maryland.

Gutierrez, R. J. and A. B. Carey, eds. 1985. *Ecology and management of the spotted owl in the Pacific Northwest*. USDA Forest Service, General Technical Report PNW-185.

Harmon, M. E., J. F. Franklin, F. J. Swanson, et al. 1986. Ecology of coarse woody debris in temperate ecosystems. In A. MacFadyen and E. D. Ford, eds., *Advances in ecological research,* Vol 15. Academic Press, New York, pp. 133–302.

Harr, R. D. 1981. Some characteristics and consequences of snowmelt during rainfall in western Oregon, *Journal of Hydrology* 53:277–304.

Harr, R. D. 1986. Effects of clearcutting on rain-on-snow runoff in western Oregon: A new look at old studies. *Water Resources Research* 22:1095–1100.

Harris, L. D. 1984. *The fragmented forest: Island biogeographic theory and the preservation of biotic diversity*. University of Chicago Press, Illinois.

Hemstrom, M. A. 1982. Fire in the forests of Mount Rainier National Park. In E. E. Starkey, J. F. Franklin, and J. W. Matthews, eds., Ecological research in national parks of the Pacific Northwest. *Proceedings of the second conference on scientific research in the national parks,* Oregon State University, Corvallis, Oregon, pp. 121–6.

Hemstrom, M. A. and J. F. Franklin. 1982. Fire and other disturbances of the forests in Mount Rainier National Park. *Quaternary Research* 18:32–51.

Kessel, M. L. 1985. Timber harvest, landslides, streams, and fish habitat on the Oregon Coast. *Journal of Forestry* 83:606–7.

Marcot, B. G., and R. Holthausen. 1987. Analyzing population viability of the spotted owl in the Pacific Northwest. *Transactions of the North American Wildlife and Natural Resources Conference* 52:333–47.

Minshall, G. W., K. W. Cummins, R. C. Petersen, C. E. Cushing, D. A. Bruns, J. R. Sedell, and R. L. Vannote. 1985. Developments in stream ecosystem theory. *Canadian Journal of Fisheries and Aquatic Science* 42:1045–55.

Morris, W. G. 1934. Forest fires in western Oregon and western Washington. *Oregon Historical Quarterly* 35(4):313–39.

Mullineaux, D. R. and D. R. Crandell 1981. The eruptive history of Mount St. Helens. In: P. W. Lipman and D. R. Mullineaux, eds. *The 1980 eruptions of Mount St. Helens, Washington*. Geological Survey Professional Paper 1250, Washington, DC.

National Audubon Society. 1986. *Report of the Advisory Panel on the spotted owl*. Audubon Conservation Report 7, National Audubon Society, New York.

Rodgers, J. D. 1986. Winter distribution, movement, and smolt transformation of juvenile coho salmon in an Oregon coastal stream. M.S. thesis, Oregon State University, Corvallis, Oregon.

Ruggerio, L. F., K.B. Aubry, R. S. Holthausen, J. W. Thomas, B. G. Marcot, and E. C. Meslow. 1988. Ecological dependency: The concept and its implications for research and management. *Transactions of the North American Wildlife and Natural Resources Conference* 53:115–26.

Sedell, J. R., J. E. Richey, and F. J. Swanson. In press. The river continuum concept: A basis for the expected ecosystem behavior of very large rivers. *Canadian Journal of Fisheries and Aquatic Science*.

Smith, D. E. 1985. Principles of silviculture, 8th ed. Wiley, New York.

Strahler, A. N. 1957. Quantitative analysis of watershed geomorphology. *American Geophysical Union Transactions* 38:913–20.

Swanson, F. J., L. E. Benda, S. H. Duncan, et al. 1987. Mass failures and other processes of sediment production in Pacific Northwest landscapes. In *Streamside management, forestry and fisheries interactions, Proceedings of College of Forest Resources, symposium*. University of Washington, Seattle, pp. 9–38.

Swanson, F. J. and C. T. Dyrness 1975. Impact of clearcutting and road construction on soil erosion by landslides in the western Cascade Range, Oregon. *Geology* 3:393–6.

Swanson, F. J., R. L. Graham, and G. E. Grant. 1985. Some effects of slope movements on river channels. In *Proceedings of the 1985 international symposium on erosion, debris flow and disaster prevention*. The Erosion-Control Engineering Society, Tsukuba, Japan, pp. 273–8.

Swanson, F. J., S. V. Gregory, J. R. Sedell, and A. G. Campbell. 1982. Land-water interactions: The riparian zone. In *Analysis of coniferous forest ecosystems in the western United States*. Hutchinson Ross, Stroudsburg, Pennsylvania, pp. 267–91

Swanson, F. J. and G. W. Lienkaemper. 1985. Geologic zoning of slope movements in western Oregon, U.S.A. In *Proceedings of the sixth international conference and field workshop on landslides, 1985*. Japan Landslide Society, Tokyo, pp 41–6.

Swanson, F. J., T. K. Kratz, N. Caine, and R. G. Woodmansee. 1988. Landform effects on ecological processes and features. BioScience 38:92–8.

Teensma, P. D. A. 1987. Fire history and fire regimes of the central western Cascades of Oregon. Ph.D. dissertation, University of Oregon, Corvallis, Oregon.

Vannote, R. L., G. W. Minshall, K. W. Cummins, J. R. Sedell, and C. E. Cushing. 1980. The river continuum concept. *Canadian Journal of Fisheries and Aquatic Science* 37:130–7.

Vest, S. B. 1988. Effects of earthflows on stream channel and valley floor morphology, western Cascade Range, Oregon. M.S. Thesis, Oregon State University, Corvallis.

Ziemer, R. R. 1981. Roots and the stability of forested slopes. In *Proceedings of a symposium on erosion and sediment transport in Pacific Rim steeplands*. Publication 132. *International Association of Hydrological Science,* Washington, DC, pp 343–61.

Part IV Planning and Management of Landscapes

Changing landscapes by design, that is, by human planning and management, is critical because the piecemeal approach focusing on each little parcel, when combined with broad-scale natural processes and population growth, is leading landscape after landscape toward degradation. Sustainable planning and management explicitly focuses on the time scale of human generations, and on the interplay between the components of ecological integrity and human aspiration.

Goals vary widely from environmental impact mitigation, maintenance of landscape diversity, and protection of rare and sensitive ecosystems to optimizing the spatial arrangement of numerous ecological characteristics and human activities. The development of alternative, iterative proposals spatially pinpoints conflicts and tradeoffs, and helps develop a harmony between ecology and economy.

Methods are also diverse. They may include delineation of theory and methods, data bases (analysis, interpretation and synthesis), terrain evaluation, sensitivity of land units to impacts, spatial linkage assessment, and hierarchical ordering. Mapping and geoinformation systems are common aids.

Some planning and management is based on general patterns common to many landscapes, while some is tailored to the uniqueness of a particular case. To enhance the former, four basic landscape types, based only on landscape structure, are identified. Structure, in turn, provides ample predictive ability on a range of processes underlying ecological integrity (e.g., plant production, rare species, erosion control, clean water) and human aspirations (e.g., food, housing, fuel, cultural cohesion). Relatively objective, rather than value-laden, variables are emphasized for sustainable environments.

Understanding the regulatory or control mechanisms in the landscape system is difficult but essential to enlightened planning and management. Mosaic stability, adaptability, and often irregularly cyclic, slowly-changing variables are key descriptors. Integrating hierarchy theory and systems analysis provides a framework for understanding landscape regulation and human decision making. In a three-tier pyramid presented, the bottom level represents a regulatory structure controlling every-day operations (inputs, flows, production, storage, etc.), as portrayed in some ecosystem or geoinformation system studies; this structure is often reasonably clear with some understanding of statistical variability. The middle level, in contrast, involves adjustments and adaptations to perturbations over a longer term; this structure has more connected feedbacks, more aggregated data bases, less predictability, and focuses on normal, but irregularly occurring events and disturbances. The top level involves the "strategic" responses of the system, with a not obvious feedback structure, unpredictability, and characteristics highly subject to external influences; this structure is critical because it focuses on the "strategy of system survival," i.e., the ability of the landscape to respond to the unexpected (e.g., certain human activities).

This hierarchical systems approach requires human judgment and local experts (especially for the top two levels), in addition to complex computer analysis. Regulation or response dynamics is portrayed visually as maps over time, with scenarios, experiments and revisions.

12. Using Landscape Ecology in Planning and Management

Wolfgang Haber

Landscape ecology deals with assemblages of ecosystems occurring in a geographically defined region (a landscape), just as ecosystem ecology deals with assemblages of plant and animal species and nonliving environmental agents occurring at a given site.

For both landscape and ecosystem ecology, the basic and critical spatial unit is the *site*. It is a small section of the earth's surface, determined to a great extent by the local geological situation (lithosphere) within the regional climate. The occurrence, quantities, and usability of site-bound substances of the earth's crust such as calcium, potassium, magnesium, phosphate, silica, and so on, together with the other abiotic and biotic factors (water, relief, plants, and animals) determine the development, productivity, and stability of the ecosystem(s) occupying, and modifiying, the site. Thus, the site can be regarded as the spatial representation of its ecosystem, a concept called *ecotope* (a term that Naveh [1984]) prefers to ecosystem (which is less exactly defined; cf. Ellenberg and Mueller-Dombois, 1974, p. 17]).

Each ecotope can be considered unique in its assemblage of living and nonliving components. However, similar ecotopes have recurring properties, allowing recognition of ecotope types (ecosystem types)—often represented by vegetation units. Such ecotope types are used for characterizing landscapes. The thorough and detailed investigation, ecological interpretation, and floristic classification of the Central European vegetation cover by the Zurich-Montpellier school of phytosociology (cf. Ellenberg, 1980) has provided landscape ecology with an unequalled, reliable basis of ecological information.

Like species diversity in an ecosystem, there is ecotope (-type) diversity in a landscape; common, frequent, or rare species have their analogy in common, frequent, or rare ecotope types. The components of an ecosystem may be grouped into ecosystem compartments such as soil, local climate, primary producers, decomposers, ecological guilds, and so on. Likewise, one may distinguish landscape compartments represented by ecotope assemblages that have certain essential features in common. Examples are the ecotopes of a hill slope, or those of a small valley bottom or floodplain. In applied landscape ecology, such landscape compartments are recognized and mapped as land use units or types. They are called ecochores by geographers (Uhlig, 1973), but this term is not in common use.

Ecotopes and ecochores can be assembled into regional natural units (RNU, in German *Naturräumliche Einheiten,* see Klink, 1973; Finke, 1986; for a U.S. approach, see Young et al., 1983), which constitute a third level in the hierarchy of spatial units. Examples of a RNU are the Munich gravel plain, or the Franconian jurassic mountains, or the Danube floodplain in southern West Germany. They are determined by common geological and geomorphological properties and a typical regional climate. Each RNU has its characteristic set of ecotopes and ecochores, often forming a pattern which is reflected in land use (Kaule et al., 1979).

Landscape Ecology and Planning

In a densely populated country such as the Federal Republic of Germany (246 inhabitants/km^2, or 637 inhabitants/mi^2), the available land has to fulfill many demands and is under intensive use, which in many areas is still being intensified. Conflicts on land use are frequent as are environmental impacts caused by land uses, and by their side- and aftereffects. The environmental policy that has arisen since about 1970 first sought solutions for conflicts and impacts by technological and economic approaches and measures, but disregarding ecological aspects and interrelations. At the end of the 1970s, increasing regional losses of plant and animal species and widespread forest decline aroused a desire for ecological solutions of environmental problems that back up, supplement, and strengthen technological measures. A holistic viewpoint began to gain a foothold; ecological planning, termed landscape planning in the Federal Republic of Germany (Olschowy, 1976; see also Turner, 1983), was called for.

Ecology-oriented environmental policy put forward three principal, interrelated objectives to be achieved by planning and management derived from landscape-ecological research (CEA, 1985; Deutscher Rat für Landespflege, 1988a,b,c):

1. Identification, reduction and/or mitigation of environmental impacts, in accordance with the sensitivity of the affected ecosystems or landscape compartments.

2. Maintenance, and where necessary enhancement, of the ecotope and landscape diversity of the country.
3. Protection of both the rarer and more sensitive ecosystem assemblages.

A holistic approach was chosen by the author and his institute (Haber, 1984; 1986a; 1987; 1988a; Haber and Burkhardt, 1986; 1988; Spandau, 1988), including the following five research steps:

Land Use Types. Identification of the principal regional land use types, conceived as compartments of regional natural units or landscapes (RNU), and arranged according to decreasing naturalness and increasing artificiality (a natural-anthropogenic gradient, Table 1). To give an example, a regional land use type would be *agriculture,* a subtype *arable land* or *pasture land;* these categories all belong to A.4 in Table 1.

Each land use type (and subtype) is then assigned a list of environmental impacts typically generated by it, classified according to (1) material or nonmaterial impacts, and (2) impact-receiving natural resources; air, water, soil or substrate, biota, and landscape (here regarded as a composite resource including physiognomy). This latter classification involves an impact-effect hierarchy: e.g., impacts on the air are more effective or dangerous than impacts on the soil, the effects of which generally remain locally restricted.

Table 1. Main ecosystem or land use types arranged according to decreasing naturalness or increasing artificiality.

A. *Bio-ecosystems*	Dominance of natural components and biological processes.
A.1 Natural ecosystems	Without direct human influence. Capable of self-regulation.
A.2 Near-Natural ecosystems	Influenced by humans but similar to A.1. Little changed after human abandonment. Capable of self-regulation.
A.3 Seminatural ecosystems	Resulting from human use of A.1 and A.2, but not (intentionally) created. Changes significantly after human abandonment. Limited capability of self-regulation. Management required.
A.4 Anthropogenic (biotic) ecosystems	Intentionally created by humans. Fully dependent on human control and management.
B. *Techno-Ecosystems* Examples: Settlements (villages, cities) Traffic systems Industrial areas	Anthropogenic (technical) systems: Dominance of technical structures (artifacts) and processes. Intentionally created by humans for industrial, economic, or cultural activities. Dependent on human control and on the surrounding and interspersed bio-ecosystems.

Spatial Pattern. Assessment and mapping of the spatial distribution pattern and area percentage of the land use types for each RNU. This yields at the same time ecotope diversity.

Most RNUs are characterized by dominance of one of the principal anthropogenic land use types, more or less interspersed with ecotopes representing the other land use types (Figure 1). This results in a characteristic pattern integrating interrelated natural and anthropogenic system components, which may be considered a new, composite "natural" resource, for example, usable for outdoor recreation.

Dominance of one anthropogenic land use type, such as agricultural land use, can be mitigated by diversification of crops promoted by crop rotations, and by maintaining small field sizes, e.g., not exceeding 8 to 10 ha (CEA, 1985; Haber, 1988b).

Impact Sensitivity. Special inventory and mapping of near-natural and seminatural ecotopes or ecotope assemblages assumed to be most sensitive to environmental impacts and worthy of protection.

While there has long been available a land register listing all types of land uses—though not considering ecological aspects at all—an inventory and registration of "natural sites" or "natural areas" (in West Germany called *biotopes*)

Figure 1. Terraced fields and hedgerows near Nideggen and the Eifel Mountains (near Aachen, Federal Republic of Germany). (Photo by G. Olschowy.)

was a new challenge. A 3-year survey of rural Bavaria, excluding settlement areas and large forests, resulted in the discovery of about 16,000 such ecotopes (category A.2 in Table 1), covering about 4.25 percent of the state's area (Haber, 1983). The percentage of these naturelike ecotopes in the different RNUs in Bavaria varied, but never exceeded 15 percent.

Spatial Linkages. Assessment of spatial interrelations among all ecotope types or ecotope assemblages of a RNU, with special emphasis on connectedness and undirectional or mutual dependences.

Urban-industrial ecosystems (category B in Table 1) depend on agri- and silvicultural ecosystems (A.4 in Table 1) for a supply of food, fiber and timber, but also on natural ecosystems (A.1 to A.3) for a supply of pure air, water, and raw materials. They do not have a self-maintaining capacity.

Agro-ecosystems (A.4) not only depend on natural ecosystems for a supply of water, raw materials, firewood, fodder for animals, and natural fertilizers, but also on urban-industrial ecosystems for a supply of technology, artificial fertilizers, pesticides, many services, and sale of agricultural produce. Agro-ecosystems have a limited self-maintaining capacity exemplified by organic agriculture, but even this depends on some external supply and, if it is not purely subsistence agriculture, on market sales.

Natural ecosystems are independent and self-maintaining, but in many cases are being included into land use practices, supplying wild food, grazing for animals, litter, firewood, or timber. In such cases the natural ecosystems will often undergo successional changes to seminatural ecosystems, which in turn may be valued as such—and maintained—because of their species composition, characteristic communities, or general appearance enhancing landscape beauty. In this case, the ecotopes become dependent on maintenance measures.

Connectedness and dependence between ecosystems or ecotypes may be direct or indirect, and may require spatial adjacency (interfaces) or not. In most cases, transport of materials is necessary, which of course involves accessibility of the ecotopes. The greater the dependence and the longer the distance, the bigger are the transport needs, which often reach beyond the boundaries of a RNU. This results in a specific transport and traffic system, a distinctive type of land use characterized by its linear or network structure stretching over large areas.

Impact Structure. Assessment of the impact structure of a RNU or a landscape, using information from working steps 1 to 4, with special emphasis on impact sensitivity and impact ranges.

Starting from the procedure of step 1, where environmental impacts have been assigned to land use types and subtypes, impact-generating ecotopes are derived from the maps of step 2. As a rule, the naturelike ecotopes (A.1 to A.3 of Table 1) do not produce impacts, but suffer from receiving impacts from anthropogenic ecotopes (A.4, B), which is one of the main reasons for their decline.

The anthropogenic ecotopes or land use systems, respectively, are both sources and receivers of impacts. This requires distinguishing within-system from between-system impacts. Soil compaction caused by heavy machinery is a within-system impact, but soil erosion is not only a within-system impact but also a between-system impact because the eroded soil going into a stream is an impact on the stream ecosystem.

Furthermore, an impact hierarchy is derived from the transportability of impacting agents. Airborne impacts rank highest in importance and overall harmfulness, followed by waterborne impacts and soil- or substrate-bound impacts. Of special importance are impacts on landscape composition or physiognomy, e.g., caused by deforestation or erection of buildings. Also of major importance are the separation impacts caused by traffic lanes and networks, which act as barriers to movement for many animal species and promote fragmentation (insularization) of contiguous populations (Mader, 1979).

Of course the information and data mass gathered during the above five working steps are computerized, and transferred into an electronically operated Geographic Information System (G.I.S.). The computers also provide a transfer of information from the regional natural units (RNU) and their subunits to administrative units (municipalities, counties, districts, etc.), so that the administration can readily make use of landscape-ecological data.

A first comprehensive planning and management strategy derived from landscape ecology and based on the working steps is the Differentiated Land Use (DLU) system, formulated by Haber (1972) following suggestions of Odum (1969) in his Compartment Model. The DLU strategy applies to rather densely populated countries with multiple and conflicting land demands causing many environmental impacts. It is formulated as follows (Haber 1979):

- Within a given RNU, the prevailing land use type (originating both from land suitability and tradition) must not become the only type present. At least 10 to 15 percent of the surface has to be reserved for the other land use or ecotope types of Table 1, taking into careful consideration environmental impact generation and sensitivity.
- Within a given RNU, or major part of it, which is under intensive agricultural or urban-industrial use (including traffic, excavation, and waste deposition), an average proportion of at least 10 percent of the surface must be reserved for "natural" ecotopes of types A.1 to A.3 (Table 1), including unmanaged pastures and forests managed by selective cutting.

  This "10 percent exigency rule" may be considered a general planning rule allowing a sufficient (though not optimum) number of wild plant and animal species to coexist with human land use. The reverse rule can be formulated for national parks where visitor installations, camping sites, parking grounds, etc. should not exceed 10 percent of the area. The 10 percent natural area has to be more or less evenly distributed across the RNU, and not concentrated in a remote corner of marginal soils. A "natural ecotope network" should be the objective of this part of the planning strategy.
- The prevailing land use must be diversified within itself. Homogeneity of large tracts of land must be avoided. Field sizes must never exceed 8 to 10 ha in densely populated regions; the same applies to urban or industrial quarters of

identical construction type. Fine-scale diversity of intensive land use provides diversification, and thus mitigation, of environmental impacts.

The DLU strategy has been favorably received by planners and conservationists (Mager, 1985; Kaule, 1986). However, its application in planning is both limited and slow, because it not only involves changes in land use but also affects property rights and land ownership, which in the Federal Republic of Germany are guaranteed by the constitution. It is primarily in rural reallotment schemes and in urban renewal that the DLU strategy can be put to work in a major way. However, it also turns out to be useful in slowing down accelerating changes in land use and land use structure in those regions where the spatial distribution of both ecotopes and landscape compartments resulting from traditional practices, is still in accordance with a differentiated land use (Figures 1 and 2).

Landscape Ecology and Land Management

Planning and nature conservation authorities responsible for what in Germany is called *Landespflege* (land stewardship or landscape husbandry) are in urgent need of ecological advice in order to exert influence, within existing legislation,

Figure 2. Hedgerow landscape matching the concept of differentiated land use (DLU) near Pinneberg, Schleswig-Holstein (near Hamburg). (Photo by G. Olschowy.) See also Figure 1.

on current and impending land use practices. These include environmental impact problems, impact sources, and impact receivers. In this context, particular challenges were posed by the rather drastic and rapid changes of the traditional German landscape brought about by agricultural intensification (CEA, 1985), by management of nature parks and national parks affected by growing visitor pressure (Haber, 1986c), and by the novel forest decline which has hit Central European forests since the late 1970s (Pitelka and Raynal, 1989).

Landscape ecologists develop many site-specific data assembled and organized in geographical information systems, which can generate and even interpret site and landscape maps. However, these systems and maps produce only a relatively static picture or model of a given area.

On the other hand, dynamic feedback models of ecosystems, based on rather well known interrelations of system components, have proven useful in demonstrating and understanding changes, both intended and unintended (e.g., side- and aftereffects or trade-offs), in ecosystems. Yet the applicability of such feedback models to a given site has proven difficult because of different data structures. Highly aggregated data in the feedback models did not match the detailed data structure characterizing the site or ecotope. Both landscape and ecosystem ecologists were challenged to bridge this gap.

Progress in this field of applied landscape ecology was brought about by Swiss and West German research work in ecosystem-based "Man and the Biosphere" (MAB) projects. Messerli (1978) designed an integrated model of a regional man–environment system consisting of three interrelated parts or subsystems:

- A "natural system" of interacting abiotic and biotic resources
- a land use system comprising a gradient of decreasing naturalness and increasing artificiality
- a socioeconomic system consisting of four sections (Figure 3).

This regional system is driven and controlled by external inputs of both ecological and economic origin. Its operation exerts decisive influences on every ecosystem and every site. Landscape ecologists thus became aware of the necessity to translate regional system dynamics into predictive statements for different spatial and temporal scales. This complicated task was tackled by adopting the "hierarchical systems method" proposed by W. D. Grossmann of IIASA (Grossmann, 1983), which relies on the theory of hierarchical multilevel systems (Mesarovic et al., 1971), and on that of orderly adaptive structure (Rappaport, 1977). Its explanation requires a short digression.

Hierarchical Systems Method

In this approach, two principal "functional domains" of ecological systems are distinguished:

1. *The process domain.* Here energy and matter are processed, and biomass is produced, consumed, and decomposed. This is the everyday reality of the

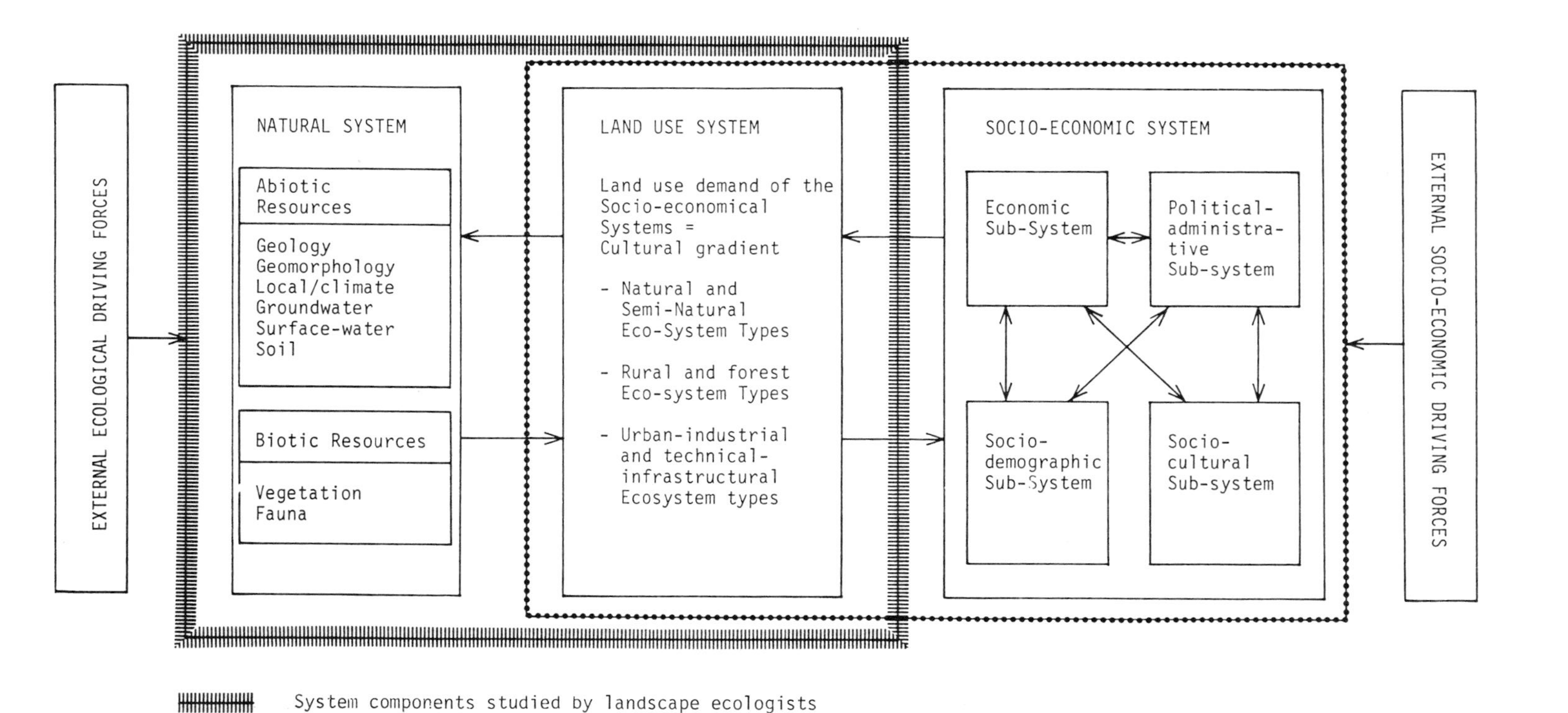

Figure 3. Integrated model of a regional ecologic-economic (or people-environment) system. (From Schaller and Spandau [1987], modified from Messerli [1978].) See Table 1 for further analysis of the Land Use System.

ecosystem, with inputs, outputs and related processes which are readily observable, measurable, and computable.

2. *The regulation and control domain.* Here information regulating inputs, flows, storage, and outputs of energy and matter is produced and processed, resulting in the steady state of the ecosystem. Contrary to the process domain, the regulating items or agents are less readily observed, measured, and/or computed; sometimes they are even intractable.

Most ecosystem research has been devoted to the process domain, with little focus on the regulation and control domain, except for some direct process-related regulation. Hence we see the prevalence of simple experimental or descriptive, and rather selective, approaches. The results of such research, however, have not proven very useful in understanding system regulation and prediction of its possible behavior. But it is just this understanding that is required in environmental planning and management to assess the effects or impacts of human activities on the landscape, on ecosystems, or on natural resources in general.

Of course landscape ecologists and ecosystem investigators have been aware of the existence and importance of the regulation and control domain, a domain which, however, has proven almost inaccessible to experimental research because of its complexity. This was the hour of ecosystem modeling stimulated by the speed of electronic data processing. But even the first modeling approaches failed because of underestimating complexity (Lee, 1973), and we gradually realized that the regulation and control domain is not completely amenable to a holistic scientific approach as was anticipated.

Therefore it is proposed to perceptionally divide the regulation and control domain into three hierarchically ordered subdomains (Figure 4), with regulatory processes within each level and between levels. The tapering of the pyramid, causing a reduction of the subdomains' dimensions from bottom to top, symbolizes decreasing internal regulation and control power, and increasing control by external influences. Accordingly, the three levels or subdomains have to be explored by using different approaches and methods, the results of which will be integrated by a special procedure called *soft coupling,* which will be described below.

Bottom Regulation Domain

The lower level or subdomain is directly and intimately connected to the process domain; thus every particular ecosystem process is rather strictly regulated and controlled. Data are relatively easily and precisely measureable, and thus can be obtained in abundance. Processes and interrelations are also abundant, but are mostly simple and linear. In spite of their multitude, their structure is transparent and their results relatively predictable. Therefore, this subdomain is often included in ecosystem investigations and even incorporated into geographical information systems to produce input-output balance sheets, and to make statistical operations possible. However, a comprehensive understanding of ecosys-

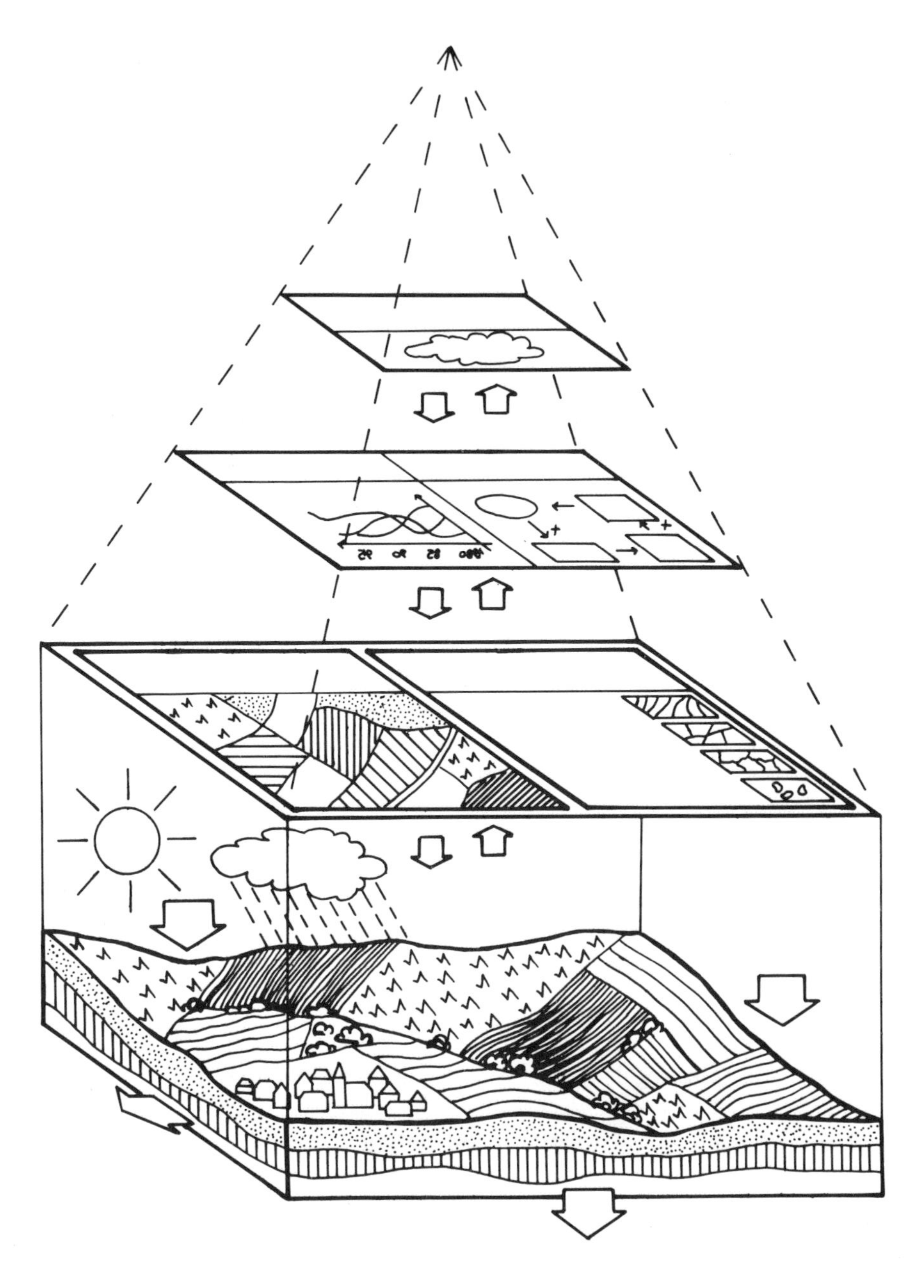

Figure 4. The hierarchical systems method applied to an ecosystem complex as process domain (bottom). The regulation and control domain is pictured as a three-storied pyramid composed of subdomains, which are conceived of as different levels of observation. From bottom to top, external control of the system increases. See text for further explanation. (Modified from Grossmann [1983].)

tem operation will not be obtained by investigation of the bottom level alone; the picture derived from it lacks the adaptive dynamics that are essential for the survival of the ecosystem.

Middle Regulation Domain

Long-term ecosystem operation depends on the regulation or buffering of influences or inputs that occur infrequently and irregularly, e.g., fire, storm, episodic frost, sudden population outbursts, temporary lack of water or availability of a nutrient. Regulation of these can be assigned to the middle level of the regulation and control domain (Figure 4), the investigation of which reveals a number of interconnected feedback loops as main components of the subdomain. The number of these units is far lower than at the bottom level, but their connectivity is much higher. Thus both the structure and the dynamics of the middle level are accessible to easily comprehensible modeling approaches. But the data used are as a rule highly aggregated, e.g., medium temperatures or average energy flows. Such data are inaccurate, even unsuitable, if applied to a given single process. On the other hand, data aggregation or averaging requires a large, extended series of measurements and observations that are expensive.

It appears that operation of the middle subdomain differs markedly from the rather strict regulation principles typical for the bottom subdomain. The middle level is less predictable and is characterized by uncertainties in the data, in spite of a plausible structure. Internal variability of components appears to be essential for regulating the irregular influences of the kind mentioned above. However, indicator organisms and processes, which can be derived from bottom level studies, may serve as keys for investigation of middle-level dynamics.

In short, regulation at the bottom level (lower subdomain) is concerned with the normal everyday operation of the ecosystem(s), whereas at the middle level, normal but irregularly occurring events or disturbances are handled and understood as far as possible.

Top Regulation Domain

The top level of the regulation and control domain is the most difficult to explore (Figure 4). Here external influences from the ecosystem's environment are very strong, unpredictable, and even erratic, endangering the very survival of the system. Its responses may be called *strategic,* but are highly uncertain and hardly predictable. At this level data are lacking and the basic structure is unknown. Thus, feedback approaches based on structure are no longer appropriate, not to mention experimental approaches. The investigator looking into this strategic subdomain can only utilize a small amount of highly aggregated information that is low in precision and of questionable value. Therefore, the dynamics of the top level are presented in scenarios and other more speculative approaches with their intrinsic subjectivity.

However, understanding of the top level is critical because it is here where the very existence of the system in the most general sense, the "strategy of system survival" or, to take Holling's (1978) term, the *viability* of the system is being determined. This includes, e.g., long-term cyclical successions and adaptations which have become known as resilience. According to Holling (1978), strong unpredictable or erratic events keep ecosystems in a kind of steady training that maintains their viability. If all events were foreseeable, the systems would have "learned" to adaptively anticipate them, and therefore would have gradually lost their capability to vigorously respond to the unexpected.

Application of the Hierarchical Systems Method

Now that the methodological background has been explained, the problem of ecologically sound management of ecotopes impacted by human actions or activities can be addressed again. This requires ecotope-specific predictions of consequences, including "normal" environmental dynamics. For this purpose, the results of the investigation of the sites or ecotopes that largely equal the process domain of the ecosystem(s) under consideration have to be integrated, or at least related, with those of the three levels of the regulation and control domain.

Whereas electronic data processing proves most useful in investigating the single levels, the interrelation of the results appears at best only partly accessible to mathematical treatment. Therefore, it should not be left to the computer alone, but should involve human judgment and even intuition. From the author's experience, it is strongly recommended that a local expert who is scientifically familiar with the ecotopes and ecosystems, particularly with their past development, performance, or treatment, execute this integration or interrelation of results.

The reason for this recommendation is the uniqueness of sites or ecotopes mentioned at the beginning. Ecotopes that appear similar, and that are represented in the GIS by the same data set, do not necessarily have the same properties, and may respond differently to certain impacts because of a different history. The response of trees to air pollution may be influenced by past silvicultural measures that differ between otherwise similar sites. The computer would not know this, but the local expert does and thus is able to correctly and reasonably interpret data, define linkages with feedback models and scenarios, and select the appropriate solution for site-specific problems.

This *soft coupling* of scenarios (top level) and feedback models (middle level) with the reality of the ecotope allows sensible predictive proposals for land management and for environmental impact statements. Regulation or response dynamics can be translated into a succession of ecotope maps ("maps over time") that show not only the present state, but also past and future developments, and allowing comparisons with the actual developments. If they are not in accordance, the scenarios or feedback models will have to be revised, so a test of hypotheses will be possible.

The procedure has been successfully worked out in the West German MAB 6 research work (Schaller and Spandau, 1987), and in several environmental impact issues in West Germany, including evaluation of forest dieback process

(Grossmann et al., 1983), assessment of environmental impacts brought about by proposed Olympic Winter Games in the Berchtesgaden regional system (Haber, 1986b), and by increasing touristic use of Berchtesgaden National Park (Spandau, 1988).

The method proved operational without knowledge of all ecotope details. In testing hypotheses of forest dieback, middle-level interdependences and feedbacks involving degrees of air pollution, soil pollution, deposition, chemical degradation of pollutants, forest area, tree density, etc., could be established in a holistic way without incorporating or even knowing all site-specific data. For example, the reduction of the filtering capacity of the forest canopy by air pollutants causing leaf fall will result in increased air pollution; fewer pollutants are filtered out, even if the emissions remain constant. More air pollution will cause more forest dieback, producing a further loss of filtering capacity, followed by more air pollutants remaining in the air. This model of positive feedback loops does not require specific knowledge of single pollutants or their effects. It can even be applied to yet unknown pollutants, and tested for validity when the pollutants have been identified. On the other hand, the method identifies the forest sites which are particularly sensitive to impacts of a certain pollutant.

Conclusion

There is ample evidence that the hierarchical systems method is a useful and valuable tool for applied landscape ecology. The author and his group are convinced of having found a solution for environmental management or impact problems involving different spatial and temporal scales, and considering both site details and general tendencies.

There is one important experience that should be emphasized—in reality the confirmation of an old experience: Even in the computer age, landscape ecologists must never ignore the site or the ecotope, its present qualities, and its history. Familiarity with the ecotopes, i.e., with the field, is a strength and sometimes also a weakness of landscape ecologists. Natural history, so long neglected or even despised as being unscientific, is gaining a new scientific importance.

Acknowledgments

I thank Lutz Spandau, Jorg Schaller, Frank Golley, I. S. Zonneveld, and Richard Forman for many valuable suggestions and for their help in preparing the text.

References

CEA (Council of Environmental Advisers). 1985. *Rat von Sachverständigen für Umweltfragen: Umweltprobleme der landwirtschaft*. Kohlhammer, Stuttgart/Mainz, Federal Republic of Germany.

Deutscher Rat für Landespflege. 1988a. Zur Entwicklung des ländlichen Raumes. *Schriftenreihe des Rates* 54:233–346.

Deutscher Rat für Landespflege. 1988b. Eingriffe in Natur und Landschaft. *Schriftenreihe des Rates* 55:353–445.
Deutscher Rat für Landespflege. 1988c. Zur Umweltverträglichkeitsprüfung. *Schriftenreihe des Rates* 56:453–550.
Ellenberg, H. 1980. Vegetation Mitteleuropas mit den Alpen. 4th ed. Ulmer, Stuttgart, Federal Republic of Germany.
Ellenberg, H. and D. Mueller-Dombois. 1974. Aims and methods of vegetation ecology. Wiley, New York.
Finke, L. 1986. *Landschaftsökologie*. Westermann, Braunschweig.
Grossmann, W. D. 1983. Systems approaches towards complex systems. Fachbeitraege Schweiz. *MAB-Information* (Bern) 19:25–57.
Grossmann, W. D., J. Schaller, and M. Sittard. 1983. "Zeitkarten": Eine neue Methodik zum Test von Hypothesen und Gegenmaßnahmen bei Waldschäden. *Allgemeine Forst-Zeitschrift* 39:834–7.
Haber, W. 1972. Grundzüge einer ökologischen Theorie der Landnutzung. *Innere Kolonisation* 21:294–8.
Haber, W. 1979. Raumordnungs-konzepte aus der Sicht der Ökosystemforschung. *Forschungs- und Sitzungsberichte Akademie f. Raumforschung und Landesplanung* 131:12–24.
Haber, W. 1983. Die Biotopkartierung in Bayern. *Schriftenreihe Deutscher Rat für Landespflege* 41:32–7.
Haber, W. 1984. Über Landschaftspflege. *Landschaft + Stadt* 16:193–9.
Haber, W. 1986a. Über die menschliche Nutzung von Ökosystemen—unter besonderer Berücksichtigung von Agrarökosystemen. *Verh. Gesellsch. für Ökologie* 14:13–24.
Haber, W., ed. 1986b. Mögliche Auswirkungen der geplanten Olympischen Winterspiele 1992 auf das regionale System Berchtesgaden. *MAB-Mitteilungen* (Bonn) 22:1–219.
Haber, W. 1986c. National parks. In A. D. Bradshaw, D. A. Goode, and E. Thorp, eds., *Ecology and design in landscape*. Blackwell, Oxford, United Kingdom, pp. 341–53
Haber, W. 1987. Zur Umsetzung ökologischer Forschungsergebnisse in politisches Handeln. *Verhandlungen der Gesellschaft für Ökolagie* 15:61–9.
Haber, W. 1988a. Über den Umweltzustand der Bundesrepublik Deutschland am Ende der 1980er Jahre. *Korrespondenz Abwasser* 35:1084–9.
Haber, W. 1988b. Anforderungen des Naturschutzes an die Landwirtschaft. *Schriftenreihe Integrierter Pflanzenbau* (Bonn) 4:26–44.
Haber, W. and I. Burkhardt. 1986. Protecting the environment by means of landscape ecology. *Universitas* (English edition) 28:233–8.
Haber, W. and I. Burkhardt. 1988. Landschaftsökologie in der Landespflege in Weihenstephan. *Berichte zur deutschen Landeskunde* 62:155–73.
Holling, C. S., ed. 1978. *Adaptive environmental assessment and management*. Wiley, New York.
Kaule, G. 1986. *Arten- und Biotopschutz*. Ulmer, Stuttgart, Federal Republic of Germany, p. 461.
Kaule, G., J. Schaller, and H. M. Schober. 1979. *Auswertung der Kartierung schutzwürdiger Biotope in Bayern*. Schutzwürdige Biotope in Bayern, Vol. 1, R. Oldenbourg, Munich/Vienna.
Klink, H J. 1973. Die naturräumliche Gliederung als ein Forschungsgegenstand der Landeskunde. In K. Paffen, ed., Das Wesen der Landschaft. Wege der Forschung, Vol. 39, Wissensch. Buchgesellschaft, Darmstadt, Federal Republic of Germany, pp 466–93.
Lee, D. B., Jr. 1973. Requiem for large-scale models. *Journal of the American Institute of Planners* 34:163–78.
Mader, H. U. 1979. Die Isolationswirkung von Verkehrsstrassen auf Tierpopulationen. *Schriftenreihe für Landschaftspflege und Naturschutz* 19:1–26.

Mager, K. D. 1985. Umwelt–Raum–Stadt. Zur Neuorientierung von Umwelt- und Raumordnungspolitik. P. Lang, Frankfurt/Bern/New York, p. 390.

Mesarovic, M., M. Macko, and T. Takahara. 1971. *Theory of hierarchical multilevel systems*. Academic Press, New York.

Messerli, B. and P. 1978. Wirtschaftliche Entwicklung und ökologische Belastbarkeit im berggebiet (MAB Schweiz). *Geographica Helvetica* 33:203–10.

Naveh, Z. 1984. Conceptual and theoretical basis of landscape ecology as a human ecosystem science. In Z. Naveh and A. S. Lieberman, eds., *Landscape ecology: Theory and application*. Springer-Verlag, New York, pp. 26–105.

Odum, E. P. 1969. The strategy of ecosystem development. *Science* 164:262–270.

Olschowy, G. 1976. The development of landscape planning in Germany. *Landscape Planning* 3:391–411.

Pitelka, L. F. and D. J. Raynal. 1989. Forest decline and acidic deposition. *Ecology* 70:2–10.

Rappaport, R. A. 1977. Maladaptation in social systems. In J. Friedman and M. J. Rowlands, eds., The evolution of social systems. Duckworth, London, pp. 49–71.

Schaller, J. and L. Spandau. 1987. MAB-Projekt 6: Der einfluss des menschen auf hochgebirgs-ökosysteme—integrierte auswertungsmethoden und modelle für die ökosystemforschung berchtesgaden. *Verh. Gesellschaft für Ökologie* 15:35–47.

Spandau, L. 1988. Angewandte ökosystemforschung im Nationalpark Berchtesgaden, dargestellt am Beispiel sommertouristischer Trittbelastung auf die Gebirgsvegetation. *Forschungsberichte Nationalpark Berchtesgaden* 16:1–88.

Turner, T. H. D. 1983. Landscape planning: A linguistic and historical analysis of the term's use. *Landscape Planning* 9:179–92.

Uhlig, H. 1973. Landschaftsökologie. In K. Paffen, ed., *Das Wesen der Landschaft*. Wissenschaftliche Buchgesellschaft, Darmstadt, Federal Republic of Germany, pp. 268–85.

Young, G., F. Steiner, K. Struckmeyer, and K. Brooks. 1983. Determining the regional context for landscape planning. *Landscape Planning* 10:269–96.

13. Basic Premises and Methods in Landscape Ecological Planning and Optimization

Milan Ruzicka and Ladislav Miklos

Several key scientific disciplines participate in the research of landscape ecology. Yet within the framework of comprehensive landscape research, the need for new theoretical and methodological approaches in each of these disciplines is ever more urgent.

All planning activity aimed at utilization, protection and development of a territory, its environment and its natural resources, should be based on knowledge of the ecological essence of a landscape. Until now efforts have focused on learning the preconditions for the ecologically optimum uses of a territory. Planning methods differ depending on the complexity and heterogeneity of a landscape as a study object as well as on the goal, such as land use optimization.

The objective of this chapter is to present a landscape ecology planning and optimization approach (LANDEP) that we and our colleagues have used successfully over the past two decades. We begin by delineating the landscape ecology research and approaches, including biotic components, most critical in landscape ecology planning. We then consider the nature of landscape ecology planning, followed by a description of the main steps of LANDEP, a scientific system of research methods directed towards landscape optimization. We conclude with some results of using LANDEP, and also consider future research developments and applications.

Foundations in Landscape Ecology

Overall landscape ecology is a spatial (chorological) expression of site specific (topological) ecological properties across a landscape. At present we distinguish three main theoretical and methodological trends in landscape ecology:

1. *Landscape ecological research on ecosystems and their spatial relationships.* Includes primary and secondary landscape structure (Appendices A and B), processes and relationships among ecosystems, and energy flow in areas and along lines.
2. *Development of new methods.* Includes computer programming, mathematical and systems approaches, remote sensing and digitization analyses, and synthesis and interpretation of landscape ecological data.
3. *Theoretical basis for landscape ecology modelling and planning.* Based on the systematic classification of landscape ecology complexes (units) and regions.

These trends indicate a robust, developing discipline. The content of landscape ecology and planning, in turn, can be divided into five areas:

1. *Theory and methods of comprehensive ecological research and landscape ecology planning (modelling).* Includes the: (a) theory of generalizing ecological information in regard to its spatial expression; (b) methods for interpreting ecological factors in landscape ecology synthesis; (c) principles for a systems approach in comprehensive landscape ecology research and planning; and (d) methods of using landscape ecology knowledge in management, planning and production.
2. *Landscape ecology data bases, data analysis, interpretation, and partial syntheses.* Includes (Appendices A, B and C) the: (a) primary landscape structure, landscape components, elements, factors and processes; (b) secondary landscape structure, landscape components, elements, factors and processes; and (c) socioeconomic landscape structure, phenomena and processes.
3. *Synthesis and evaluation of landscape ecology data.* Includes the: (a) classification of the landscape ecological types (LETs), that is, the ecologically homogeneous spatial units in the landscape; (b) determination of limits for ecological functions for each land use; (c) determination of the suitability of a LET for a particular land use function; and (d) the combinations of LETs into whole landscapes or regions.
4. *Proposed ecologically optimum land use.* Includes the: (a) proposed ecological data for a land use based on ecologically functional limits; (b) alternative land use proposals based on the order of land use suitability for ecological functions; (c) ecologically optimum proposals for land use, that is, a theoretically ideal proposal; and (d) proposal that relates the economic and social (governmental, etc.) requirements for a land use with ecological conditions.

5. *Protection and shaping of the landscape*. Includes the: (a) steps in land use planning resulting from the landscape ecology models of LANDEP (described below) and other simplified methods; (b) spatial analysis, synthesis and classification of environmental problems in the landscape; (c) interests in landscape protection, nature conservation, and natural resource protection; and (d) steps for relating the development interests of society with landscape ecological conditions.

Biotic Components in Landscape Ecology Research

The biotic components are the basic parts of a landscape that reflect, and to a certain degree are indexes of, the ecological properties and processes taking place in the landscape. Applied and simplified methods of botanical and zoological research are needed to permit rapid understanding of the spatial expression of vegetation and animal communities. This would also aid our understanding of these organisms as functioning components of ecosystems and landscapes.

The significance of the question of spatial position and roles of biotic components in a landscape system emanates from the worldwide, but especially European, interest in ecologically optimizing the use of nature and its resources. Two basic natural vegetation formations, forests and perennial grasslands, are of special interest. For example, within these formations the roles of swamps and scattered landscape greenery or natural vegetation patches is being studied. Cultural vegetation, such as agroecosystems, is also of special interest in the study of landscape structure and function. Indeed, cultural vegetation is a key integrated component in the ecological evaluation and use of landscape components.

Ecosystem science has focused on the production functions of the vegetation formations under the aegis of varied projects and programs (e.g., UNESCO-MAB, UNEP, and IUCN). Increasingly scientific studies are considering the quality, spatial distribution, and share of natural elements in a vegetation formation or specific territory. These aspects are especially critical to understanding landscape ecology stability, landscape-creative dimensions of biotic elements and components, carrying capacity, and ecological corridors and barriers in the landscape. However, these aspects at present are poorly understood.

General Landscape Ecology Planning Considerations

Driven by society's needs for development and by increasing problems associated with society-nature interactions, landscape ecology planning has become one of the most significant directions of landscape ecology research (Figure 1).

We now see that national economies, the world economy, and civilization itself are not immune from global environmental change. In addressing this problem, scientific and technical cooperation in the socialist countries (of the

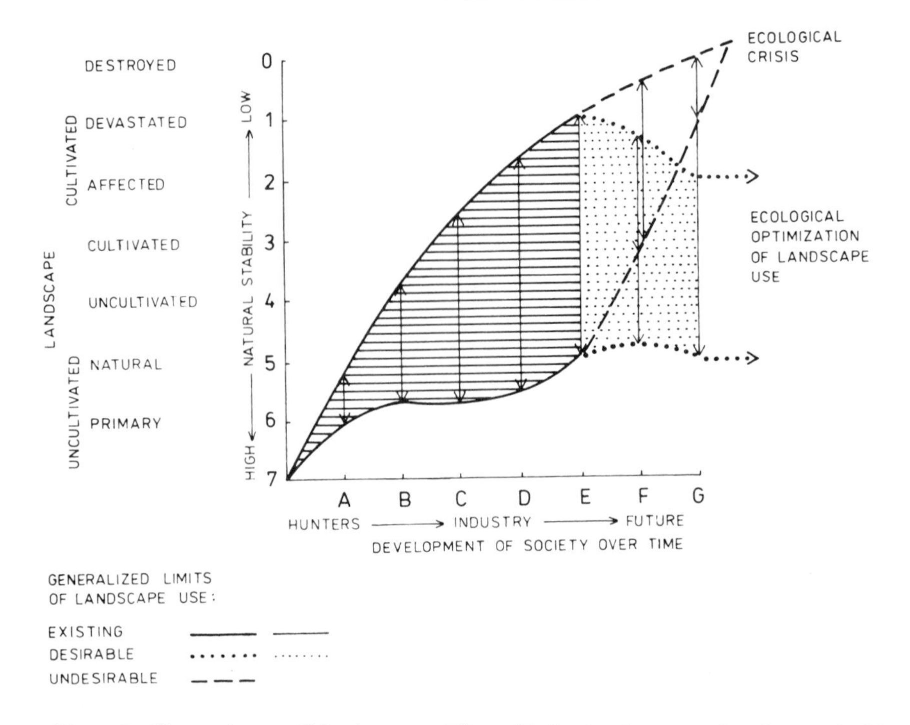

Figure 1. Change in overall landscape stability with the development of society, and with landscape-ecological planning. Vertical arrows indicate the range of widespread land uses present. Two future scenarios beginning at point E (the present) are indicated on the right. Without landscape-ecological planning the expected range of land uses narrows and natural stability decreases, ending with an ecological crisis. With landscape-ecological planning the expected results differ markedly.

CMEA) is coordinated through the Council for Protection and Improvement of the Environment. Among the 14 activities of the Council is "Protection of Ecosystems and the Landscape" dealing with landscape ecology problems. Also landscape ecology planning is included here within the framework of a special theme.

The following five landscape ecology problems deserve particular attention:

1 rational utilization of natural resources;
2 creation of ecologically optimum landscape structure and ecological data for territorial planning (landscape management);
3 creation of favorable living conditions for the inhabitants of towns and settlements, and harmonization of the urbanization process with ecological conditions;
4 transformation of nature consistent with the development needs of different branches of the national economy that affect ecological conditions;
5 nature conservation, including maintaining the natural gene pool of living organisms.

International research and cooperation in this area are developing within the framework of international, nongovernmental and governmental programs. For example, programs of the United Nations (UNEP and UNESCO-MAB) and the International Union for the Conservation of Nature (IUCN) involve the environment, nature and culture. The International Association for Landscape Ecology (IALE), established in 1982 in the CSSR (Czechoslovakia) at the 6th International Symposium on the Problems of Ecological Research of Landscapes, specifically deals with problems of ecological planning.

The LANDEP System for Landscape Optimization

LANDEP is a complex system of applied scientific activities, which includes biological, ecological, geographical, agricultural, silvicultural and other research methods. These methods are united by an integrated modelling process directed towards landscape optimization (Ruzicka and Miklos, 1981).

The LANDEP system sequentially includes a comprehensive landscape ecological analysis, a synthesis component, a landscape ecological evaluation of the territory, and a proposal for optimum land uses. Individual aspects of such a method appear in all planning approaches, but with the possible exception of the McHarg (1969) approach (cf. Junea 1974), LANDEP appears to be the only comprehensive landscape ecology planning system yet available.

The theory and methods of LANDEP have been worked out and tested in Czechoslovakia over the past two decades. The results obtained so far (Ruzicka 1970; 1973a; 1973b; 1976; 1979; 1982; 1985; Ruzicka et al., 1988) permit an unusually broad range of ecological perspectives in a process that leads to landscape management, regional planning, and projecting. The results also open possibilities for a new branch of basic research. In principle, LANDEP permits planning the optimal use of ecological properties of the landscape, as well as creating conditions for a harmony between humanity and the landscape. In territorial planning practice, LANDEP has a simplified form for the ecological proposal of territory (Figure 2).

The LANDEP concept stresses the need for evaluation of the landscape as a territory in which human and societal activities develop on the basis of natural phenomena and processes. LANDEP contains two basic parts (Ruzicka and Miklos, 1979; 1981; 1982a; 1982b) (Figure 3):

1. *Landscape ecology data*. The core of this part is as follows: inventories and assessment of the abiotic and biotic components, the contemporary landscape structure, ecological phenomena and processes, and effects and consequences of human activities upon the landscape (Figure 4). Analysis, interpretation and synthesis (typification and regionalization) complete this part.
2. *Ecological optimization of landscape use*. This part relies on the landscape ecology data, particularly for the ecologically homogeneous spatial units. Thus, the spatial units are compared with the requirements and development

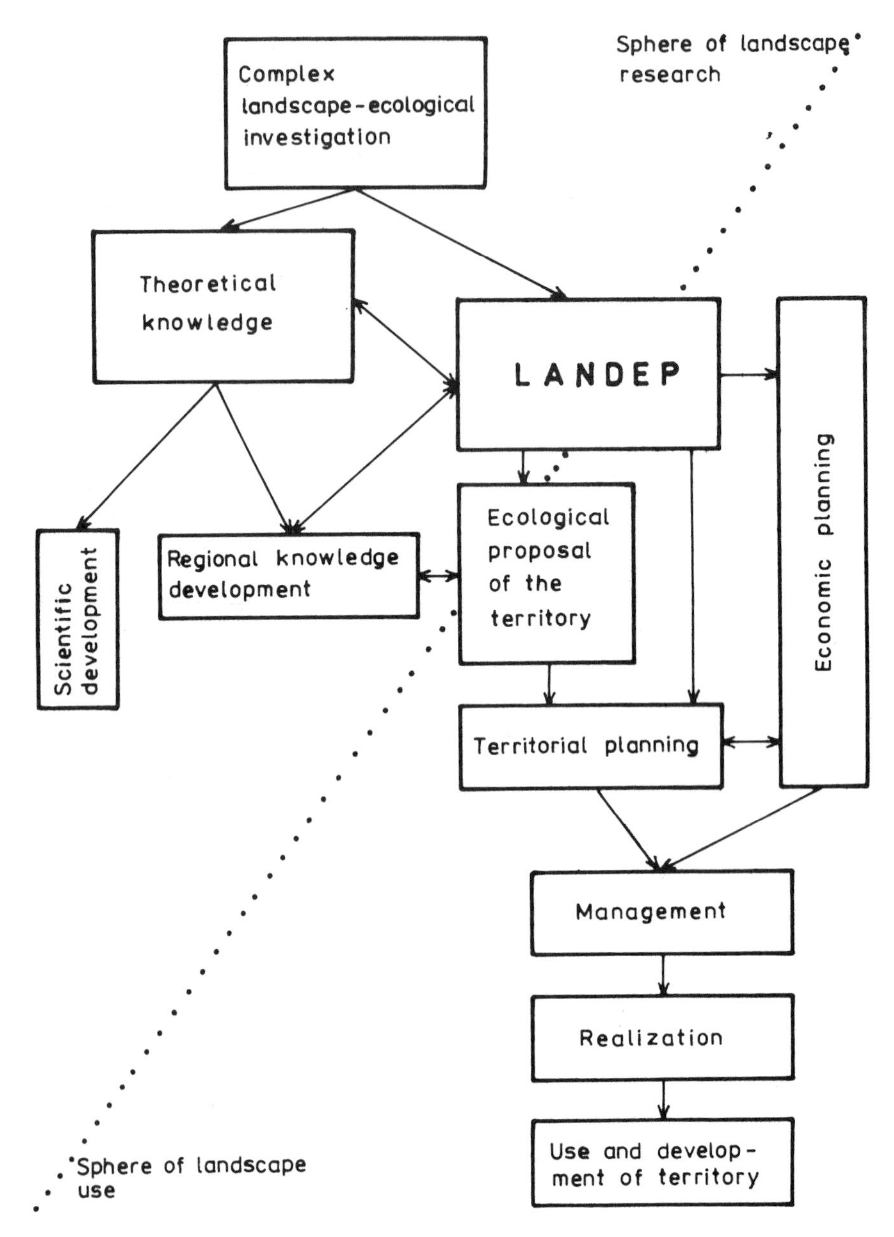

Figure 2. The position of LANDEP, a comprehensive set of methods for landscape planning and optimization, in science and practice.

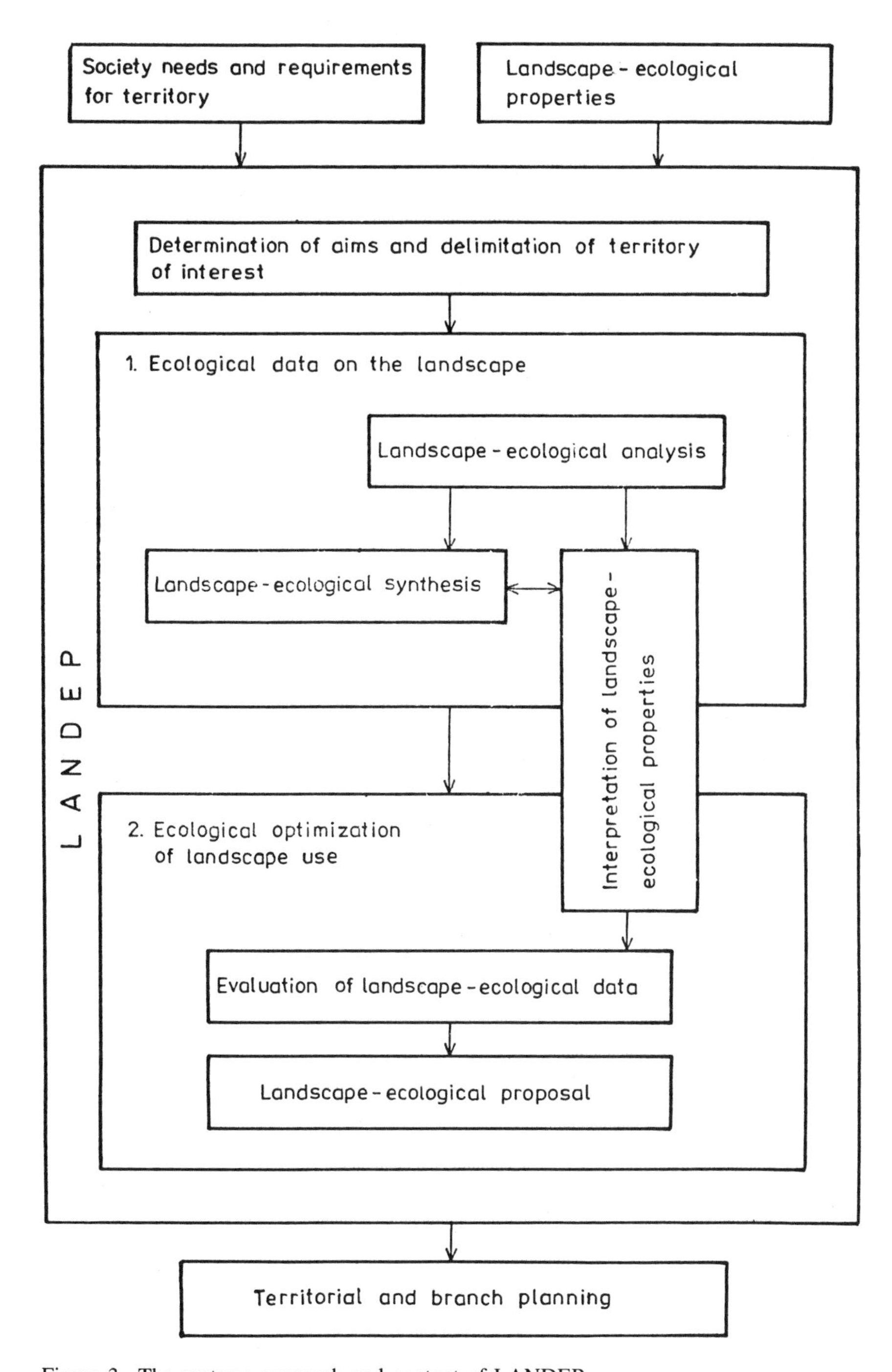

Figure 3. The systems approach and content of LANDEP.

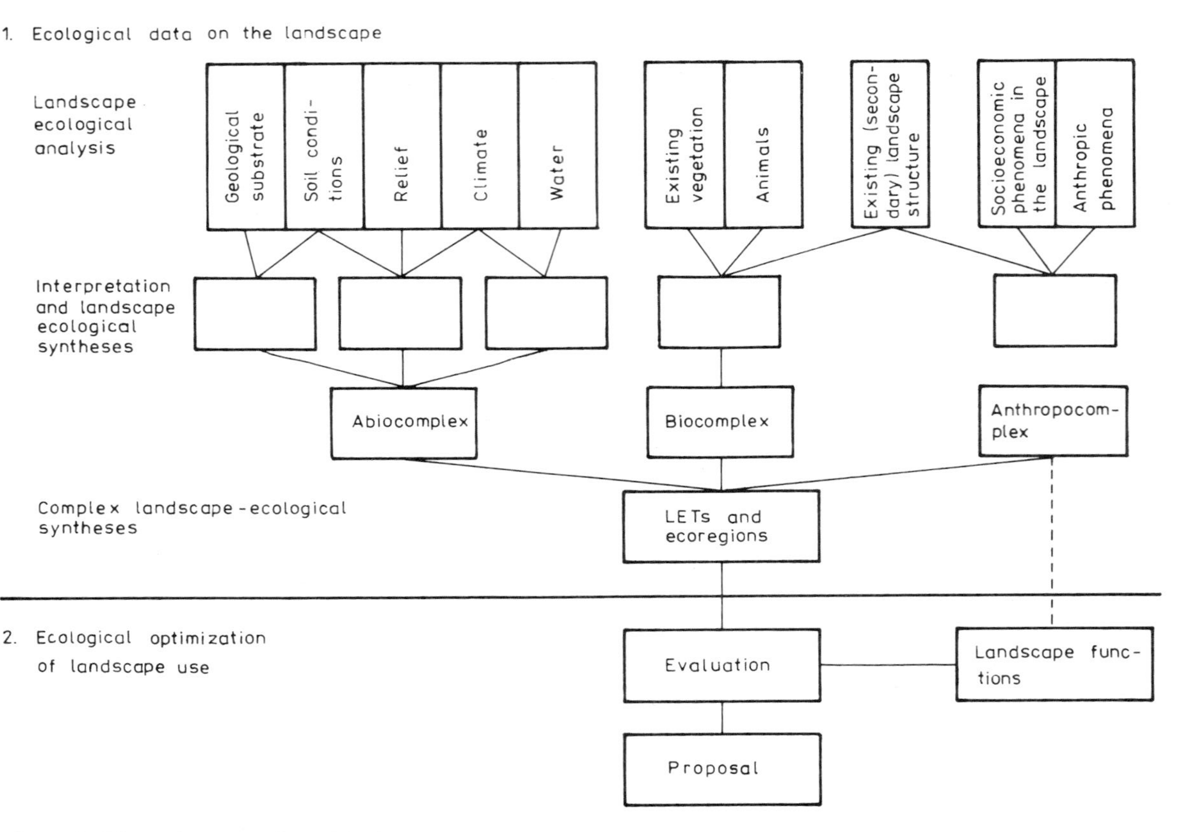

Figure 4. The main steps of LANDEP.

needs of society for a particular territory. Following evaluation of the degree of appropriateness of each spatial unit for a particular human activity or land use, a proposal is made on the most suitable location of the activity in the landscape based on landscape ecology criteria (Figure 4).

Optimization in LANDEP may result in proposing the most suitable location of a planned land use (landscape utilization type) in the landscape based on landscape ecology criteria. However, because landscape optimization should preferably not retard economic development too much, optimization here means choosing the site causing the least evil. That is, a location is determined where a given human activity will be in least conflict with natural conditions. This goal is attained by the complex processes of LANDEP (Ruzicka and Miklos, 1981; 1982a; 1982b).

The LANDEP method is based on confronting land use requirements with the ecological ability of a given territory to support the projected use. In this process three essential questions are answered (Ruzicka and Miklos, 1982a):

1. How is a given set of ecological properties of the landscape adapted to the functional demands of land uses, that is, to what extent can some activity be developed in a given area?
2. What effects have locating a particular activity had on the ecological characteristics of a given area in the past?
3. What is the present state of natural and human conditioned processes and properties of the landscape (e.g., stability, balance, and resistance)?

By the gradual reevaluation and combination of these partial evaluations, we may find: (a) what activity has the best functional potential to be performed in a given area; (b) what activity is most suitable from a combined ecological and economic perspective; and (c) what danger a particular activity poses to the landscape, as well as the most suitable measures available to mediate the threat.

The optimization process in LANDEP is wholly systematized and partly automated. It simulates the management process occurring in the human mind, aimed at the best planning of space. We have tried to keep this process objective on two important points: (a) consistency during the whole decision-making process according to predetermined accurate principles, and (b) consistency over the whole area of concern.

We must, therefore, divide this process into analytical phases that logically arrange the results and state the underlying systematic procedure. We have also tried to automate this system to the greatest extent with computer techniques (Figure 5).

The systematic LANDEP process permits many ways to simplify, without loss of the main logical principles of the methods. During simplification, the ecological analysis is narrowed to the most important ecological properties of the landscape. It is then checked mostly by reevaluation of already worked up data, supplemented by informational investigation in the field.

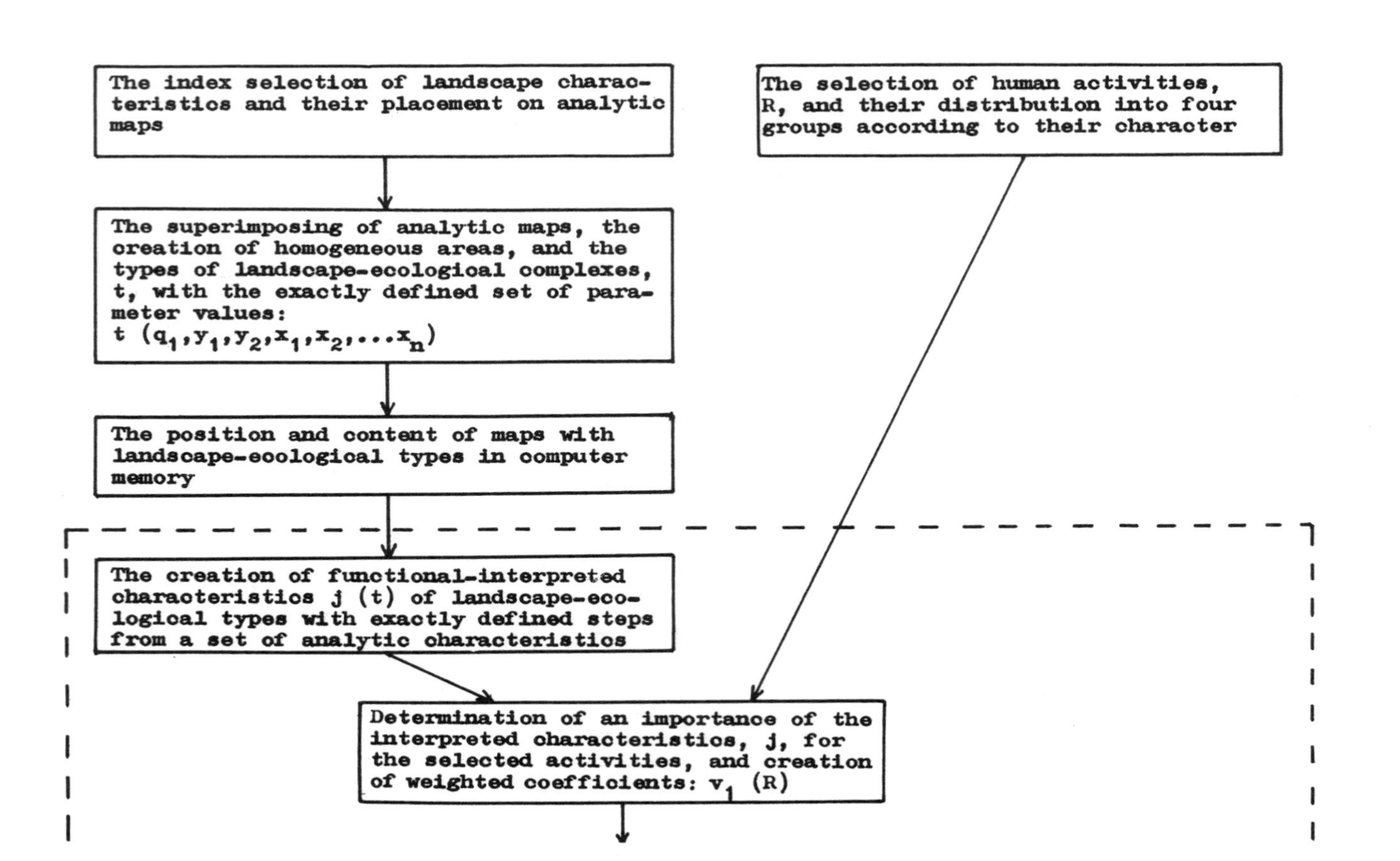

The index selection of landscape characteristics and their placement on analytic maps
The selection of human activities, R, and their distribution into four groups according to their character
The superimposing of analytic maps, the creation of homogeneous areas, and the types of landscape-ecological complexes, t, with the exactly defined set of parameter values:
t $(q_1, y_1, y_2, x_1, x_2, \ldots x_n)$
The position and content of maps with landscape-ecological types in computer memory
The creation of functional-interpreted characteristics j (t) of landscape-ecological types with exactly defined steps from a set of analytic characteristics
Determination of an importance of the interpreted characteristics, j, for the selected activities, and creation of weighted coefficients: v_1 (R)

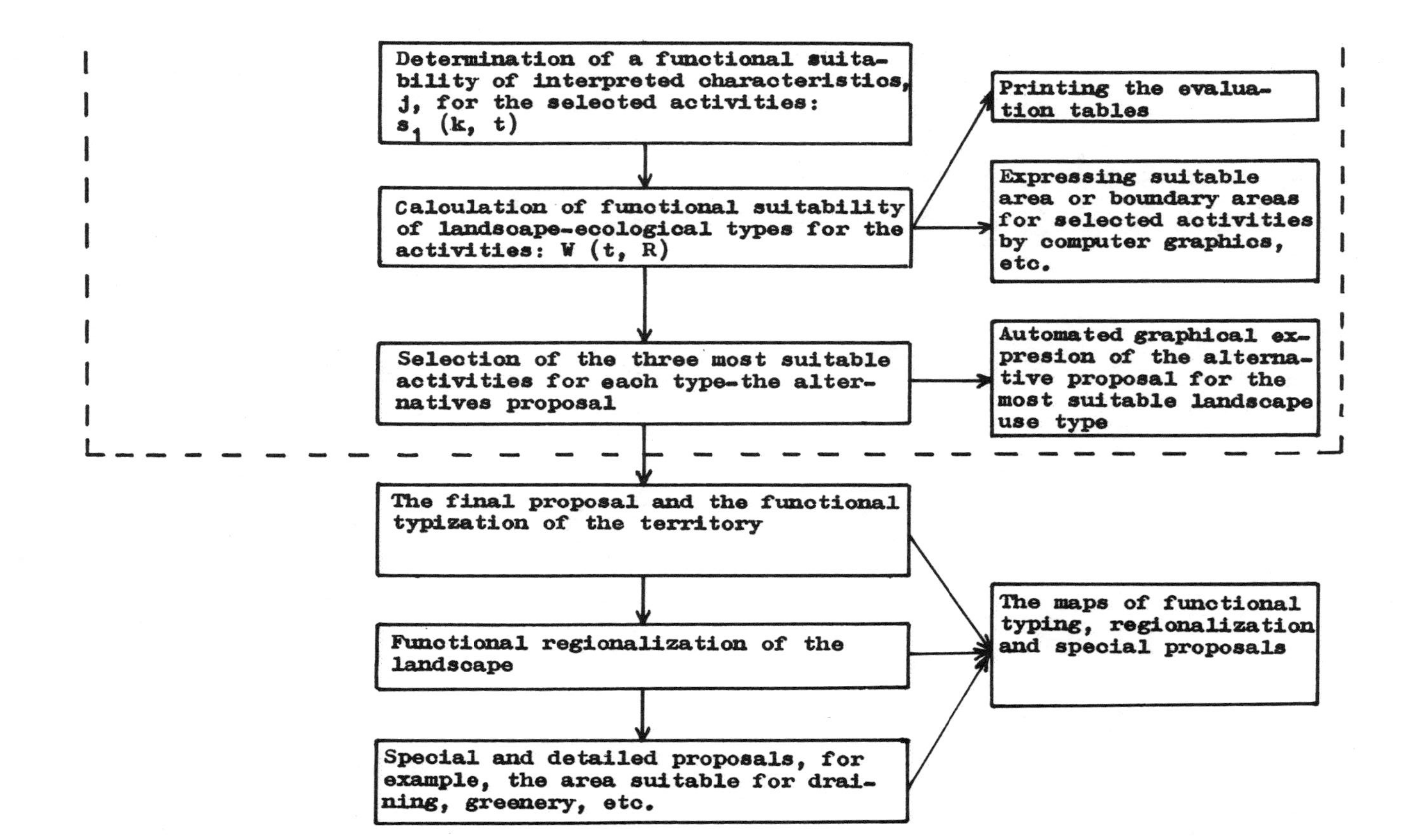

Figure 5. The systematization and automation scheme of the LANDEP method. Steps within the dashed line are accomplished by computer.

The Main Steps of LANDEP

Landscape Ecological and Socioeconomic Data on the Territory

Analytical Portion

The most frequently used analyses are (Figure 4 and Appendices A, B and C):

1. *Delimitation of the territory of interest,* where both administrative boundaries and natural boundaries are recognized (Figure 3).
2. *Geological basis,* from the standpoint of its resistance, carrying capacity and tectonics (from engineering, geology and hydrology).
3. *Soil-forming substrate, soils and ground water,* on the basis of Quaternary sediments and of soil-ecological properties (from soil and stand reconnaissance).
4. *Morphometry of relief,* mainly the inclination, orientation, curvature, and forms (from evaluation of topographic maps).
5. *Hydrography,* the size and shape of a partial catchment area.
6. *Climatic conditions,* based on climatic regions and wind conditions (from atlases), and interpretation of topographic relief forms for insolation and shading.
7. *Potential and actual vegetation,* evaluated on the basis of physiognomic-ecological formations.
8. *Animal biotopes,* analysis and interpretation.
9. *Contemporary landscape structures,* resulting from human economic activity and natural factors (Fig. 6 and Appendix B). Mapping is based on classification with six groups of elements (forests and scattered natural greenery, grasslands, arable lands, denuded substrate, water areas and flows, and built structures and settlements), and its application to individually determined spatial units.
10. *Socioeconomic phenomena,* connected with: (a) industrialization, urbanization and traffic; (b) agriculture (related to soil use intensity, chemical fertilization, amelioration, etc.); (c) recreation and housing; and (d) natural resources and nature protection (Appendix C).

Synthetic Portion

The aim of synthesis (Figures 3 and 4) is to create ecologically homogeneous areal units that in turn are of vital importance in the LANDEP process. These units are of different order and content. They form landscape-ecological types (LETs) differing in landscape properties both vertically and horizontally. LETs have clearly defined content (represented by codes), and can be arranged into a logical and tabular form. The areas of homogeneous LETs form the elements of synthetic maps (Figure 7).

A second aim of synthesis is to provide a clear statement or representation of spatial structure, using analytical indexes for LETs and for regions.

Synthetic maps are the deliberate databases for the subsequent LANDEP process. The contents of LETs put into the computer memory are: (a) all

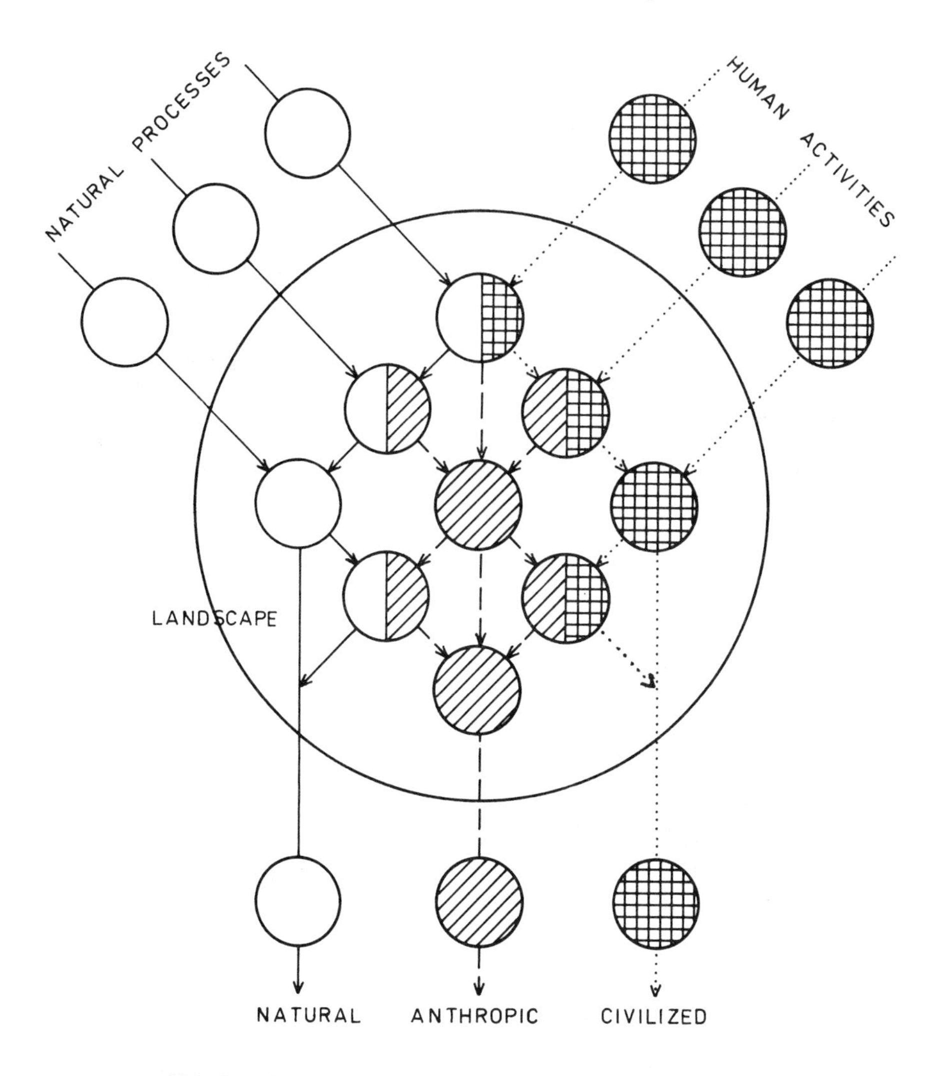

Figure 6. Components of landscape structure resulting from interactions between human economic activity and natural forces.

necessary values for the individual landscape properties or indices; and (b) the areal coordinates for the boundaries of LETs. This is a key synthesis step for LANDEP.

The "overlay" map is the most used methodological portion of the spatial synthesis. In the subsequent steps of LANDEP, the LETs serve as a basis for the optimization process, and the landscape properties, represented as a code set for each LET, enter into the decision-making process.

Clusters of LETs are then recognized in a regionalization process that spatial-

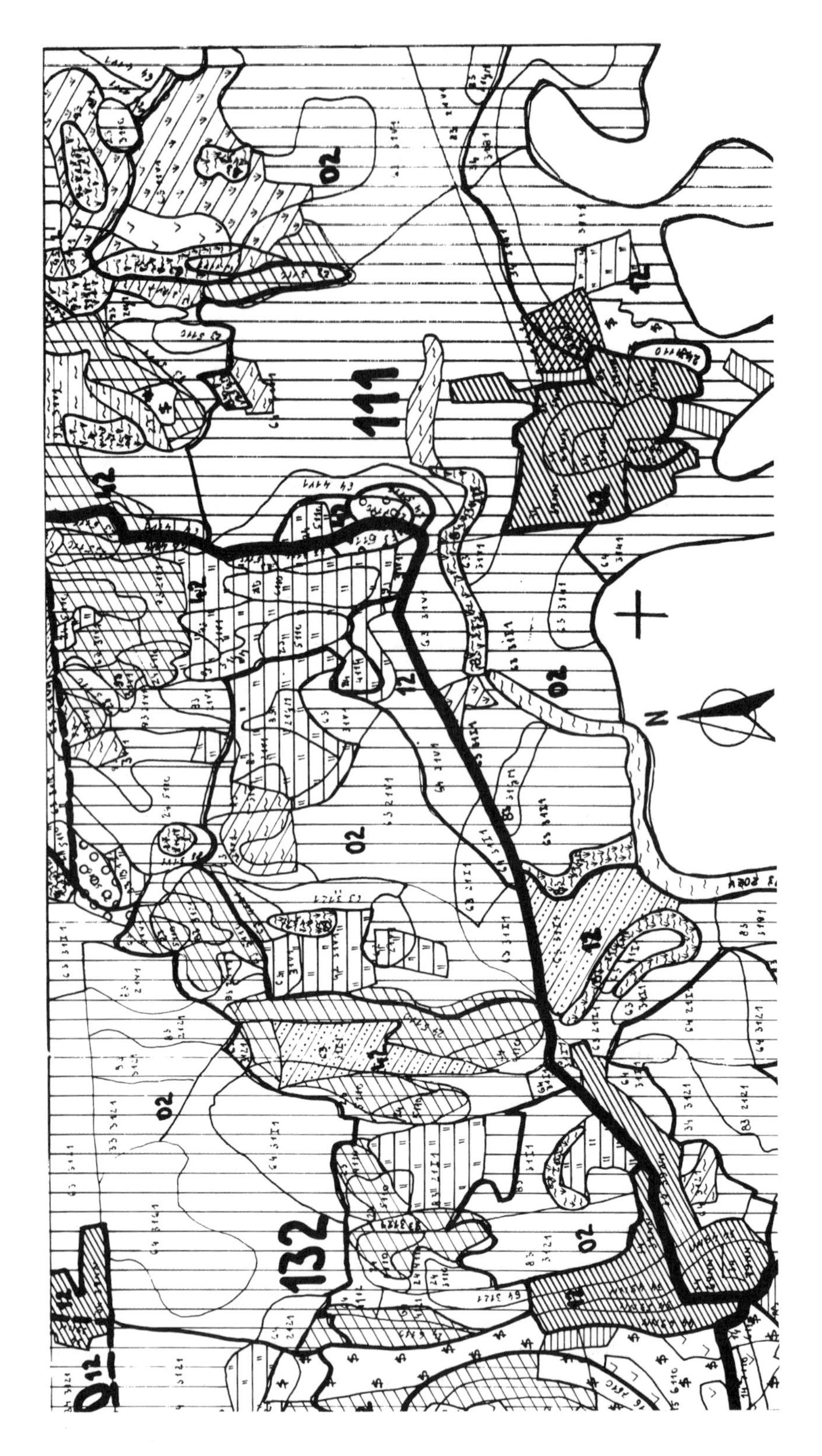

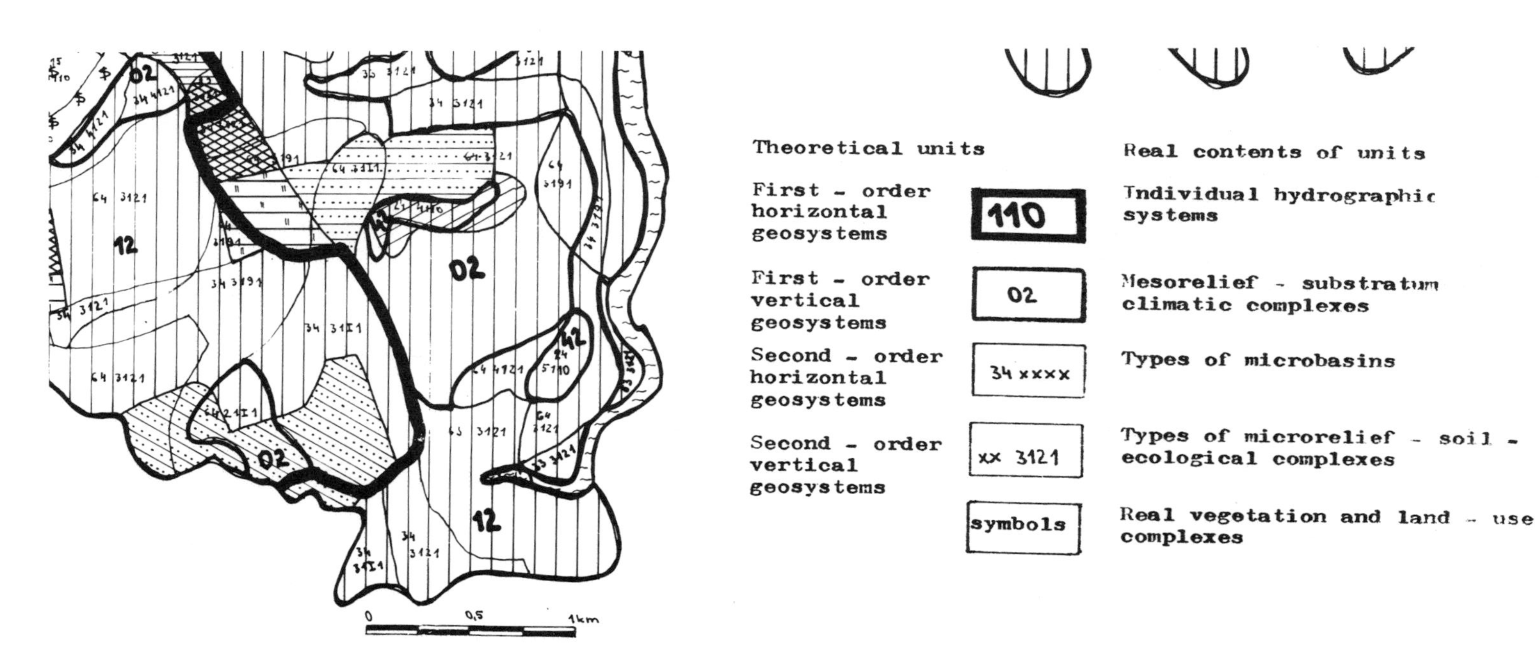

Figure 7. Example of a synthetic map with integrated landscape-ecological types (LETs) in the East Slovakian Lowland.

ly differentiates larger territorial units of the landscape, that is, regions. These regions serve for an overall description of the territory, as well as the basis for separating functional units.

Interpretation of Landscape-Ecological Data

The primary purpose of "interpretation" (Figure 3) is to provide indications of ecological uniquenesses in the landscape that are often not otherwise revealed in the field. Interpretation, as defined here, is a process of transforming basic landscape ecological indices into a form amenable to the process of optimization.

Interpretation of the basic landscape ecology properties makes it possible to obtain from analytic methods a range of functional characteristics, j, that we call interpreted properties. These include availability, arability, waterlogging, soil trophism (physical), carrying capacity of the substrate, insolation from relief, material transport dynamics, anthropogenic vegetation change, degree of "synantropization" of the landscape, suitability for housing, etc. Detailed individual methods are used for determining each of these. The essence of these methods is to state what the combination of analytic indexes (based on LETs) is, and how it influences the interpreted properties, j, of landscapes.

Landscape Ecology Optimization

Optimization is the core of LANDEP methodology. Here, the landscape indexes are compared with selected human activities according to the following steps.

Evaluation

The objective of the evaluation process (Figure 3) is to state (a) the suitability of geosystems (mainly LETs) for human activities, and (b) the limits of landscape properties for a human activity. Again the process is systematized and automated, with two primary inputs: (1) the interpreted functional properties, j, which depend on the original analytic data set (of LETs); and (2) the selected human activities, R. Four human activity groups are of key importance in the decision-making process: (1) "ecological" (forests, natural greenery, reservoirs, etc.); (2) agricultural (arable land, pasture, etc.); (3) permanent culture and recreation (orchard, garden, vinyard, cotagering, etc.); and (4) investment (varied building activities).

The distribution of human activities into classes reflects similarity in the way a function is accomplished. It also reflects similarity in the requirements for a function, "physical" stability in the landscape (that is, if the activities can be changed frequently or not), and importance for the protection and creation of the natural landscape (hence, their "ecological importance").

The three step process of evaluation involves: (1) the determination of weighting coefficients, (2) functional suitability of interpreted properties, and (3) total suitability of LETs for human activities.

Weighting Coefficients

Naturally the different interpreted properties do not influence our decision about the suitability for a given activity in the same way. On the contrary, a property has different importance for different activities. For example, trophism has a different importance for forage cultivation than for building houses. This circumstance is solved by forming weighted coefficients of the interpreted properties for each human activity evaluated, V_j^R.

Functional Suitability of Interpreted Properties

In this step the *functional suitability* of each class of interpreted property, j, for each activity, R, is stated (Figure 5) as $-S_j^{R,LET}$. The limit values of the interpreted properties for each R ($S_j^{R,LET}$ = limit), and the limit sizes in particular, are determined. These values are initially excluded from the proposal for a given activity, based on predetermined principles that limit realization of the activity. Then the zero values ($S_j^{R,LET} = 0$) are determined (the values making a given activity impossible).

Naturally, we are unable to express the individual functions in precise mathematical terms. Therefore, the values $S_j^{R,LET}$ are determined in the order of suitability of each value, j, for an activity, R, on a relative scale.

A team of "experts" decides upon the functional suitability, $S_j^{R,LET}$, for a selected activity. The basic groups of evaluation criteria are: (a) the technical feasibility for realizing a given activity, that is, the practical suitability of a given property for R; (b) the prognosis of a localized influence of the given activity on a locality (we should emphasize the roles and inclusion of landscape-ecology prognosis in this step of the LANDEP process); (c) the influence of the localized activity on specific landscape ecology properties, such as the biological balance or ecological stability; and (d) the expectation that a given localized activity can be realized, based on economic and geographic criteria such as distance and position.

Total Suitability

We assume that the degree of suitability of a particular LET for a particular human activity, is determined by the cumulative effect of its properties, R, being evaluated by their theoretically interpreted properties, j. Thus the total suitability, W (R,LET), is the sum of these partial suitabilities, expressed as a percent of the maximum possible suitability of a given LET and a given activity:

$$W(R, LET) = \sum_{j=1}^{n} S_j^{R,LET} \cdot V_j^R$$

Proposals

Proposals (Figure 3) are aimed at harmonizing the ecological properties of the landscape with its development use for humans and society. The proposal process is divided into four steps.

Initial Selection of Alternative Proposals

The foregoing procedure results in a decision concerning the suitability of a given LET for a given human activity. In the following procedure, in contrast, a decision about the best activity for a given type is made. This phase of the management process results in three or more alternative proposals for the most suitable activity for each LET (Figure 5). The following characteristics are considered in selecting the activities in this alternative proposal:

1. The suitability of the existing land use, that is, how appropriate is the present activity on a given site.
2. The character of the present land use and the distribution (with or without priority) of the land uses into classes and into human activities.
3. The suitability of other human activities for a given type, that is, other evaluation values expressed in percent.
4. The possibility, need and intention of looking for various alternatives.
5. The physical stability of the existing land use as a limiting factor in the selection of alternatives. This determines whether a change from the present use is possible, and if so, how it is technically possible.

The algorithms in this overall alternative selection process determine whether to maintain or change the land use, and in the latter case, what land use should replace the existing one. In essence, for each LET, a selection is made based on simultaneous calculations that determine the human activity with the highest suitability, and the activity group with the highest priority (Miklos et al. 1986). In this process, the present land use is maintained if it falls into the high priority activity group, performs a stabilizing function, or can be changed only with difficulty.

Final Proposal Selection

The next step (Figure 5) of the decision-making process selects from among the alternative proposals. The ecologically most suitable activities are first graphically portrayed on maps (e.g., by color). In this manner, the most suitable human activity or *function* is illustrated for every LET. This *functional typing of territory,* whereby the original LETs are spatially replaced by new functional types, is the basic result of ecological optimization.

The management principles underlying the final selection of the most suitable human activity result primarily from space conditions. Most important are the size of a homogeneous area, properties of adjacent areas, similarity of proposals for adjacent areas, and spatial configuration of surrounding areas.

In larger territories certain groupings of the same or similar functional types can be recognized and correlated with overall natural conditions. This phase of the optimization process is effectively a regionalization of the territory, whereby *functional regions*, that is, characteristic groups of functional types, are delineated (Figure 5).

Functional typing and regionalization serve as the basis for management, decision making and planning in territory development.

Protection and Management of the Environment

This step represents a further stage of the proposal process, in which the proposed ecologically optimal landscape use is compared or confronted with any existing or valid documents of territorial planning. This emanates from the fact that all human activities required for social and economic development must be located somewhere in this or another landscape.

Graphic Expression of the Management Process

The basic results of the evaluation and initial selection of alternative proposals are presented in tabular form. The results are also expressed on hand-made or computer-drawn maps (Figure 8).

Besides presenting basic results the automatization process allows a graphic expression of the (a) boundaries appropriate for selected human activities, (b) areas of minimum values for selected activities, (c) areas of critical values based on individual ecological indexes, and (d) optimum locations for selected human activities, etc.

Results and Use of LANDEP

LANDEP as part of a landscape ecology research program can only be developed on a team basis, with composition of the team reflecting the LANDEP content. Each member of the team must also be skillful in obtaining all necessary published and unpublished data, and in elaborating his or her topic inventively for its use in the LANDEP program. The amount and quality of landscape data used in LANDEP must be modified according to their significance and usage for final theoretical and practical objectives.

To verify and modify the LANDEP methods in practice, approximately 100 projects of small to large territories (scales from 1:500 through 1:500,000) have been done in collaboration with other institutes. Close collaboration with regional planners and practitioners made it possible to develop simplified methods of LANDEP that cooperating institutions could use. To work out and use these simplified methods, additional ecological teams were included in the collaborating institutes. These teams were involved in regional planning and in planning for development of agricultural production.

In Czechoslovakia, LANDEP is in the process of being incorporated into

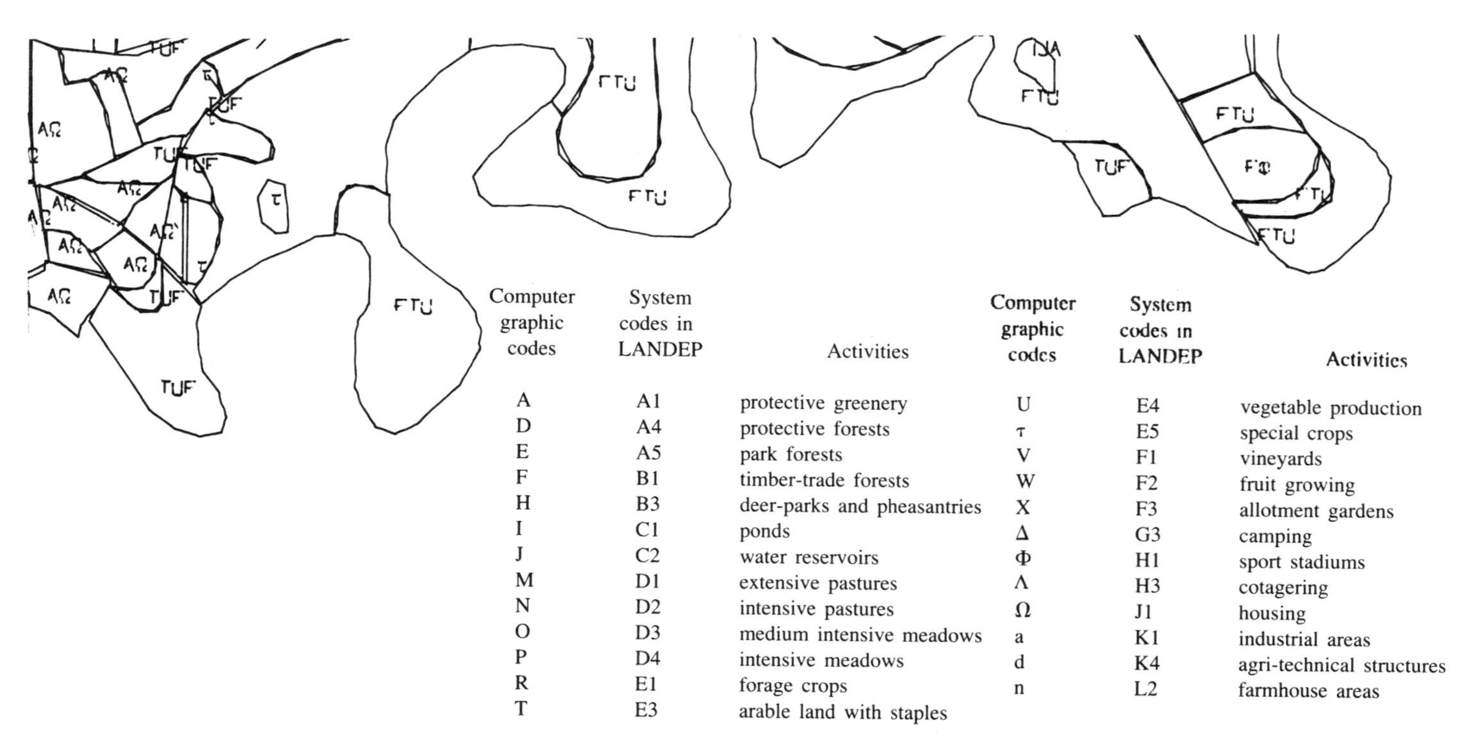

Computer graphic codes	System codes in LANDEP	Activities	Computer graphic codes	System codes in LANDEP	Activities
A	A1	protective greenery	U	E4	vegetable production
D	A4	protective forests	τ	E5	special crops
E	A5	park forests	V	F1	vineyards
F	B1	timber-trade forests	W	F2	fruit growing
H	B3	deer-parks and pheasantries	X	F3	allotment gardens
I	C1	ponds	Δ	G3	camping
J	C2	water reservoirs	Φ	H1	sport stadiums
M	D1	extensive pastures	Λ	H3	cotagering
N	D2	intensive pastures	Ω	J1	housing
O	D3	medium intensive meadows	a	K1	industrial areas
P	D4	intensive meadows	d	K4	agri-technical structures
R	E1	forage crops	n	L2	farmhouse areas
T	E3	arable land with staples			

Figure 8. Initial proposal map showing the highest priority optimal land uses for each homogeneous unit of the landscape (1 : 25,000 scale). Produced by computer graphics. Note that this area overlaps the easternmost portion of Figure 7.

methods of regional planning. Landscape ecology data are becoming a part of routine projects where there is little time available for long-term planning.

LANDEP results in a specific proposal, including a map, for the optimal localization of human activities in the landscape. Several important aspects of LANDEP recommend use of the method in practice:

1. The choice, extent and mode of elaborating often detailed data on natural conditions is not left to good will, professional knowledge or the sense of an urban planner, but rather is incorporated into an integrated spatially-organized proposal. The proposal is developed from ecological data and contains clear cut criteria and procedures as to the what, where and why of the ecological proposal.
2. In principle, LANDEP is not in disagreement with economic development of the territory, because it respects all categories of landscape use required by society. The role of LANDEP is to provide the optimum ecological arrangement of such requirements in a given territory, and to pinpoint ecological problems for society caused by poor spatial arrangements.
3. Though LANDEP is applied to socioeconomic categories of territorial development, natural indexes play a crucial role in deciding the localization of these categories. This concern is motivated by an effort to preserve the "life" of the landscape in a harmony between economy and ecology.

These aspects also guarantee a close collaboration among urbanists and ecologists during further stages of territorial planning, because: (a) the urban planner or decision maker becomes aware of conflicts, especially of ecological losses when the ecological proposal is not taken into account; (b) the kind and significance of the conflicts in a given territory can be directly compared on maps based on ecological versus traditional urban planning approaches; and (c) the character of the conflicts forces urbanists and ecologists to decide jointly on a final proposal for territorial development, and on a mode for minimizing associated negative consequences.

These aspects are missing in many other methods that have the same goals as the LANDEP program. A few brief examples might be mentioned. A high degree of comprehensiveness and systematization is achieved by applied landscape ecology methods based on so called *potential concepts* (e.g., Haase, 1978; Nieman, 1977; Jager and Hrabowski, 1976; Junea, 1974; Mannsfeld, 1979; McHarg, 1969). This is also the case for the methods for determining agroecopotential in Hungary (Goczan 1980). However, these concepts are missing the explicitly expressed proposal level as described above. This is illustrated in the methods of landscape planning in the Federal Republic of Germany (Olschowy 1975) and in Holland (van der Maarel and Vellema, 1975) which are based on modelling systems. Systematization of the contents of individual steps of these models is at a different level. Of course, many theoretical models of this type exist. Most often practically oriented models use a hierarchical branching type of evaluation of the landscape. These are particularly evident in evaluations for recreation, nature conservation, human effects on the landscape, etc. (e.g.,

Lebedeva, 1983; FAO, 1976; Zee, 1984; Hrabowski, 1980; Preobrazenskij et al., 1974; Aleksandrova et al., 1985).

Comparison of the systematized LANDEP method with other methods is complex. Applied methods are highly dependent on local landscape conditions and on particularities of the objectives given (Miklos, 1982).

Further Research Development and Objectives

When addressing questions of ecological evaluation and use of biotic components of the landscape, simplified and applied research methods for vegetation and animals should be developed. Such methods should be aimed at a step by step understanding of their ecological structure and spatial expression. In essense, biotic phenomena and processes should be investigated as part of a landscape complex, as well as indicators of ecological landscape characteristics and use.

The principles and methods of working up landscape-ecology plans should focus on harmonizing land uses with ecological conditions (Fig. 6), and on pinpointing the conflicting land use interests of society. The process should also create prerequisites for conservation of natural resources and for rational ecologically optimum use of the landscape, focused on long-term societal interests and development of humankind.

The systems approach should be applied to understanding the landscape with a major focus on spatial configuration. Further development of the methods of using remotely sensed data, mathematical methods of evaluating spatial relationships, and computer systems, such as geographic information systems, are needed.

The development of specific principles for determining ecologically optimum land uses, landscape conservation, and the creation of new landscape structure would be particularly valuable. Basic research on the structure of models that link general plans to detailed solutions should be enhanced. Finally, for regional and economic planning, we need to develop a simplified method of ecologically evaluating a territory and apply this evaluation in practice.

Application for Society

Knowledge of the ecological properties and biotic components of a landscape can be applied in practice through landscape ecology planning (Fig. 1). Direct application is possible when solving the problems of conservation and rational soil use, optimization of agricultural production, landscape protection against erosion, and increase in ecological stability of the landscape.

The practical focus of landscape ecology planning and the systems approach enable us to play socially important roles. At present there is considerable pressure to satisfy the social order to solve important tasks connected with the national economy, development of individual regions, and society at large. The applicability and progressive character of the results achieved to date have stimulated a number of institutions into collaboration or other forms of using the results.

The widest application potential has been within regional planning where clear linkages among ecological data, analytic procedures, and presentation of territorial plans are especially critical. The second major use of this landscape ecology planning and optimization is in branch planning. Here the search is for ecologically optimum solutions on how to use land for agriculture, water systems, dislocation of industry, power engineering, transportation development, and recreation.

The ecological dimensions of using nature and natural resources should become government policy, as well as a prerequisite in the further development of society.

References

Aleksandrova, T. D., V. S. Preobrazhensky, P. G. Shischenko, eds. 1985. Geoecological accesses to the projecting of natural-technical geosystems. *Proceedings of the landscape-ecological summer-school*. Krym, USSR, AN SSSR, Institut geografii, Moskva, p. 235.

FAO. 1976. A framework for land evaluation. *Soils Bulletin* 32:72.

Góczán, L. 1980. Research, typisation and evaluation of agricultural territory. *Akedémiai Kiadó,* Budapest, p. 125.

Haase, G. 1978. Zur Ableitung und Kennzeichnung von Naturpotentialen. *Petermanns Geographische Mitteilungen* 122:113–125.

Hrabowski, K. 1980. Evaluation of potential and mapping of resources; on the example of building-up potential. In: *Structura, dinamika i rozvitije landšaftov*. AN SSSR, Institut geografii, Moskva, pp. 178–88.

Integrated Ecological Studies or Human Settlements. 1975. The Hong-Kong Human Ecology Programme. *Urban Ecology* 1:81–85.

Jäger, K. D., K. Hrabowski. 1976. Zur Strukturanalyse von Anforderungen den Gesellschaft an der Naturraum, dargestellt am Beispiel des Bebauungspotentials. *Petermanns Geographische Mitteilungen* 120:29–37.

Junea, N. 1974. Medford. Landscape Architecture, Univ. of Pennsylvania, p. 64.

Lebedeva, N. Ja. 1983. Analysis of the documents of the complex planning process. In: *Ochrana landsaftov i projektirovanie*. Akademia nauk SSSR, Institut geografii, Moskva, pp. 72–78.

Maarel, van der E., K. Vellema. 1975. Towards an ecological model for physical planning in the Netherlands. In: *Ecological Aspects of Economic Development Planning Report*. Seminar United Nations Economic Commision for Europe, Rotterdam. Geneva.

Mannsfeld, K. 1979. Die Beurteilung von Naturraumpotentialen als Aufgabe der geographischen Landschaftsforschung. *Pettermanns Georgraphische Mitt.* 123:2–6.

McCoy, K. 1975. Landscape planning for a new australian town. *Urban Ecology* 1:129–271.

McHarg, I. L. 1969. *Design with Nature*. natural History Press, New York.

Miklós, L. 1986. Spatial arrangement of landscape in landscape ecological planning. *Ekológia* (CSSR) 5:49–70.

Miklós, L. 1982. Conceptions of applied landscape-ecological research in foreign countries. *Acta ecologica* 9, 26:76–122.
Miklós, L., D. Miklisová, Z. Reháková. 1986. Systematization and automatization of decision-making process in LANDEP method. *Ekológia* (CSSR) 5:203–32.
Nieman, E. 1977. Eine methode zur rrarbeitung der funktions-leistungsgrade von landschaftselementen. *Arch. Naturschutz u. Landschattforsch* 17:119–157.
Olschowy, G. 1975. Ecological landscape inventories and evaluation. *Landscape planning* 37–44.
Preobraženskij, V. S. et al. 1974. System—approach by the research of recreational activity. *Izvestija AN SSSR, ser. geogr.* 1:18–27.
Ružička, M., ed. 1970. Theoretical problems of the biological landscape research. *Quaestiones geobiologicae* 7:188.
Ružička, M., ed. 1973a. Problems of applying landscape ecology in the practice. *Quaestiones geobiologicae* 11:286.
Ružička, M., ed. 1973b. Content and object of the complex landscape research in the protection and formation of human environment. *Collection of papers for Third International Symposium on Problems of Ecological Landscape Research*. Smolenice, ČSSR.
Ružička, M., ed. 1976. Ecological data for optimal landscape utilization. *Collection of papers for Fifth International Symposium on Problems of Landscape Research*. Smolenice, ČSSR, pp. 430.
Ružička, M., ed. 1979. Ecological stability, resistance, diversity, potentiality, productivity and equilibrium of landscape. *Collection of papers for Fifth International Symposium on Problems of Ecological Landscape Research*. Stará Lesná High Tatras, ČSSR, pp. 553.
Ružička, M., ed. 1982. Ecosystem approach to the (agricultural) landscape. *Collection of papers for Fifth International Symposium on Problems of Landscape Ecological Research*. Piešt'any, ČSSR.
Ružička, M. et al., ed. 1985. The topical problems of landscape ecological research and planning. *Collection of papers for Seventh International Symposium on Problems of Landscape Ecological Research*. Pezinok, ČSSR.
Ružička, M., et al., ed. 1988. Spatial and functional relationships in landscape ecology. *Proceedings of the Eighth International Symposium on Problems of Landscape Ecological Research*. Zemplínska Šírava, ČSSR.
Ružička, M., et al. 1982. Ecological viewpoints in the solution of relations between projected housing estate and its recreational background in Bratislava. *Ekológia* (CSSR) 1:157–192.
Ružička, M., et al. 1983. Ecological evaluation of the prerequisites for agricultural development in the catchment area of a water reservoir. *Ekológia* 2:199–210.
Ružička, M., L. Miklós. 1981. Methodology of ecological landscape evaluation for optimal development of territory. *Proceedings International Congress of the Netherlands Society of Landscape Ecology*, 1981. Pudoc, Wageningen 1981:99–107.
Ružička, M., L. Miklós. 1982a. Landscape—ecological planning (LANDEP) in the process of territorial planning. *Ekológia* (CSSR) 1:297–312.
Ružička, M., L. Miklós. 1982b. Example of the simplified method of landscape—ecological planning (LANDEP) of the settlement formation. *Ekológia* 1:395–424.
Ružička, M., H. Ružičková. 1986. Ecological assumption for the differentiated development of agricultural enterprises (on the example of the district of Trebišov). *Ekológia ČSSR* 5:161–186.
Zee, van der, D. 1984. Monitoring Monaregala, a landscape under pressure. In: *Methodology in landscape ecological research and planning*. IALE, Vol. II. Roskilde University Centre, Denmark, pp. 85–96.

Appendix A. Landscape components and primary landscape structure

Components	Ecological analysis and interpretation	Landscape-ecological synthesis	Landscape-ecological planning
1 Geological substrate		Interpretation of geological substrate for LE synthesis	Regional aspects of geological substrate interpretation for LANDEP expressed on a map
2 Soil and soil-forming bedrock	Soil biology, soil and vegetation	Analysis and interpretation of soil-ecological units in terms of physical and chemical properties	Regional aspects of analysis and interpretation of soil-ecological units for LANDEP expressed on a map
3 Relief		Evaluation of relief forms, morphometry and spatial relationships	Regional characteristics of relief forms, morphometry and spatial relationships
4 Water	Water and vegetation	Microbasins and hydrological systems, water flux in a landscape	Regional hydrological systems and microbasins
5 Climate	Phenological aspects of microclimate and local climate	Meso- and macroclimate, insolation	Meso- and macroclimate
6 Vegetation	Potential reconstructed vegetation, real vegetation, forests and scattered vegetation	Potential and real vegetation	Vegetation map
7 Animals	Selected animal groups with regard to their bond to the environment		Map of distribution of selected animal groups
8 Anthropogenic phenomena and processes	Synanthropization of vegetation and animals	Evaluation of anthropogenic relief forms and technogenic phenomena and processes	Regional characteristics of anthropogenic phenomena and processes

Appendix B. Landscape element types and secondary landscape structure

	Ecological analysis and interpretation								Landscape-ecological synthesis		L E P*
Type of elements	Study Subject	Purpose of study									
		Quality and quantity		Spatial structure		Function		Interpretation of purpose			
1 Forest and scattered vegetation	Physiognomic-ecological formation, communities, forest types	Mapping units and their hierarchy	Homogeneity, diversity, vertical structure	Point, line, area	Size, shape, arrangement of the area, edges	Biological, ecological, socioeconomic	In a landscape: corridors, barriers, borders, etc.	Indicator properties of vegetation and animals	Precision and detailing of mapping units of secondary landscape structure, and their use in partial syntheses and in landscape-ecological complex typing. Landscape dynamics in terms of human-land relationships.		Use of mapping units of secondary landscape structure in the elaboration of landscape ecological designs
2 Grassland and permanent grassland	Ecological formation and types of meadows and pastures, communities, cultivated grassland										
3 Crops	Weedy vegetation, permanent crops, character and types of crop rotations										
4 Water flows and basins	Aquatic vegetation and animals									Microbasins and hydrologic systems	
5 Rocks	Colonization and succession of vegetation and animals									Natural and anthropogen. elements	
6 Technical objects and built-up areas	Character and types of technical areas, ruderal vegetation and animals									Economic-geographical characteristics and zoning	

*Landscape-ecological planning

Appendix C. Socioeconomic phenomena and processes

	Ecological analysis and interpretation	Landscape-ecological synthesis	Landscape-ecological planning
1 Nature protection	Interests of nature protection		Synthesis of interests of nature and natural resource protection on regional scale
2 Natural resource protection	Protection of soil and recreation resources	Protection of soil, water resources and mineral resources	
3 Anthropogenic elements and technical phenomena		Interests of urbanization, industrialization and recreation	
4 Anthropogenic elements and phenomena with seminatural character	Interests of agriculture, water and forest management		
5 Overlap of interests in a landscape		Synthesis and evaluation of interests of particular economic branches and of landscape protection	Regional projection of interests in a landscape

14. Ecologically Sustainable Landscapes: The Role of Spatial Configuration

Richard T.T. Forman

People attempt to improve their well-being. The environment provides materials, but also constrains the effort. This interplay between human aspiration and ecological integrity is an underlying theme of sustainable development and of this article. Alternating changes over a long time span is another theme. At times, technology and organization have provided breakthroughs in sustainable societal development, whereas at other times, environmental constraints have caused social stagnation and human suffering (Clark and Munn, 1986; Jacobs and Munro, 1987).

Spatial scale is yet another basic theme. Individual local ecosystems are sometimes enhanced, but often degraded by humans. Such local ecosystems can change rapidly and markedly, and may be poor candidates to plan for sustainability. At the other end of our spatial scale, the biosphere exhibits considerable stability (Lovelock and Margulis, 1974), but also recently manifests significant degradation. Planet Earth must be analyzed and must be carefully tended for sustainability. However, is that enough? Or is there another spatial scale that should receive planning and management for a sustainable environment? The landscape as a mosaic of local ecosystems, and usually containing people and their activities, has promising characteristics and will be evaluated in this article.

Ecological spatial theory focuses on (a) scale (e.g., Allen and Starr, 1982; O'Neill et al., 1986; Milne et al., 1989); (b) pattern or dispersion (random, regular, and aggregated) (e.g., Greig-Smith, 1964; Pielou, 1974; Gardner et al., 1987), and (c) patch dynamics (appearance, persistence, and disappearance) (e.g., Pickett and White, 1985; Bormann and Likens, 1979; Levin, 1978; Paine and Levin, 1981). In contrast, the present analysis focuses on spatial configuration—that is, the adjacency, connection or juxtaposition of, for example, patches (Leopold, 1933; Harris, 1984; Forman and Godron, 1986; Merriam, 1984; Davis, 1986; Forman, 1987a).

In a decade, sustainable development (and similar terms, sustainability or sustainable environments) has attained a wide range of definitions, perhaps reflecting the many fields necessary for informed policy and action (Repetto, 1985a; Clark and Munn, 1986; Jacobs and Munro, 1987). A United Nations committee (World Committee on Environment and Development, 1987) summarizes the general tone of the concept in stating: "A sustainable condition for this planet is one in which there is stability for both social and physical systems, achieved through meeting the needs of the present without compromising the ability of future generations to meet their own needs." Yet another definition will not be proposed here, but both important strengths and significant shortcomings of this concept should become clear in this article.

The objectives are to: (1) delineate key characteristics of sustainable development, (2) evaluate the applicability of the concept to the landscape scale, and (3) examine the regulatory role of spatial configuration on key variables underlying sustainability. Human demography and direct economic considerations, both important to sustainable development, are widely discussed elsewhere and not analyzed here. However, each is indirectly mirrored in the present focus on the major ecological and human dimensions controlling sustainability.

The general approach is to first consider the concept of sustainable development in terms of time and change, variables and values, and spatial scales. Then landscape ecology (Neef, 1967; Risser et al., 1984; Naveh and Lieberman, 1984; Forman and Godron, 1986; Merriam, 1984; Turner, 1987; Forman and Moore, 1989), and especially spatial configuration, will be used in considering the basic types of landscapes and their promise as sustainable environments.

Sustainable Development

Time and Change

A constant world is impossible and cannot be an objective of sustainable development. Dreams of constancy have led inexorably to revolutionary upheavals, the demise of utopias, and obsolescent industries (C. S. Holling, personal communication).

What then is the nature of stability in sustainable development? It is not physical system stability, where, with negligible biomass, the rock outcrop or concrete runway is essentially the same year in and year out (Forman and Godron, 1986). Nor is it recovery (or resilient) stability, where, with low biomass, the system is readily disturbed but recovers rapidly. Nor is it resistance stability, where the system, commonly with high biomass, resists alteration but when altered recovers slowly. Rather it is *mosaic stability,* a shifting mosaic (Bormann and Likens, 1979) or patch dynamics (Levin, 1978; Paine and Levin, 1981; Pickett and White, 1985), where the system is heterogeneous and may change gradually or remain in steady state, while the component spatial units change at varying rates and intensities. It is like looking down on a city at night where lights blink on and off, but the total amount of light remains nearly constant.

A mosaic is the most conspicuous characteristic of the planet, a continent, a region, or a landscape. All ecological and human processes are spatially differentiated in the mosaic. Thus, mosaic stability, which includes changes, even radical changes, within specific spatial units, is a key element of sustainable development.

But more important than the simple recognition of spatial heterogeneity or overall mosaic stability, is understanding the role of spatial configuration. That is, the specific juxtaposition, adjacency, and connection of spatial units has manifold effects on the system, including regulatory processes.

A second key element of sustainable development is that adaptability and change are inevitable and important. Biological organisms and humans both adapt to and create change. For the ecological or physical system this includes climatic change and biological evolution. Social system change, broadly including economics and culture, includes technology and changing social and territorial organization (Brooks, 1986). All these changes operate over many time scales, but especially critical here are human generations, centuries, and even millennia.

A third important characteristic of sustainable development is the time period involved. At least several human generations, or more than a century, seems to be the appropriate time scale (Clark and Munn, 1986). Planning and political decisions by social and territorial institutions are primarily short-term, from hours to a few years, rarely more than a decade. Individuals and families may extend that range at times to give a higher priority to slightly longer-term decisions. But sustainability implies a much longer time frame. Indeed, the Iroquois Indians of North America are said to have considered only things that last at least seven generations (ca. 110 years) to be important. Learning from at least two centuries of history provides some insight into change related to ecological and human interactions (Clark and Munn, 1986). Perhaps historical analyses of cycles of climatic change, biological evolution pulses, major technological innovations, and so on, could make the time period of sustainable development more precise than "over human generations."

Finally it is essential to consider rates. Some ecological and human variables change slowly while others change rapidly. Some are cyclic, returning to previous states, whereas others are not, or at least do not return to precisely the previous state. Cycles may be regular or highly irregular in frequency and amplitude. In the time frame of human generations or centuries the slow variables, typically with irregular cycles, are of particular interest to sustainable development (Holling, 1986; Brooks, 1986; Clark, 1985). These slowly changing, usually cyclic variables or *foundation variables* are the underlying regulatory foundation determining whether a development is sustainable or not, and they will be examined in more detail later in this article.

The expansion of the human system may reflect a certain phase in the cycle of a slow variable, when, for example, innovation (Haggett et al., 1977), technology, and environmental resources mesh in a new way. This phase usually alternates with a phase of maintenance or contraction, when, for example, famine, war, pestilence, or atmospheric degradation predominates. Development usually implies expansion of the human system or built environment, and therefore primarily relates to only one of the alternating temporal phases of a sustained environment.

Consequently, I prefer the term *sustainable environment* (or sustainable biosphere, sustainable landscape, sustainability, etc.), which encompasses both the expansion or development phases and the contraction or stagnation phases. Nevertheless, here all the terms will be used essentially synonymously. Humankind presumably wants to plan for or establish a sustainable environment; this may or may not permit sustainable development in the narrow sense, where permanent expansion is the goal. The laws of thermodynamics and the lessons of ecology suggest the futility of the latter.

In summary, a sustainable environment (or sustainable development) includes four key characteristics: a time period of several human generations; adaptability and change in ecological and human systems; slowly changing (foundation) variables usually with irregular cycles; and mosaic stability, permitting ongoing rapid fluctuations within component spatial units.

Variables and Values

It is not possible to eliminate values from sustainable development. The values associated with a variable vary from person to person, political group to group, and time to time. Consequently, it seems best to minimize values in the concept and to use an operational concept of sustainable development. In this manner, a particular case of development may be tested objectively to determine whether it was sustainable or not, presumably with general agreement on the results by persons or groups with different values. It also avoids the inappropriate assigning of economic or monetary values to many variables, such as friendship, rare species, and aesthetics, that at times, and to some people, have more value than bread or gold.

The test of sustainability must include both ecological and human dimensions (Bugnicourt, 1987; Sunkel, 1987; Gadgil, 1987). Is ecological integrity maintained or attained? Are human aspirations maintained or achieved? Note that at this broad general level values are explicit: It is good to maintain or achieve ecological integrity and human aspirations.

As mentioned above, it is the set of slowly changing or foundation variables, rather than rapidly changing variables, that are of primary interest in the time frame of sustainable development (Holling, 1986; personal communication). It is convenient and useful to divide the foundation variables into ecological and human categories (Table 1), recognizing that there are important feedbacks among variables within a category and between categories.

Socioeconomic variables, such as equitable control over resources, security, material progress, social and educational institutions, and gross economic product, might be added to those underlying human aspirations (Table 1). However, caution is warranted because generally these are heavily value-laden, and change significantly in short time frames.

These are variables (Table 1) the levels of which determine whether an environment is sustainable or not. Each variable in turn is controlled by a complex of more specific regulatory processes. For example, several processes described in the soil loss equation (Jenny, 1980) determine the rate of soil erosion, and hence the amount and fertility of soil present. Similarly, many factors and processes affect the abundance of housing, including raw materials, transportation, soil or substrate suitability, and nearby employment.

Many of the foundation variables will change together, emphasizing the interactions or feedbacks present. For example, widespread soil loss is typically associated with major decreases in atmospheric quality, biological diversity, and cultural cohesion. Severe famines are commonly associated with losses in biological productivity, in fuel, and in health. Long-term gains in variables are also generally correlated. These gains and losses produce the slow irregular patterns of change in foundation variables, and underlie the manifest alternating phases of human development and stagnation or degradation.

High qualities or levels of the variables maintain or achieve ecological integrity and human aspirations. Uninterrupted noncyclic development or expansion of the built environment is unsustainable. It can be expected to lead inevitably to crash, usually a rapid acute degradation. Striking examples of unsustained environments with human causes are the massive deforestation of central China ca. 3000 B.C., resulting in centuries of siltation and floods; the overgrazing of many Mediterranean lands, resulting in loss of soil and biological productivity; and the 1930s dust bowl of the North American Great Plains,

Table 1. Slowly changing foundation variables that regulate sustainable development.

I. Variables underlying ecological integrity
  a. Soil
  b. Biological productivity
  c. Biological diversity
  d. Fresh water
  e. Oceans
  f. Atmosphere
II. Variables underlying human aspirations
  a. Basic human needs of food, water, health, and housing
  b. Fuel
  c. Cultural cohesion and diversity

resulting in desertification of an area the size of France or Thailand (Worster, 1979). Thus, in a sustainable environment the achievement of ecological integrity and human aspirations continues through both the development or expansion phases, and the stagnation or modest degradation phases.

Finally, we must consider the special role of recovery time for a variable (Forman, 1987b; Repetto, 1985b). Fortunately, few changes are permanent or irreversible (such as species extinction). Most characteristics can recover rapidly. For example, a farmer can turn a forest into field overnight by cutting, a desert into rice culture by irrigation, or rice culture into desert by turning off the water. However, it is the group of nearly irreversible or long-recovery-time characteristics that is of special interest in sustainability. Primeval forest cutting, suburban spread, severe wind erosion, siltation of key surface-water impoundments, and heavy-metal smelter establishment are examples. Here the recovery time is often measured in human generations or centuries, exactly the time scale critical in sustainable environments.

Spatial Scale

Much of the thought on sustainable development focuses on the planet or biosphere (Repetto, 1985a; Clark and Munn, 1986). Such a focus is essential, because if humanity does not design a sustainable planet, there is no habitable place to go. Furthermore, sustainability at the biosphere scale has important effects on sustainability at finer spatial scales, as hierarchy theory demonstrates (Allen and Starr, 1982; Clark, 1985; O'Neill et al., 1986).

Nevertheless, biological and human survival probably depends also on sustainability at finer scales. Both the effectiveness of planning and management and the simple probability of success in attaining sustainability differ sharply according to spatial scale. This issue will be examined below, but first it is important to consider the options for spatial scale, and explore what controls the foundation variables at each scale.

The basic options, in addition to the biosphere, are the continent, biome, region, landscape, and local ecosystem. The continent and biome (McNeely, 1987) usually have distinct boundaries, but in most cases are only loosely tied together by transportation and economics, and encompass extremely dissimilar areas of human land use. The region (such as the southwestern USA, southwestern Australia, the Loire valley, the Andes of Venezuela, and the maritime provinces of Canada) often has diffuse boundaries, determined by a complex of physiographic, cultural, economic, political, and climatic factors. It is tied together relatively tightly by transportation, communication, and culture, but is extremely diverse ecologically. An example differentiating the region from a landscape is instructive.

New England in the US is a relatively distinct, widely recognized region (Figure 1). Its boundaries include an ocean, a long lake, remoteness from early centers of European settlement, and the results of historical political ac-

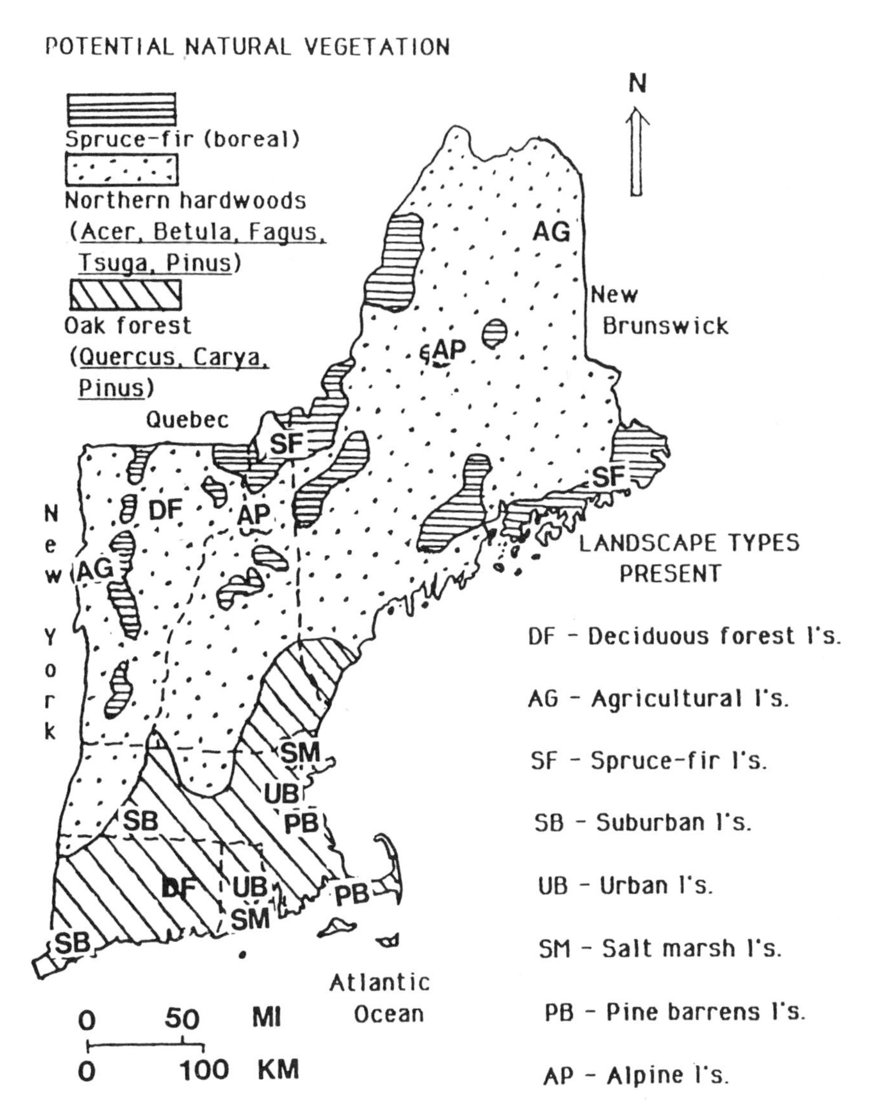

Figure 1. The New England region, with potential natural vegetation mapped, and with examples of present landscape types located. Potential natural vegetation, modified from Kuchler (1964), is in the hypothetical absence of human effects. The eight landscape types present in the region are not mapped; rather, the locations of two examples of each are marked. The first four landscape types (deciduous forest, agriculture, spruce–fir, and suburbia) are widespread, while the latter four are small and scattered.

comodation. New England is tied together by a cool climate, a tradition of governing by town meeting, a transportation network, and cultural nuances including architecture, religion, and language. However, different portions of the region differ markedly in their ecology—e.g., from the wild spruce–fir (boreal) forests of the high mountains to the houses and exotic species of suburbia.

This region is composed of at least eight landscape types, each landscape being a coherent repetitive land mosaic extending for kilometers (Forman and Godron, 1981; 1986). Four types are widespread (deciduous forest landscapes, suburban landscapes, agricultural landscapes, and spruce–fir landscapes), and at least four more (urban, salt marsh, pine barren, and alpine landscapes) are scattered within these (Figure 1). Each type could be subdivided for special purposes—e.g., cultivated and pasture landscapes instead of agriculture, or oak and northern hardwoods landscapes in lieu of deciduous forest. In the region, two or three alpine landscapes and about six urban landscapes are present. Overall, the New England region is composed of several dozen landscapes.

Each landscape is a mosaic where the local ecosystems or land uses are repeated in similar form throughout (Forman and Godron, 1986). Thus, whereas different sections of a region are quite dissimilar ecologically, the landscape manifests an ecological unity with similar ecological conditions found in all sections.

At a still finer scale the local relatively homogeneous ecosystem, such as a marsh, a cornfield, a woodlot, or a pond, might also be managed for sustainability. In areas remote from human activity, some individual ecosystems remain in similar form for generations or centuries. However, in such areas, major natural disturbances significantly alter many of the ecosystems in this time frame (Pickett and White, 1985; Runkle, 1982; Mooney and Godron, 1983). Moreover, in most landscapes today human population or activity is pronounced. Here, few local ecosystems escape major and frequent alterations over several human generations. In short, the local ecosystem in a landscape mosaic should be planned, managed, and cared for, but overall it is not a promising spatial scale for planning a sustainable environment.

In summary, an operational concept of sustainable development that minimizes values is recommended. The key slowly changing or foundation variables underlying ecological integrity are soil, biological production, biological diversity, fresh water, oceans, and air, and those underlying human aspirations are basic human needs of food, health and housing, fuel, and cultural cohesion and diversity. Many of the variables are interlinked and change together, producing the slow cycles of change expected in sustainable environments. Certain nearly irreversible variables with long recovery times are of special concern in sustainable environments. While most sustainability literature has focused on the biosphere, a landscape within a region is a highly promising spatial scale for planning a sustainable environment.

Role of Landscape Ecology

Deeper insight into landscapes and their ecology is now required. A brief introduction to landscape ecology precedes a more detailed analysis of landscape types. Four fundamental landscape types are identified. For each, the key spatial structures or configuration are pinpointed, and in turn, their effects on the foundation variables of sustainable environments are illustrated.

Landscape Ecology in Brief

Landscape ecology focuses on the spatial relationships, fluxes, and changes in species, energy, and materials across large land mosaics (Forman and Godron, 1981; 1986; Risser et al., 1984; Brandt and Agger, 1984; Turner, 1987). A structural approach to landscape ecology elucidates how these objects (species, energy, and materials) are distributed in relation to the sizes, shapes, numbers, kinds, and configuration of the ecosystems or landscape elements present. Patch, corridor, and background matrix analyses have been particularly fruitful.

A functional approach builds on this and explores the interactions among the landscape elements, that is, the flows of objects between adjacent ecosystems or through the mosaic. Edge and stream corridor studies, forest–field interactions, and vertebrate radiotracking studies have provided especially rich insights.

A dynamic or change approach focuses on the alteration in structure and function of the ecological mosaic over time. Geographic-information-system and satellite-image technology, landscape logging patterns, and quantitative modeling have contributed significantly here.

Almost all the principles and theory emerging at the landscape scale appear applicable to a region or any other spatially heterogeneous ecological system.

Basic Landscape Types

Landscapes are often differentiated according to vegetation, physiography, agricultural practices, human populations, and the like. However, a fundamental classification based on the preceding structural, functional, and dynamic (origin and developmental) characteristics does not yet exist (Forman and Godron, 1986). For convenience, therefore, all landscapes will be separated into four categories based on structural characteristics alone: (1) scattered patch landscapes, (2) network landscapes, (3) interdigitated landscapes, and (4) checkerboard landscapes. Rather than being mutually exclusive, the categories should be thought of as the four tips or points of a tetrahedron. All specific landscapes are located in the tetrahedral volume, and each contains varying proportions of the structure represented by each tip.

Scattered Patch Landscapes

This landscape has a predominant background matrix of one ecosystem or landscape-element type, in which patches of one or more other types are enmeshed (Figure 2). Examples are suburbia with scattered school yards, desert with scattered oases, and rangeland with patches of woods. The key spatial characteristics of the scattered patch landscape are: (a) relative area of the matrix, (b) patch sizes, (c) interpatch distances, and (d) patch dispersion (aggregation, regularity, or randomness).

These spatial configurations in a landscape in turn exert regulatory controls on many of the foundation variables. For example, relative area has a major effect on the source and sink functions of the matrix. Thus, dust, nitrogen oxides, and smoke from an extensive matrix will significantly alter atmospheric quality. An extensive matrix may also saturate or alter the enmeshed patches (examples are the oasis effect where heat from dry surroundings desiccates a moist patch, or where high human populations overexploit the fuelwood in scattered woods). Interpatch distance affects the spread of many disturbances, species, and pests from patch to patch (Johnson, 1988). It also may regulate pest outbreaks in the matrix by providing stepping stones for the movement of controlling predators.

Thus the scattered-patch landscape has unique spatial configurations of local ecosystems. These configurations exert major controls on the levels of slowly changing variables that determine sustainable development.

Network Landscapes

These are characterized by prominent intersecting corridors throughout the landscape (Figure 2). Examples are hedgerow grids in pastureland, logging roads in forest, and dendritic irrigation or stream systems in grassland. The key spatial characters are: (a) corridor width, (b) connectivity, (c) network circuitry, (d) mesh size, (e) node size, and (f) node distribution.

Numerous effects on foundation variables are evident. Food-crop production, soil desiccation and erosion in some areas depend heavily on the width and connectivity of windbreak corridors (Caborn, 1965; Les Bocages, 1976; Forman and Baudry, 1984; Baudry, 1984; Ryszkowski and Kedziora, 1987). The movement of wide-ranging, often rare mammals is doubtless strongly affected by connectivity and circuitry. Flooding and water quality depend on stream corridor or riparian systems (Davenport et al., 1976; Gorham et al., 1979; Schlosser and Karr, 1981; Verry and Timmons, 1982; Lowrance et al., 1984; Decamps et al., 1988); indeed, many coastal marine fisheries, nutrient levels, and delta formations depend on river and hedgerow corridor systems that inhibit erosion. Community and cultural cohesion, as well as diversity, are significantly enhanced by the separating effect of corridors—such as greenbelts between suburban neighborhoods, or large wildlife corridors that prevent strip (or ribbon) development and maintain the integrity of towns.

Interdigitated Landscapes

Here two or more continuous landscape-element types are prominent and mesh or interfinger along their common boundary (Figure 2). Examples are housing development along roads interfingering with the unbuilt surroundings, and agri-

a. Scattered patch landscapes

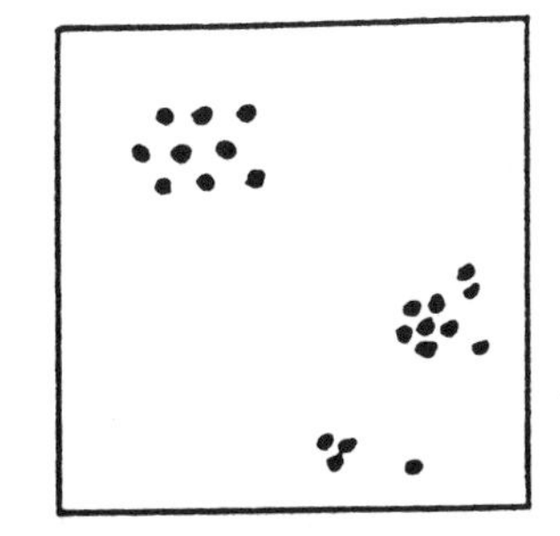

b. Network landscapes

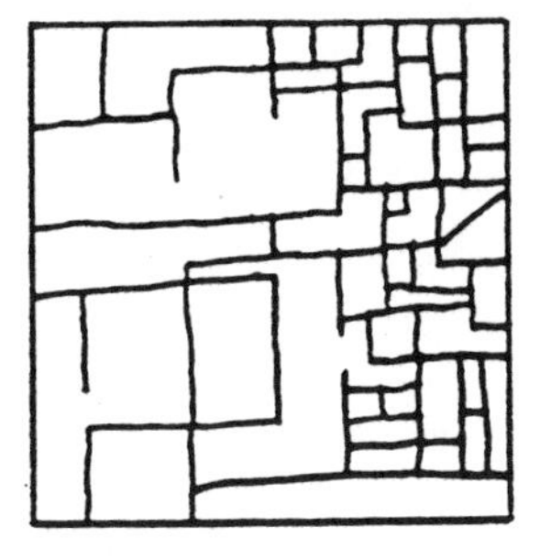

c. Interdigitated landscapes

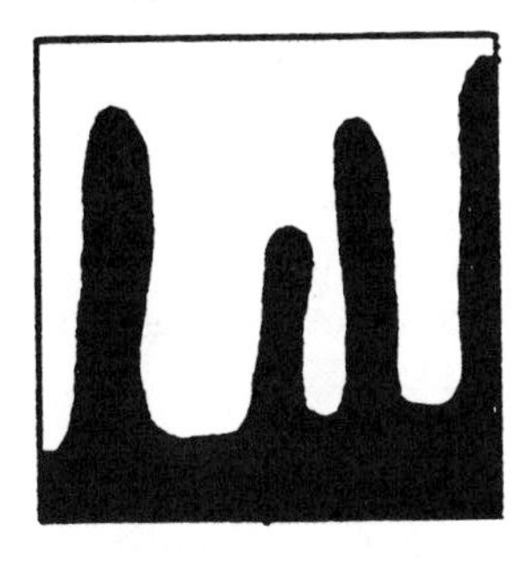

d. Checkerboard landscapes

Figure 2. Four basic landscape types characterized by structure. In each landscape only two types of landscape elements (ecosystems or land uses) are included, indicated by black and white. The dendritic example combines characteristics of two landscape type, network and scattered patch.

culture and forest interdigitating in a ridge and valley area. The predominant spatial characteristics are the: (a) relative areas of each element type; (b) abundance and orientation of peninsulas; and (c) length and width of peninsulas.

Peninsular orientation significantly affects wind penetration and crop production, and width constrains housing developments and biological diversity (Milne and Forman, 1986; Forman and Godron, 1986). Total boundary length may be considerable in this landscape, resulting in high densities of edge species as well as animals that require two or more ecosystems in proximity. Interactions between ecosystems are rampant in interdigitated landscapes, where, for example, herds of field herbivores inhibit forest regeneration in woods, and forest herbivore herds ravish adjacent agricultural plantings.

Checkerboard Landscapes

These landscapes have a grid with two or more landscape-element types in alternating cells of the grid (Figure 2). Examples are some systematically managed logging patterns, and some highly regular agricultural fields with alternating crops. The salient spatial characteristics here are: (a) grain size of the landscape (due to the average area or diameter of component patches); (b) the regularity or completeness of the grid; and (c) total boundary length (or amount of edge).

Grain size of the landscape determines the abundance of interior species and biological diversity, because a fine-grained landscape contains primarily generalist species such as weeds and edge species (Forman and Godron, 1986; Franklin and Forman, 1987). Regularity and completeness of the grid control the movement and colonization of many objects such as crop pollinators, disease vectors, and people (O'Brien, 1984; Hudson, 1985). The regeneration of trees for forest production is often enhanced in a logged checkerboard, and boundary-related phenomena such as tree blowdowns are widespread (Franklin and Forman, 1987). But the highly dissected nature of the checkerboard landscape means that extensive stretches of matrix or patch are absent, thus minimizing, for example, atmospheric dust pollution in dry areas or the buildup of extensive fires. Checkerboards may enhance human culture by providing proximity for kinship with nature (Bugnicourt, 1987; Gadgil, 1987).

In summary, the four landscape types presented exhibit sharply different spatial configuration of local ecosystems. These spatial configurations exert manifold and major regulatory controls on the slow cyclic variables of sustainable environments.

Landscapes and Sustainability

The biosphere is an important scale for sustainability, not only because we have no other place to live, but because in a hierarchical system the conditions at a broad scale affect those at finer scales.

Concurrently, conditions in a smaller unit affect the broader scale, as well as conditions in neighboring comparable small units (O'Neill et al., 1986; Hall,

1981). Therefore it is critical to identify the most appropriate fine-scale unit to plan and manage for sustainability.

The farther we get away from an individual caring for his or her own garden, the less effective planning and management decisions are. Thus managing a local ecosystem may be easy, but managing the planet very difficult (ignoring the political question of whether it is good for humankind to have one or a few persons managing the world).

Conversely, the probability of achieving sustainability decreases at finer scales. Large rapid fluctuations in individual ecosystems are normal, whereas the broad-scale natural regulatory processes provide considerable stability, as suggested by the Gaia hypothesis or empirical result (Lovelock and Margulis, 1974; McElroy, 1986; Ryszkowski and Kedziora, 1987). Again our attention is drawn to identifying a scale most appropriate for human planning and management for sustainable development.

As noted above, the landscape has significant advantages over the region or continent for sustainable development. Its relatively distinct boundaries and the commonality of ecological process over its area, combined with the developing scientific understanding from landscape ecology, point to the landscape as a highly promising scale for a sustained environment.

Unfortunately case studies from which to draw lessons are few. A lucid analysis of a New Brunswick, Canada, landscape (Regier and Baskerville, 1986; Wynn, 1980; Hall, 1981) showed a two-century period of relative stability for the human community, concurrent with changing resource use due to overexploitation, and with a continuously shrinking resource base. Eventually, economic stagnation and contraction arrived. This temporal analysis demonstrated the essential requirement for both ecological and human dimensions in sustainable development.

More detailed spatially explicit information on the ecological and human components is available for the Pine Barrens landscape of the state of New Jersey, USA over three centuries (Forman, 1979; Collins and Russell, 1988), and would be promising for an evaluation of the characteristics of sustainability. It appears that some population and cultural stability, along with some overall resource stability, resulted from: spatially differentiated resource type, availability, and use; a low overall resource base that inhibited city formation; and some constancy in linkages with four types of surrounding landscapes.

Despite the paucity of case studies at the landscape scale, spatial configuration is a simple and concrete handle for planners and managers. An example is useful. A farmer will maximize grain production on his or her field—a local ecosystem. The farmer's town or county will balance grain production, clean water, recreation, and so on, in its area, not by requiring one tent site and one small wetland in each grain field, but by spatial differentiation (zoning, government incentives, etc.). The state will balance grain production, clean water, recreation, industrial areas, rare and endangered species, transportation systems, and so on, in a similar manner. The spatial configuration of the mosaic of good soils, stream corridors, industrial areas and the like controls in a major way the levels of the variables attained. *Adjacency,* the effect an adjoining system has on a landscape element, is especially critical here (Forman 1987), and is readily incorporated into planning and management.

Thus a planner generally can consider many possible spatial configurations in a landscape to achieve a particular level of a variable, such as grain production, housing density, or available fresh water. Similarly, a particular level of many variables, such as those underlying ecological integrity or achievement of human aspiration, can be produced with a certain spatial configuration of landscape elements. Rearranging the configuration should almost always increase or decrease ecological integrity.

This leads to a provocative hypothesis. I suspect that for any landscape, or major portion of a landscape, there exists an optimal spatial configuration of ecosystems and land uses to maximize ecological integrity, achievement of human aspirations, or sustainability of an environment. If so, the challenge is to find it. The development of theory and principles at the landscape scale will enhance our ability to meet this difficult, but tractable, challenge.

Finally, let us return to mosaic stability. We can tie landscapes and the biosphere together, using both the time frames and foundation variables of sustainable development. At any time, individual landscapes are in different phases of their irregular cycles of slowly changing foundation variables. Yet when combined into a global mosaic of landscapes, sustained stability of the biosphere may be possible. Why don't we tend, tenderly and sustainably, each garden of the mosaic?

Summary and Conclusion

We have explored the interface between sustainable development and landscape ecology to identify key characteristics of sustainable environments, examine their applicability to landscapes, at a scale finer than that of the biosphere, and evaluate the role of spatial configuration in regulating variables critical to a sustainable environment. An operational concept of sustainable environments that minimizes values has been recommended. Key sustainability characteristics include a time frame of several human generations (more than a century), slow regulatory foundation variables with irregular cycles, adaptability and change in ecological and human systems, a mosaic stability that allows ongoing fluctuations within individual spatial units. At least six foundation variables are required for attaining ecological integrity and three for achieving human aspiration. Most are linked and change slowly together. Four basic landscape types in the biosphere were identified based on spatial structure alone: (1) scattered patch landscapes, (2) network landscapes, (3) interdigitated landscapes, and (4) checkerboard landscapes. The key spatial characteristics of each were delineated based on landscape-ecology theory, and examples of their effect on foundation variables presented. Evidence points to the landscape as an optimal spatial scale for planning and managing for a sustainable environment, and this should go hand in hand with sustainability of the planet.

Acknowledgments

It is a pleasure to thank Peter Jacobs and William C. Clark, whose writing and discussions on sustainable development have stimulated me. Michael Binford, Katharine Poole, Kristina Hill, and Harvey Brooks have also contributed to my thinking on sustainability, and Elgene Box aided with comments on the manuscript. This paper is based on a talk presented at the 1988 meeting of the International Federation of Landscape Architects.

References

Allen, T. F. H. and T. B. Starr. 1982. *Hierarchy: Perspectives for ecological complexity*. University of Chicago Press, Illinois.

Baudry, J. 1984. Effects of landscape structure on biological communities: The case of hedgerow network landscapes. In J. Brandt and P. Agger, eds, *Proceedings of the first international seminar on methodology in landscape ecological research and planning*. Roskilde University Center, Denmark, Vol. 1, pp.55–65.

Bormann, F. H. and G. E. Likens. 1979. Catastrophic disturbance and the steady state in the northern hardwood forest. *American Scientist* 67:660–9.

Brandt, J. and P. Agger, eds. 1984. *Proceedings of the first international seminar on methodology in landscape ecological research and planning*. 5 vol. Roskilde University Center, Denmark.

Brooks, H. 1986. The typology of surprises in technology, institutions, and development. In W. C. Clark and R. E. Munn, eds., *Sustainable development of the biosphere*. Cambridge University Press, United Kingdom, pp.325–48.

Bugnicourt, J. 1987. Culture and environment. In P. Jacobs and D. A. Munro, eds., *Conservation with equity. Strategies for sustainable development*. International Union for the Conservation of Nature and Natural Resources, Gland, Switzerland, pp.95–106.

Caborn, J. M. 1965. *Shelterbelts and Windbreaks*. Faber and Faber, London.

Clark, W. C. 1985. Scales of climate impacts. *Climatic Change* 7:5–27.

Clark, W. C. and R. E. Munn, eds. 1986. *Sustainable development of the biosphere*. Cambridge University Press, United Kingdom.

Collins, B. R. and E. W. B. Russell, eds. 1988. *Protecting the New Jersey pinelands: A new direction in land-use management*. Rutgers University Press, New Brunswick, New Jersey.

Davenport, D. C., P. E. Martin, E. B. Roberts, and R. M. Hagan. 1976. Conserving water by antitranspirant treatment of phreatophytes. *Water Resources Research* 12:985–90.

Davis, J. C. 1986. *Statistics and data analysis in geology*. Wiley, New York.

Decamps, H., M. Fortune, H. Gazelle, and G. Pautou. 1988. Historical influence of man on the riparian dynamics of a fluvial landscape. *Landscape Ecology* 1:163–74.

Forman, R. T. T., ed. 1979. *Pine barrens: Ecosystem and landscape*. Academic Press, New York.

Forman, R. T. T. 1987a. The ethics of isolation, the spread of disturbance, and landscape ecology. In M. Turner, ed. *Landscape heterogeneity and disturbance*. Springer-Verlag, New York, pp.213–229.

Forman, R. T. T. 1987b. Emerging directions in landscape ecology and applications in natural resource management. In R. Herrmann and T. Bostedt-Craig, eds, *Proceedings of the conference on science in the national parks*. U.S. National Park Service and the George Wright Society, Fort Collins, Colorado, pp.59–88.

Forman, R. T. T. and J. Baudry. 1984. Hedgerows and hedgerow networks in landscape ecology. *Environmental Management* 8:495–510.

Forman, R. T. T. and M. Godron. 1981. Patches and structural components for landscape ecology. *BioScience* 31:733–40.

Forman, R. T. T. and M. Godron. 1986. *Landscape ecology*. Wiley, New York.

Forman, R. T. T. and P. N. Moore. 1989. Patch boundary theory in landscape mosaics. In Di Castri, F. and A. Hansen, eds., *Landscape boundaries*: Consequences for biotic diversity and ecological flows. Kluwer, Amsterdam.

Franklin, J. F. and R. T. T. Forman. 1987. Creating landscape patterns by forest cutting: Ecological consequences and principles. *Landscape Ecology* 1:5–18.

Gadgil, M. 1987. Culture, perceptions and attitudes to the environment. In P. Jacobs and D. A. Munro, eds. *Conservation with equity: Strategies for sustainable development*. International Union for the Conservation of Nature and Natural Resources, Gland, Switzerland, pp.85–94.

Gardner, R. H., B. T. Milne, M. G. Turner, and R. V. O'Neill. 1987. Neutral models for the analysis of broad-scale landscape patterns. *Landscape Ecology* 1:19–28.

Gorham, E., P. M. Vitousek, and W. A. Reiners. 1979. The regulation of chemical budgets over the course of terrestrial ecosystem succession. *Annual Review of Ecology and Systematics* 10:53–84.

Greig-Smith, P. 1964. *Quantitative plant ecology*. 2nd ed. Butterworth, London.

Haggett, P., A. D. Cliff, and A. Frey. 1977. *Locational analysis in human geography*. Wiley, New York.

Hall, T. H. 1981. Forest management decision making, art or science? *Forestry Chronicle* 57:233–8.

Harris, L. D. 1984. The fragmented forest. Island biogeography theory and the preservation of biological diversity. University of Chicago Press, Illinois.

Holling, C. S. 1986. The resilience of terrestrial ecosystems: Local surprise and global change. In W. C. Clark and R. E. Munn, eds. *Sustainable development of the biosphere*. Cambridge University Press, United Kingdom, pp.292–317.

Hudson, J. C. 1985. *Plains country towns*. University of Minnesota Press, Minneapolis.

Jacobs, P. and D. A. Munro, eds. 1987. *Conservation with equity: Strategies for sustainable development*. International Union for the Conservation of Nature and Natural Resources, Gland, Switzerland.

Jenny, H. 1980. *The soil resource: Origin and behavior*. Springer-Verlag, New York.

Johnson, W. C. 1988. Estimating dispersibility of *Acer, Fraxinus* and *Tilia* in fragmented landscapes from patterns of seedling establishment. *Landscape Ecology* 1:175–88.

Kuchler, A. W. 1964. Potential natural vegetation of the conterminous United States. American Geographical Society, Special Publication 36, New York.

Leopold, A. 1933. *Game management*. Scribners, New York.

Les Bocages: Histoire, ecologie, economie. 1976. I.N.R.A., C.N.R.S., E.N.S.A., et Université de Rennes, France.

Levin, S. A. 1978. Pattern formation in ecological communities. In J. A. Steele, ed. *Spatial pattern in plankton communities*. Plenum Press, New York, pp.433–70.

Lovelock, J. E. and L. Margulis. 1974. Atmospheric homeostasis by and for the biosphere. The Gaia hypothesis. *Tellus* 26:1–10.

Lowrance, R., R. Todd, J. Fail, Jr., O. Hendrickson, Jr., R. Leonard, and L. Asmussen. 1984. Riparian forests as nutrient filters in agricultural watersheds. *BioScience* 34:374–7.

McElroy, M. B. 1986. Change in the natural environment of the Earth: the historical record. In W. C. Clark and R. E. Munn, eds., *Sustainable development of the biosphere*. Cambridge University Press, United Kingdom, pp.199–211.

McNeely, J. A. 1987. The biome approach to sustainable development. In P. Jacobs and D. A. Munro, eds., *Conservation with equity: Strategies for sustainable development*. International Union for the Conservation of Nature and Natural Resources. Gland, Switzerland, pp.251–74.

Merriam, G. 1984. Connectivity: A fundamental characteristic of landscape pattern. In J. Brandt and P. Agger, eds., *Proceedings of the first international seminar on methodology in landscape ecological research and planning*. Roskilde University Center, Denmark, Vol. 1, pp.5–15.

Milne, B. T. and R. T. T. Forman. 1986. Peninsulas in Maine: Woody plant diversity, distance, and environmental pattern. *Ecology* 67:967–74.

Milne, B. T., K. M. Johnston and R. T. T. Forman. 1989. Scale-dependent aggregation of wildlife habitat in a landscape using a spatially-neutral Bayesian model. *Landscape Ecology* 2:101–110.

Mooney, H. and M. Godron, eds. 1983. *Disturbance and ecosystems*. Springer-Verlag, New York.

Naveh, Z. and A. S. Lieberman. 1984. *Landscape ecology: Theory and application*. Springer-Verlag, New York.

Neef, E. 1967. Die theoretischen Grundlagen der Landschaftslehre. Haack, Geographische Anstalt Gotha/Leipzig.

O'Brien, M. J. 1984. *Grassland, forest, and historical settlement: An analysis of dynamics in northeast Missouri*. University of Nebraska Press, Lincoln, Nebraska.

O'Neill, R. V., D. L. DeAngelis, J. B. Waide and T. F. H. Allen. 1986. *A hierarchical concept of ecosystems*. Princeton University Press, New Jersey.

Paine, R. T. and S. A. Levin. 1981. Intertidal landscapes: Disturbance and the dynamics of pattern. *Ecological Monographs* 51:145–78.

Pickett, S. T. A. and P. S. White, eds. 1985. *The ecology of natural disturbance and patch dynamics*. Academic Press, New York.

Pielou, E. C. 1974. *Population and community ecology*. Gordon and Breach, New York.

Regier, H. A. and G. L. Baskerville. 1986. Sustainable development of regional ecosystems degraded by exploitive development. In W. C. Clark and R. E. Munn, eds., *Sustainable development of the biosphere*. Cambridge University Press, United Kingdom, pp.75–101.

Repetto, R., ed. 1985a. *The global possible: Resources, development and the new century*. Yale University Press, New Haven, Connecticut.

Repetto, R. 1985b. Overview. In R. Repetto, ed. *The global possible: Resources, development and the new century*. Yale University Press, New Haven, Connecticut.

Risser, P. G., J. R. Karr, and R. T. T. Forman. 1984. *Landscape ecology—directions and approaches*. Illinois Natural History Survey, Special Publication Number 2. Champaign, Illinois.

Runkle, J. R. 1982. Patterns of disturbance in some old-growth mesic forests of eastern North America. *Ecology* 63:1533–46.

Ryszkowski, L. and A. Kedziora. 1987. Impact of agricultural landscape structure on energy flow and water cycling. *Landscape Ecology* 1:85–94.

Schlosser, I. J. and J. R. Karr. 1981. Riparian vegetation and channel morphology impact on spatial patterns of water quality in agricultural watersheds. *Environmental Management* 5:233–40.

Sunkel, O. 1987. Beyond the World Conservation Strategy. Integrating conservation and development in Latin America and the Caribbean. In P. Jacobs and D. A. Munro, eds., *Conservation with equity: Strategies for sustainable development*. International Union for the Conservation of Nature and Natural Resources, Gland, Switzerland, pp.35–54.

Turner, M. G., ed. 1987. *Landscape heterogeneity and disturbance*. Ecological Studies 64. Springer-Verlag, New York.
Verry, E. S. and D. R. Timmons. 1982. Waterborne nutrient flow through an upland-peatland watershed in Minnesota. *Ecology* 63:1456–7.
World Committee on Environment and Development. 1987. *Our common future*. Oxford University Press, United Kingdom.
Worster, D. 1979. *Dust bowl: The southern plains in the 1930s*. Oxford University Press, New York.
Wynn, G. 1980. *Timber colony: A historical geography of early nineteenth century New Brunswick*. University of Toronto Press, Canada.

Concluding Remarks

Rather than only tying down neat packages of knowledge, the sequence of chapters here is designed to forge new linkages and explore dimensions of an emerging frontier. Beginning with a historical background, the book progresses from natural processes in short-term change, to interacting natural and human processes in unplanned long-term change, to change by design, that is, human planning and management. This distinctive integration of natural processes and human activities is increasingly a hallmark of landscape ecology.

A second distinguishing feature is the focus on spatial heterogeneity and its ecological consequences. We can describe and model, albeit primitively, the structure, fluxes and changes in a landscape system of patches, corridors and matrix. But understanding with predictive ability how a species, including *Homo sapiens,* reacts to, uses, and depends on spatial heterogeneity remains an important challenge.

Planning and management for a single ecological or human characteristic, such as crop production, biodiversity, housing, or erosion control, on a small parcel within a landscape is commonplace and short-term. But planning and management that balances the range of characteristics of ecological integrity and human aspiration in optimum spatial arrangements over the landscape is the major objective. Such a goal should lead to greater short-term value, as well as a sustainable environment. To accomplish this we must look beyond methodology in planning and management, and articulate and build on underlying theory, especially the regulatory processes.

Understanding for these challenges, as the basic goal of human inquiry, will depend on our ability to develop explicit concepts and principles at the landscape scale. Such a body of theory will be robust when directly linked to or derived from "first principles," including thermodynamics and molecular movement.

As citizens become increasingly aware of the ever expanding examples of desertification, urbanization, overgrazing, irrigation, deforestation and reforestation, the world's attention is inexorably drawn to the ecology of changing landscapes.

Index